THE FISCHER–TROPSCH SYNTHESIS

PROFESSOR FRANZ FISCHER
19 March 1877–1 December 1947

THE FISCHER–TROPSCH SYNTHESIS

Robert Bernard Anderson
Department of Chemical Engineering
McMaster University
Hamilton, Ontario, Canada

With a chapter on the Kölbel–Engelhardt Synthesis by H. Kölbel and M. Ralek
Institute of Technical Chemistry
Technical University
Berlin (West), Germany

1984

ACADEMIC PRESS, INC.
(Harcourt Brace Jovanovich, Publishers)
Orlando San Diego San Francisco New York London
Toronto Montreal Sydney Tokyo São Paulo

ACADEMIC PRESS, INC.
Orlando, Florida 32887

United Kingdom Edition published by
ACADEMIC PRESS, INC. (LONDON) LTD.
24/28 Oval Road, London NW1 7DX

Library of Congress Cataloging in Publication Data

Anderson, Robert Bernard, Date
The Fischer-Tropsch synthesis.

Includes index.
1. Fischer-Tropsch process. 2. Coal liquefaction.
I. Kölbel, Herbert. II. Rálek, M. (Milǒs), 1928- .
III. Title.
TP352.A53 1983 662'.6623 83-15762
ISBN 0-12-058460-3 (alk. paper)

PRINTED IN THE UNITED STATES OF AMERICA

84 85 86 87 9 8 7 6 5 4 3 2 1

To the second generation of researchers on the Fischer–Tropsch Synthesis, including Professor Paul H. Emmett, Dr. C. Charles Hall, Professor Herbert Kölbel, Professor Helmut Pichler, and Dr. Henry H. Storch

Contents

Preface

The author chose to write this book as a retirement project to keep him busy; it has. Papers based on the researches spawned by the OPEC oil embargo have been appearing in large numbers since about 1979. An attempt has been made to include pertinent papers through 1982 and a few from 1983. The current global overabundance of oil may again diminish research on coal-to-oil processes. The possibility of new plants producing fluid fuels from coal before 2000 seems unlikely. Material covered in the book begins in 1955 and continues from the major reviews of the Fischer–Tropsch synthesis published by authors from the United States Bureau of Mines at Bruceton, Pennsylvania.

I wish to acknowledge the assistance of many in preparing this book: D. S. Hodgson, McMaster University, for preparing all of the graphical material; Professors H. Kölbel and M. Ralek, Technical University, Berlin, for contributing Chapter 7 on the Kölbel–Engelhardt synthesis; Dr. R. C. McLane, Chemical Abstracts Services, Columbus, Ohio, for aiding in the translation of Chapter 7 and for preparing the "Quest for a Homogeneous Fischer–Tropsch Synthesis" in Chapter 5; Professor Günther Wilke, Max-Planck Institute for Coal Research, Mülheim an der Ruhr, West Germany, for furnishing biographical material and the excellent photograph of Franz Fischer; Dr. Jacques Monnier, McMaster, and Energy, Mines and Resources, Ottawa, for technical editing of the manuscript and checking the thermodynamic data; Dr. Masa Saito, National Research Institute for Pollution and Resources, Japan, for reviewing the manuscript; Ms. E. M. Tooke, McMaster, for a literature review; Amy Stott, Laura Honda, and Jan Reid, Engineering, Word Processing Center, McMaster, for making the task easier; graduate student Henry Meng and Dr. Kevin S. Smith, SASOL 1, South Africa, for technical editing; and my wife Jane, for putting up with another major project, even in retirement.

Discussions and useful reports, manuscripts, reprints, and theses were kindly furnished by C. N. Bartholomew, A. Deluzarche, M. E. Dry, P. H. Emmett, J. Happel, A. Kiennemann, H. Kölbel, M. J. McGlinchey, R. M. Madon, C. B. Murchison, M. Ralek, W. M. H. Sachtler, C. N. Satterfield, H. Schulz, and others.

THE FISCHER–TROPSCH SYNTHESIS

CHAPTER 1

Introduction

1.1 Scope of the Book

A definition of the Fischer–Tropsch synthesis (FTS) that delineates the material covered here is as follows: a hydrogenation of oxides of carbon producing higher hydrocarbons and/or alcohols, the carbon chains of the molecules being predominantly straight in the range C_4 to C_{10}. The FTS is a practical way of converting coal to gasoline, diesel oil, wax, and alcohols. Coal is gasified to furnish the hydrogen plus carbon monoxide. The FTS is currently used commercially in three large plants in South Africa, and during World War II about one-fifth of the gasoline used in Germany was obtained in this way. When coal-to-oil processes are required in North America, the FTS will be a prime contender. In addition to liquid fuels, the FTS can provide largely straight-chain higher alcohols and olefins for the chemical industry.

The early German and other research on FTS to about 1950 was summarized by the author and his colleagues at the United States Bureau of Mines (*1, 6, 12, 13, 23*). The present volume brings this account up to date, through 1982. Usually, engineering aspects of FTS are not covered in this book, and the reader is referred to references *5, 8, 20,* and *22* and older work (*9, 18, 23*).

Chapter 1 begins with a short history of the FTS and related reactions, followed by a brief appraisal of the current situation. Chapter 2 covers the thermodynamics of the hydrogenations of oxides of carbon, as well as reactions that catalytic materials may undergo. Chapter 3 describes the preparation of iron carbides, nitrides, and carbonitrides and their properties as catalysts for the FTS; Bureau of Mines work from 1950 to 1960 is summarized. Then, Chapter 4 reports on tests of all types of catalysts, including a number of metals not usually regarded as catalysts for the FTS. The mechanism of the FTS is the subject of Chapter 5. Despite the use of modern research tools, many aspects of the reaction are still the subject of polemics. Chapter 6 describes the poisoning of catalysts in the

FTS by sulfur compounds. Apparently, it is still best to remove virtually all of these molecules from the feed. In the final chapter, Professors H. Kölbel and M. Ralek report on the Kölbel–Engelhardt synthesis. Professor Herbert Kölbel was a codiscoverer of this process in 1949, and earlier, a co-worker of Franz Fischer at the Kaiser-Wilhelm Institute at Mülheim.

1.2 A Short Historical Sketch

1902 P. Sabatier and J. D. Senderens hydrogenated CO over Ni to produce CH_4 (*21*).

1910 A. M. Mittasch, C. Bosch, and F. Haber developed promoted Fe catalysts for the synthesis of NH_3 (*15*).

1912–1913 The Kaiser-Wilhelm Association for the Advancement of Knowledge established its third institute in Mülheim, dedicated to coal research, the Kaiser-Wilhelm-Institut für Kohleforschung (KWIK). Franz Fischer, Professor of Electrochemistry at the Technical University, Berlin, was appointed Director of the new institute.[1]

1913 Badische Anilin und Soda Fabrik (BASF) received patents on the preparation of hydrocarbons and oxygenates by the hydrogenation of CO at high pressure, usually on oxide catalysts.

1920 Brennstoff-Chemie was started by Fischer. KWIK also had been publishing *Abhandlung zur Kenntnis der Kohle* since 1917.

1922–1923 Fischer and Hans Tropsch obtained Synthol, mostly oxygenates, from H_2 + CO on alkalized Fe and other catalysts.

1923 BASF and I. G. Farben (IG) obtained patents on Zn-based, Fe-free catalysts for the methanol synthesis.

1925 Fischer and Tropsch announced the synthesis of higher hydrocarbons at atmospheric pressure; Co and Ni seemed to be promising catalysts. A reason for working at 1 atm was that BASF had patents covering higher pressures. The idea developed that hydrocarbons are produced at 1 atm and oxygenated molecules at higher pressure; this notion may have delayed the development of useful Fe catalysts for a decade.

1926–1930 Studies were made on the new process in England, Japan, and the United States. Workers at the U.S. Bureau of Mines incorporated ethylene in the synthesis on Co catalysts.

[1] The early history of the Kaiser-Wilhelm-Institut für Kohleforschung and biographical material on Franz Fischer are largely taken from references *10, 11, 18, 19,* and *23*.

1928 Tropsch left KWIK to become director of a coal research institute in Prague. In 1931, he went to Chicago to work at Universal Oil Products and to be Adjunct Professor at the Armour Institute and later at the University of Chicago.

1928–1930 P. K. Frohlich and W. K. Lewis studied the synthesis of higher alcohols on alkalized $ZnO–Cr_2O_3$ and other catalysts.

1931–1932 Fischer and K. Meyers developed $Ni–ThO_2$–kieselguhr and $Co–ThO_2$–kieselguhr, which were relatively sophisticated catalysts designed for the primitive reactors of that time. Initial studies were on Ni because of the short supply of Co; however, the Ni catalysts did not operate properly in large reactors. The yields of CH_4 were too large. Pilot plant work was started at KWIK in 1930.

1933–1939 Pilot-plant tests and further development of Co catalysts were made by Ruhrchemie AG.

1935 Tropsch returned to Germany critically ill; he died on 8 October.

1935–1945 The FTS was operated commercially in Germany using Co catalysts.

1936 Fischer and Helmut Pichler found that Co catalysts also operated satisfactorily in the middle pressure range, 5–20 atm.

1936 The first commercial catalytic cracking unit, developed by Eugene Houdry, was installed (*14*).

1937 Fischer and Pichler discovered that synthesis on Fe was greatly improved by operating at 5–20 atm. At about the same time, Herbert Kölbel at Rheinpreussen also performed medium-pressure tests of Fe catalysts. Alkalized Fe in this pressure range became a possible replacement for Co in the German commercial plants.

1938 Fischer and Pichler produced high-melting-point waxes in high-pressure synthesis on Ru.

1939–1944 German industry joined in the development of an Fe catalyst to replace Co in the existing plants. Improved reactors were also studied, including fixed-bed reactors with recycle, hot-gas recycle, oil recycle, and slurry reactors.

1939–1944 Fischer and Pichler developed the Isosynthesis, and Otto Roelen the OXO process.

1942–1950 Fears of an impending shortage of petroleum in the United States caused wide interest in coal-to-oil processes. A large research and development program was authorized for the U.S. Bureau of Mines, and the American petroleum and related industries initiated studies of the FTS, often on Fe in fluidized or entrained reactors.

1943 Fischer retired from KWIK and moved to Munich. Karl Ziegler became the second Director of KWIK at Mülheim.

1943–1944 Kölbel initiated and directed the comparative tests I and II held at Schwarzheide to select an Fe catalyst to replace Co in existing reactors; KWIK and five companies participated (*18*). Although the results were remarkably good, no replacement catalyst was selected, probably because of the overall disarray of activity that attended the approach of the end of the war.

1945 World War II ended, and the operation of FTS plants in Germany stopped. Some plants in what is now East Germany (GDR) were confiscated by the Soviet Union.

1947 Fischer died in Munich on 1 December.

1948 P. H. Emmett used molecules tagged with ^{14}C to study the mechanism of the FTS.

1948 The Kaiser-Wilhelm Gesellschaft becomes the Max-Planck Gesellschaft, and the institute at Mülheim became a Max-Planck Institute.

1949 Kölbel and F. Engelhardt discovered that H_2O and CO react on FTS catalysts to yield typical FTS products.

1950 A fluidized–fixed bed process developed by Hydrocarbon Research, Trenton, New Jersey was installed in Brownsville, Texas. This plant, called Carthage Hydrocol, Inc., used reformed natural gas. Severe operating difficulties required designing a new reactor, which was installed in 1953. The new reactor operated properly, but the plant was promptly shut down, sold, and dismantled. By the time the plant was operating correctly, the price of natural gas had more than doubled. Merely selling the gas was more profitable than converting it to gasoline and chemicals. At the same time, in South Africa, the SASOL FTS plant using coal was constructed and opened in 1955. Lurgi gas generators and Rectisol gas-cleaning units were employed. Two types of FTS reactors, both with Fe catalysts, were used: a fixed-bed with recycle unit designed by Ruhrchemie in West Germany, having long tubes with an internal diameter of ~2 in., and an entrained-solids reactor by M. W. Kellogg in New Jersey. After a year or more of start-up problems, these units have operated successfully up to the present (*7*).

1953 R. B. Anderson and J. F. Shultz found that Fe nitrides are unique, durable catalysts for the FTS that produce large yields of alcohols (*3*). Anderson developed equations that predict the isomer and carbon-number distributions of FTS products (*4*). Kölbel and P. Ackermann successfully operated a large slurry reactor in a 10-ton-per-day demonstration plant.

1954 The abundance of petroleum in the world and its availability at very low prices caused the decrease or termination of most of the programs of research and development on coal-to-oil processes.

1957 G. Natta and co-workers (*16, 17*) studied the methanol and higher alcohol syntheses.

1959 Pichler resumed work on FTS on Ru, which led to the polymethylene synthesis.

1961 H. H. Storch died; he was the architect of the useful scientific and engineering research programs on coal-to-oil processes at the U.S. Bureau of Mines in Pittsburgh and Bruceton, Pennsylvania (*2, 4*).

1966 Imperial Chemical Industries (ICI) developed a moderate-pressure methanol synthesis employing copper–zinc oxide catalysts in a gas-recycle reactor.

1967 U.S. Bureau of Mines workers developed methods for flame spraying catalysts onto metals for the platelet assemblies of hot-gas-recycle reactors. These techniques led to the development of a new FTS unit, the tube-wall reactor (*5*).

1969 Domestic production of petroleum in the United States fell below demand; the United States became an importer of oil.

1973 The oil embargo by the Organization of Petroleum Exporting Countries (OPEC), followed by a fourfold increase in the price of petroleum, accelerated research on coal-to-oil processes in the United States and many other countries. With the demise of the U.S. Bureau of Mines, federally sponsored coal-to-oil research became part of the Energy Research and Development Administration (ERDA) and subsequently the Department of Energy.

1974 H. Pichler died in Karlsruhe on 13 October.

1975 In South Africa, the decision was made to build SASOL II, scheduled for operation in 1980, and in 1979, plans were made for SASOL III, to begin operation in 1982 (*7*). The new plants are similar to the initial plant, except that fixed-bed FTS reactors were not included in the new installations.

1976 Mobil announced a process for converting methanol to an aromatic gasoline and C_2 to C_4 olefins on the shape-selective catalyst ZSM5. This process may be a serious competitor for FTS.

1981 In New Zealand a methanol plant operating on natural gas plus the Mobil process was planned, scheduled for completion in 1985.

1982 The worldwide recession sharply decreased demand for petroleum. OPEC was unable to control the production and price of oil among its members. The relatively low cost and abundance of petroleum has discouraged new coal-to-oil ventures.

1.3 A Brief Appraisal of the Present Status of the Fischer–Tropsch Synthesis

In the decade following the end of World War II, research and development on the FTS was pursued vigorously in the United States, England, and West Germany, but this activity decreased abruptly in the 1950s. SASOL maintained a modest continuing research and development program devoted largely to practical problems, and the U.S. Bureau of Mines continued a small research activity.

The renewed interest in and funding for FTS research, starting in 1970, has involved academic workers to a very much larger extent than in the previous period. In most instances, the research teams of the 1950s had been dispersed, and the new researchers attacked the FTS with new ideas and excellent new research tools. However, neophytes in FTS research require 2–3 years to digest the voluminous literature. For this reason, most of the useful contributions have appeared since 1978.

Capillary gas chromatography, often coupled with mass spectrometry, has been an important research tool that has made the determination of carbon number and isomer distribution almost routine. Mössbauer spectroscopy has proved to be the ideal way of obtaining quantitative analyses of phases present in Fe catalysts. However, studies of Fe comprise only a small part of the current total research effort. Precipitated Fe containing as much as 10 Mn to 1 Fe has been shown to be highly selective, producing large yields of α-olefins.

Many workers have chosen supported Ru as an ideal catalyst for investigation, because Ru is neither oxidized nor carburized during synthesis. Rhodium has been shown to have interesting properties in the FTS and probably should be included with Fe, Co, Ni, and Ru as FTS catalysts. Nickel has been studied widely, but often under conditions for which CH_4 is the principal product. As catalysts, Mo and W also produce higher hydrocarbons, but their activity is low, possibly because these metals oxidize easily. Studies are in progress to test the idea that the pore structure of supports or the particle size of metal crystallites may alter the carbon-number distribution of FTS products. Other research has been examining the possibility that shape-selective catalysis may drastically alter product distribution. Zeolites, including ZSM5, have been used as catalyst supports and/or as subsequent refining catalyst in an additional bed. Not enough data have been reported to pass judgement on these ideas. Sufficient research has been done on the "quest for a homogeneous FTS" to permit the statement that the discovery of this process seems highly *unlikely*.

Major contributions have not yet been made on the FTS by the sophisticated methods of the surface physicist, involving ultrahigh-vacuum apparatus. Methods have been devised for treating the catalyst under usual industrial conditions and then transferring it to the ultrahigh-vacuum system without exposure to air.

Tagged molecules and atoms have been used as tracers in numerous mechanistic studies. The postulated intermediates in the FTS have changed from carbides to oxygenated species and back to carbides, methylene, methyl, etc. The jury is still out on trying to select the most appropriate mechanism. The old chain-growth schemes from 1950 are still valid, but the formation of branched carbon chains is still a subject of debate. More recent studies of poisoning by sulfur compounds confirm the older ideas; removal of virtually all sulfur-containing molecules seems the best procedure.

Engineering studies have included examination of fluid-bed, slurry, hot-gas-recycle, and tube-wall reactors. For most purposes, significantly improved catalysts and reactors have not emerged. For the production of gasoline, the entrained iron process of SASOL seems to be the best choice. To obtain a wider distribution products, including wax, fixed-bed units of the type used in SASOL I or the slurry reactor, both using Fe catalysts, seem appropriate. Excellent diesel fuel may be produced by hydrorefining the heavy fractions.

References

1. Anderson, R. B., *in* "Catalysis" (P. H. Emmett, ed.), Vol. 4, pp. 1–372. Van Nostrand-Reinhold, Princeton, New Jersey, 1956.
2. Anderson, R. B., *Fuel* **41,** 295 (1962).
3. Anderson, R. B., *Catal. Rev.—Sci. Eng.* **21**(1), 53 (1980).
4. Anderson, R. B., *ACS Symp. Ser.* **222,** 389 (1983).
5. Baird, M. J., Schehl, R. R., Haynes, W. P., and Cobb, J. T., Jr., *Ind. Eng. Chem. Prod. Res. Dev.* **19,** 175 (1980).

6. Cohn, E. M., *in* "Catalysis" (P. H. Emmett, ed.), Vol. 4, pp. 443–473. Van Nostrand-Reinhold, Princeton, New Jersey, 1956.
7. Dry, M. E., CHEMTECH **12,** 744 (1982).
8. Falbe, J., ed., "Chemierohstoffe aus Kohle." Thieme, Stuttgart, 1977.
9. Field, J. H., Benson, H. E., and Anderson, R. B., *Chem. Eng. Prog.* **56,** 44 (1960).
10. Fischer, F., *Ber. Dtsch. Chem. Ges. A* **68,** 169 (1935).
11. Fischer, F., "Leben und Forschung," p. 110. Max-Planck-Institut für Kohleforschung, Mülheim–Ruhr, 1957.
12. Greyson, M., *in* "Catalysis" (P. H. Emmett, ed.), Vol. 4, pp. 473–513. Van Nostrand-Reinhold, Princeton, New Jersey, 1956.
13. Hofer, L. J. E., *in* "Catalysis" (P. H. Emmett, ed.), Vol. 4, pp. 372–443. Van Nostrand-Reinhold, Princeton, New Jersey, 1956.
14. Houdry, E., Burt, W. F., Pew, A. E., Jr., and Peters, W. A., Jr., *Pet. Refiner* **17,** 574 (1938).
15. Mittasch, A., *Adv. Catal.* **2,** 81 (1950).
16. Natta, G., *in* "Catalysis" (P. H. Emmett, ed.), Vol. 3, p. 349. Van Nostrand-Reinhold, Princeton, New Jersey, 1955.
17. Natta, G., Colombo, U., and Pasquon, I., *in* "Catalysis" (P. H. Emmett, ed.), Vol. 5, p. 131. Van Nostrand-Reinhold, Princeton, New Jersey, 1957.
18. Pichler, H., *Adv. Catal.* **4,** 271 (1952).
19. Pichler, H., *Ber. Dtsch. Chem. Ges. A* **100,** 127 (1967).
20. Riekena, M. L., Vickers, A. G., Haun, E. C., and Koltz, R. C., *Chem. Eng. Prog.* **78**(4), 86 (1982).
21. Sabatier, P., and Senderens, J. D., *C.R. Hebd. Seance, Acad. Sci.* **134,** 514, 680 (1902).
22. Schulz, H., *in* "Future Sources of Organic Raw Materials. Chemrawn I" (L. E. St. Pierre and G. R. Brown, eds.), pp. 167–183. Pergamon, Oxford, 1980.
23. Storch, H. H., Golumbic, N., and Anderson, R. B., "The Fischer–Tropsch and Related Syntheses." Wiley, New York, 1951.

CHAPTER 2

Thermodynamics of the Fischer–Tropsch Synthesis

The thermodynamics of the Fischer–Tropsch synthesis (FTS) and relevant reactions of catalytic material has been described by Anderson (*1*, *10*). That work is reviewed briefly here, and more recent work is described.

2.1 Thermodynamics of Producing Organic Molecules

For each hydrocarbon or oxygenated organic molecule, a family of equations such as those shown in Table 1 for olefins are obtained by adding the water–gas shift [Eq. (5)] successively to Eq. (1) (*3*). Equation (2) is typical of the primary synthesis reactions; with some catalysts, particularly iron, the water–gas shift may occur as rapidly as Eq. (2). Table 1 also presents the standard-state Gibbs free energy and enthalpy changes, ΔG^0 and ΔH^0, for the production of 1-hexene by Eqs. (1) to (4) at 427°C. Both the free energy and enthalpy changes become more negative in proceeding from Eq. (1) to Eq. (4), that is, the reactions become more favorable thermodynamically and more exothermic passing down the table. Because synthesis reactions usually involve a decrease in number of moles, equilibrium conversions increase substantially with increasing operating pressure. At moderate to high pressures, large conversions may be obtained even when the value of ΔG^0 is positive.

Figures 1 and 2 present the standard-state heats of reaction and Gibbs free energy changes as a function of temperature for reactions typical of Eq. (2) of Table 1, that is, with water as a product. Formaldehyde, methanol, and acetic acid are produced according to Eqs. (6) to (8), respectively. To keep all of the curves on the same scale, the enthalpy and free

Table 1
Equations for the Production of Olefins and Thermodynamic Data for Production of 1-Hexene[a]

Equation	ΔG^0	ΔH^0
(1) $3\ H_2 + 1\ CO_2 = 2\ H_2O + \frac{1}{n}(C_nH_{2n})$	5.68	−25.96
(2) $2\ H_2 + 1\ CO = H_2O + \frac{1}{n}(C_nH_{2n})$	2.63	−35.01
(3) $H_2 + 2\ CO = CO_2 + \frac{1}{n}(C_nH_{2n})$	−0.42	−44.06
(4) $H_2O + 3\ CO = 2\ CO_2 + \frac{1}{n}(C_nH_{2n})$	−3.47	58.52
(5) $H_2O + CO = H_2 + CO_2$	−3.04	−9.06

[a] Data for 1-hexene at 427°C given in kilocalories.

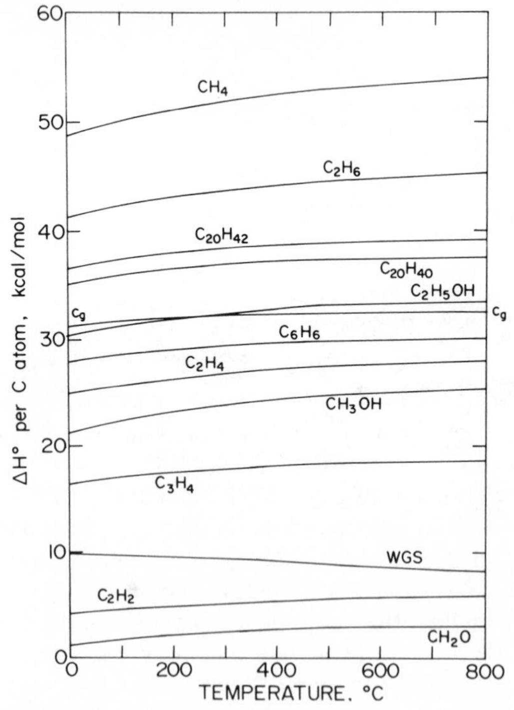

Fig. 1 Standard heats of reaction per carbon atom in the organic product for equations of type (2) in Table 1, where C_3H_4 is methyl acetylene, C_6H_6 is benzene, and C_g is graphite. WGS, Water–gas shift reaction.

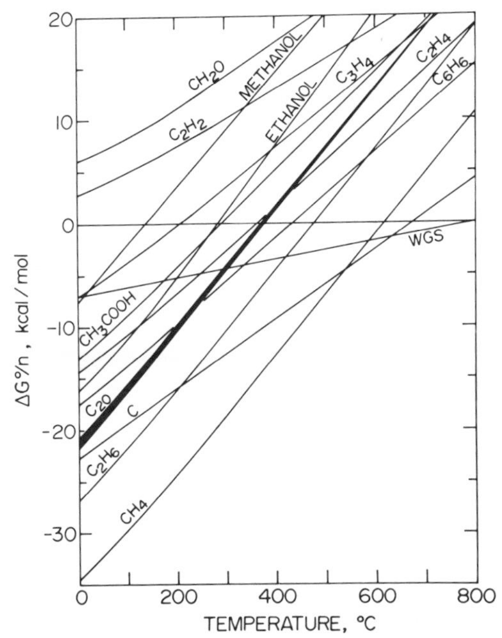

Fig. 2 Standard free energies of reaction per carbon atom of the organic product for equations of type (2) in Table 1, where C_3H_4 is methyl acetylene, C is graphite, and the top of the C_{20} band represents the 1-olefin and the bottom the *n*-paraffin. $-\Delta G^0 = RT \ln K$. WGS, Water–gas shift. Reproduced with permission from Anderson *et al.* (*3*).

energy changes are divided by the number of carbon atoms in the product. Data for the water–gas shift are also given in Figs. 1 and 2. For reactions typical of Eq. (1), data may be obtained by subtracting the contribution of the water–gas shift from the value shown; for reactions of the type of Eqs. (3) and (4), the water–gas shift is added once or twice, respectively. With increasing carbon number *n*, the heats of reactions tend toward the band between the C_{20} paraffin and C_{20} olefin. The same trend holds for Gibbs free energy changes, but the curves for olefins appear to rotate as well as being displaced with increasing *n*.

$$H_2 + CO = HCHO \tag{6}$$

$$2\ H_2 + CO = CH_3OH \tag{7}$$

$$2\ H_2 + 2\ CO = C_2H_4O_2 \tag{8}$$

Thus, the ordinates of Figs. 1 and 2 are, respectively, $\Delta H^0/n$ and $\Delta G^0/n$, and the equilibrium constant for the reaction is given by

$$K = \exp[-n(\Delta G^0/n)]/RT. \tag{9}$$

Table 2
Thermodynamics of the Synthesis of Acetylene at 227°C[a]

	Reaction type	ΔH^0	ΔG^0
2	$\frac{3}{2} H_2 + CO = \frac{1}{2} C_2H_2 + H_2O$	−5.0	8.4
5	$H_2O + CO = H_2 + CO_2$	−9.5	−4.8
1	$\frac{5}{2} H_2 + CO_2 = \frac{1}{2} C_2H_2 + 2\ H_2O$	4.5	13.2
3	$\frac{1}{2} H_2 + 2\ CO = \frac{1}{2} C_2H_2 + CO_2$	−14.5	3.6
4	$\frac{5}{2} CO + \frac{1}{2} H_2O = \frac{1}{2} C_2H_2 + \frac{3}{2} CO_2$	−19.21	1.2

[a] Data given in kilocalories per mole for the reaction as written, for example, for $\frac{1}{2}$ C_2H_2 in the top reaction.

Of the molecules shown in Fig. 2, only formaldehyde, acetylene, and methanol have positive ΔG^0 values, even below 200°C, and hence the conversion at equilibrium may be small. However, sizeable yields of methanol are obtained at 250–300°C and 100 atm in the current commercial processes.

The thermodynamics of reactions producing acetylene from carbon monoxide at 227°C are given in Table 2. The ΔG^0 values become less positive in proceeding from reaction type 1 to 4. Although to the author's knowledge acetylene has never been produced in the FTS, at least reaction type 4 seems possible thermodynamically.

Compared with the formation of paraffins of the same carbon number, reactions producing molecules containing a hydroxyl group or double bond have more positive ΔG^0 values. Two or more of these structures increase the ΔG^0 values further; here, an acetylenic structure has about the same effect as two olefin bonds. Gibbs free energy changes for some reactions producing molecules containing hydroxyl, olefinic, and acetylenic groups are shown in Table 3.

In addition, molecules with branched carbon chains usually have more negative ΔG^0 values than those with straight carbon chains; however, the differences are small and would be hardly discernible on the scale of Fig. 2. Similar statements can be made for internal and terminal olefins, internal olefins usually being favored thermodynamically.

Thus, on the basis of thermodynamics, a vast variety of molecules can be produced in synthesis reactions up to 400°C and for some up to 500°C, particularly at elevated pressures, including acetaldehyde and higher aldehydes, ketones, esters, and naphthenes, which are not shown in Fig. 2. Only in the methanol synthesis is a single product obtained. Usually a broad spectrum of types of molecules, carbon numbers, and carbon-chain

Table 3
Changes of Gibbs Free Energy for Formation of C_2 and C_3 Compounds[a]

Compound		ΔG^0, *kcal/mol*	
		227°C	*427°C*
Ethane	C_2H_6	−29.18	−5.88
Ethylene	C_2H_4	−11.09	5.90
Acetylene	C_2H_2	16.85	27.72
Acetaldehyde	CH_3CHO	−4.05	13.36
Ethanol	C_2H_5OH	−6.78	16.41
Ethylene glycol[b]	$C_2H_4(OH)_2$	15.77	38.94
Methyl ether	$(CH_3)_2O$	7.21	31.20
Propane	C_3H_8	−37.28	−2.24
Propylene	C_3H_6	−23.06	5.43
Cyclopropane	C_3H_6	−11.60	18.38
Methylacetylene	C_3H_4	2.71	25.19
Allene	C_3H_4	4.83	27.51
n-Propanol	C_3H_7OH	−15.94	18.96
Allyl alcohol	$H_2C{=}CHCH_2OH$	0.28	29.27
Acetone	$(CH_3)_2CO$	−18.63	11.06

[a] For reactions producing H_2O as product except as noted. Calculated from data of Stull *et al.* (*11*).
[b] $3\ H_2 + 2\ CO = C_2H_4(OH)_2$.

structures is produced, and the distribution of molecules depends on the selectivity of the catalyst employed. Generally the reaction products are not in thermodynamic equilibrium with other product molecules or reactants; therefore, valuable clues regarding the reaction mechanism may be obtained from detailed analysis of products (*2, 8, 13*). We now consider the products that would be obtained if all of the reactants and products were in equilibrium, for a limited number of typical molecules (*3, 12*).

To obtain the equilibrium situation for systems involving two or more reactions, the Gibbs free energy for the system must be minimized while maintaining an atom balance. Except for some trivial simple cases, these calculations involve an "educated" trial-and-error search and are generally beyond the scope of hand calculations, because of the nonlinear simultaneous equations that must be solved. In the present work the procedure of White, Johnson, and Dantzig (*14*) was used. The computer program generally converged on the final result in ~15 iterations. In all cases the equilibrium constant for the water–gas shift was calculated from product compositions, and in a few cases similar calculations were made

Table 4
Compounds Included in Equilibrium Calculations for Figs. 2–5, in addition to H_2, CO, CO_2, and H_2O

Group	*Compounds*[a]
1	Methane
2	Ethane, propane, *n*-butane, isobutane, *n*-pentane, isopentane, neopentane
3	Ethylene, propylene, acetone
4	Formaldehyde, acetaldehyde, formic acid, acetic acid, methanol, ethanol, acetylene

[a] Note: group 1 includes all molecules shown; group 2, all molecules but methane; etc.

for other reactions. The calculated equilibrium constants were always correct to four or more significant figures. Free energy data were obtained from free energy functions tabulated by Pitzer and Brewer (*9*), interpolated numerically to desired temperatures. Calculations were made in terms of partial pressures, rather than fugacities, and small errors were therefore introduced in the calculated equilibrium compositions at 200 and 2000 atm.

Molecules used in the calculations in addition to H_2, CO, H_2O, and CO_2 are listed in Table 4. In the first calculation, involving group 1, all of the molecules in the table were included. In group 2, methane was deleted from the calculation. For group 3, all paraffins were excluded, and for group 4, paraffins, olefins, and acetone were deleted. Temperatures used were 262, 350, and 500°C; the two lower temperatures correspond to temperatures that may be expected for operation of iron catalysts in fixed-bed and fluidized- or transported-bed reactors, respectively, and 500°C is the highest temperature at which substantial conversion may be expected. Feed gases were 3 H_2 + 1 CO_2, 2 H_2 + 1 CO, 1 H_2 + 1 CO, 1 H_2 + 2 CO, and 1 H_2O + 3 CO. The 1 H_2 + 1 CO feed may be considered as 1.5 H_2 + 1.5 CO in the present context.

The calculated data in terms of mole percentage of the product exclusive of H_2O, CO_2, H_2, and CO are given in Figs. 3 to 6 for groups 1 to 4, respectively. Not all of the molecules are plotted, often because their concentrations are too small, but in some cases because the graphs would become too complicated. For example, in the last two columns of Figs. 3 and 4 a number of C_3 to C_5 paraffins has been omitted; however, ethyl-

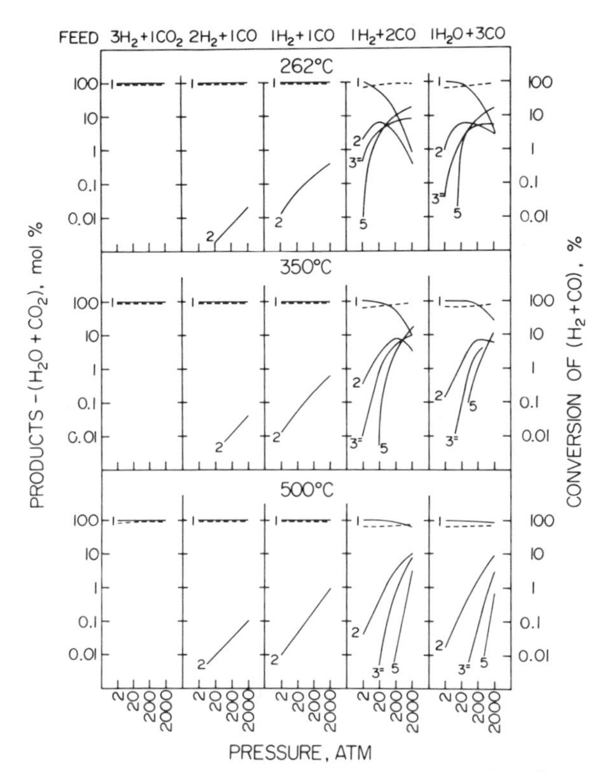

Fig. 3 Equilibrium compositions and conversions for molecules of group 1 of Table 2. ---, Percentage conversion of H_2 + CO; 1, methane; 2, ethane; 5, *n*-pentane; $3^=$, propylene. Reproduced with permission from Anderson *et al.* (*3*).

ene is off scale. Also shown on the graphs as a broken curve is the conversion of H_2 + CO, defined as moles of (H_2 + CO) consumed divided by moles of (H_2 + CO) fed. This quantity is not changed by the water–gas shift.

For group 1 (Fig. 3), methane is the predominant product, and other hydrocarbons are produced only for feeds lacking enough hydrogen to permit complete conversion to CH_4. Nearly complete conversions of H_2 + CO are obtained. If CH_4 is deleted from the calculations (group 2), the results shown in Fig. 4 are obtained. Ethane becomes the dominant product, except again for the feed compositions on the right side of the chart, where hydrogen is deficient for producing all C_2H_6. At 500°C and lower pressures, conversions are substantially less than 100%.

For group 3 (Fig. 5), propylene is the dominant product, the yields of ethylene, not shown, varied from 0.2 to 50%. Acetone appears in substantial amounts along with ethanol, acetic acid, and acetaldehyde. At 500°C

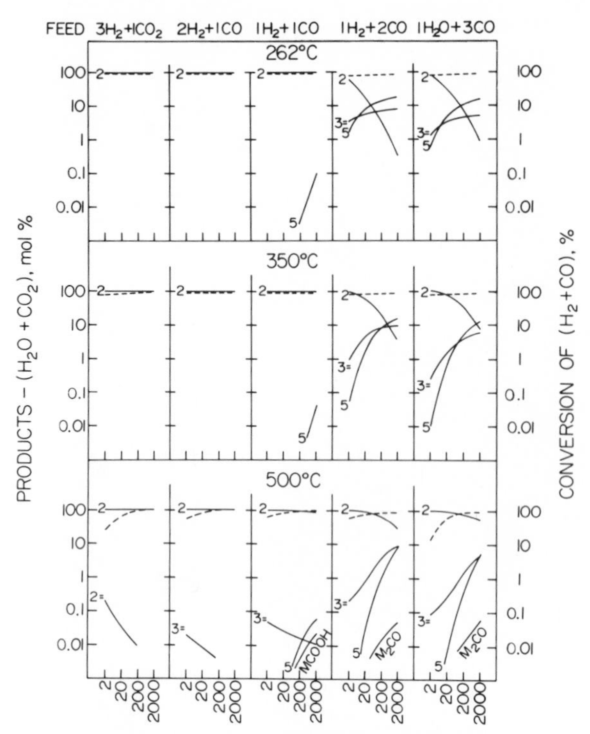

Fig. 4 Equilibrium compositions and conversions for molecules of group 2 of Table 2. ---, Percentage conversion of H_2 + CO; 2, ethane; 5, *n*-pentane; $2^=$, ethylene; $3^=$, propylene; MCOOH, acetic acid; M_2CO, acetone. Reproduced with permission from Anderson *et al.* (*3*).

and 2 atm, the conversion is low for some feed mixtures, particularly 3 H_2 + 1 CO_2. For group 4 (Fig. 6), with all paraffins, olefins, and acetone deleted from the calculation, ethanol is the major product if enough hydrogen is present (left side of graph); otherwise acetaldehyde (right side of graph). Acetic acid is also an important product. Acetylene and methanol appear on some portions of the graph, but only small amounts of these molecules are present. Conversions of the feed at low pressures are often small at 350°C, and always small at 500°C.

The lesson to be learned from Figs. 3–6 is that the catalyst used must be selective; otherwise the product will be mainly methane. For example, high yields of methanol may be obtained at 250–300°C and elevated pressures, but only on special catalysts containing combinations of oxides of copper, chromium, and zinc and essentially free of group VIII metals.

The thermodynamics of reactions of products from the FTS are considered in detail in reference *1*, and the conclusions are presented briefly

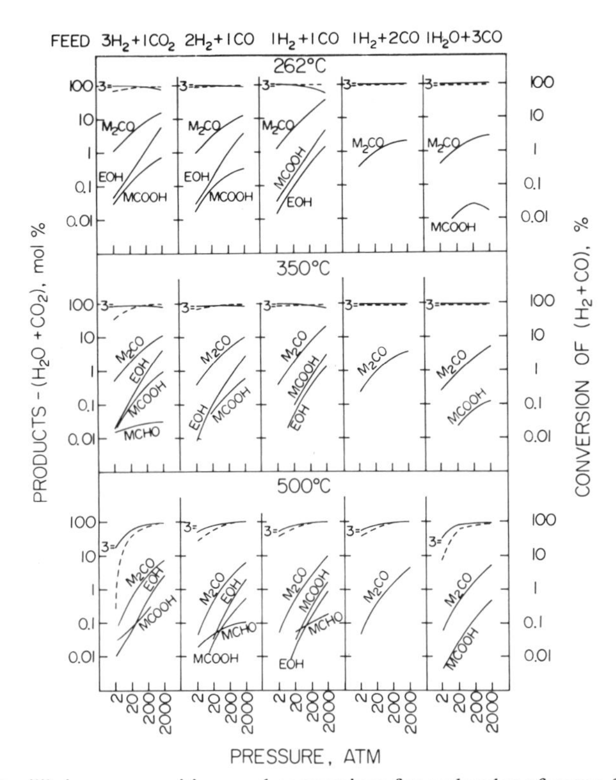

Fig. 5 Equilibrium compositions and conversions for molecules of group 3 of Table 2. ---, Percentage conversion of H_2 + CO; $3^=$, propylene; M_2CO, acetone; EOH, ethanol; MCOOH, acetic acid; MCHO, acetaldehyde. Reproduced with permission from Anderson *et al.* (*3*).

here. Hydrogenation of olefins and dehydration of alcohols are possible thermodynamically under usual synthesis conditions. The hydrogenolysis of paraffins is possible at usual FTS temperatures, but the combination of two paraffins to produce one larger paraffin molecule plus H_2 is impossible. Most reactions of two olefins or one olefin and one paraffin are usually possible thermodynamically. The term "incorporation" has been used to denote the building-in of an organic molecule together with H_2 + CO in the FTS. The incorporation of ethylene, ethanol, or methanol in any amount is thermodynamically feasible. Higher olefins and alcohols have a smaller thermodynamic tendency to react in this way. Incorporation of most paraffins is possible only to a limited extent; for example, at 300°C it is thermodynamically possible to add CH_4 to H_2 + CO only to the extent that less than ~30% of the carbon in the paraffin produced comes from CH_4.

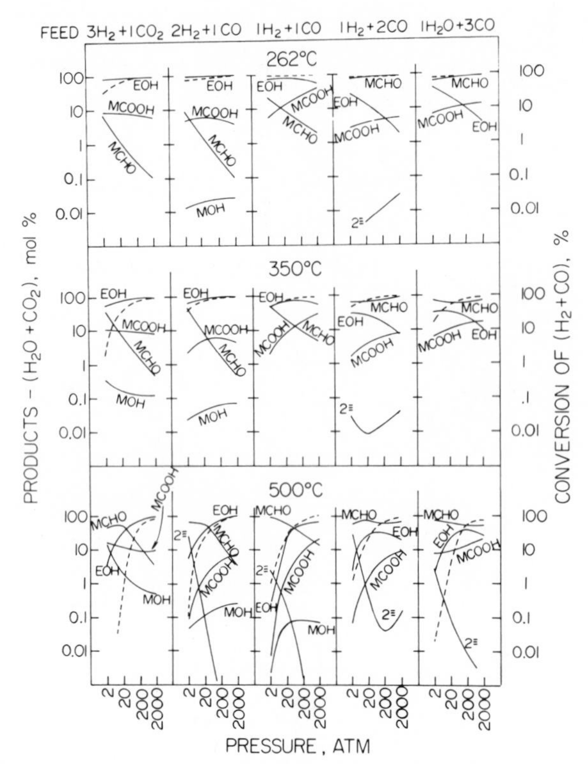

Fig. 6 Equilibrium compositions and conversions for molecules of group 4 of Table 2. ---, Percentage conversion of H_2 + CO; EOH, ethanol; MCOOH, acetic acid; MCHO, acetaldehyde; MOH, methanol; 2≡, acetylene. Reproduced with permission from Anderson *et al.* (*3*).

2.2 Thermodynamics of Reactions of Catalytic Materials

Thermodynamic data for inorganic substances pertinent to the FTS have largely been taken from Barin and Knacke, *Thermochemical Properties of Inorganic Substances* (*4*) and the *Supplement* (coauthored by Kubaschewski) (*5*). Tables 5–11 present the available thermodynamic data at 400–700 K in the form of equilibrium constants for pertinent reactions of the active metals for the FTS and for copper, zinc, molybdenum, rhodium, rhenium, tungsten, and iridium, which are similar in some respects to the active metals. The data presented in the author's previous accounts for the thermodynamics of the FTS (*1, 10*) generally agree with those of the newer compilation.

Tables 5 and 6 give the equilibrium constants for the reduction of oxides

Table 5
Reduction of Oxides

Reaction	*Equilibrium constant* $K = p_{H_2O}/p_{H_2}$, *at*			
	400 K	*500 K*	*600 K*	*700 K*
$\frac{1}{4}$ $Fe_3O_4 + H_2 = \frac{3}{4}$ $Fe + H_2O$	0.00170	0.0143	0.0550	0.1350
$CoO + H_2 = Co + H_2O$	150.0	104.2	75.7	57.5
$NiO + H_2 = Ni + H_2O$	550.8	439.5[a]	343.6	278.0
$Cu_2O + H_2 = 2\ Cu + H_2O$	9.1×10^{10}	1.1×10^{9}	5.6×10^{7}	6.4×10^{6}
$CuO + H_2 = Cu + H_2O$	5.5×10^{13}	2.7×10^{11}	7.4×10^{9}	5.35×10^{8}
$ZnO + H_2 = Zn + H_2O$	1.1×10^{-11}	5.75×10^{-9}	3.6×10^{-7}[b]	6.65×10^{-6}
$\frac{1}{2}$ $MoO_2 + H_2 = \frac{1}{2}$ $Mo + H_2O$	4.45×10^{-5}	0.00091	0.00646	0.0251
$\frac{1}{2}$ $RuO_2 + H_2 = \frac{1}{2}$ $Ru + H_2O$	7.3×10^{13}	3.0×10^{11}	7.1×10^{9}	4.7×10^{8}
$\frac{1}{3}$ $Rh_2O_3 + H_2 = \frac{2}{3}$ $Rh + H_2O$	2.6×10^{17}	2.4×10^{14}	2.2×10^{12}	7.1×10^{10}
$\frac{1}{2}$ $WO_2 + H_2 = \frac{1}{2}$ $W + H_2O$	4.2×10^{-5}	0.00090	0.00652	0.0257
$\frac{1}{2}$ $ReO_2 + H_2 = \frac{1}{2}$ $Re + H_2O$	4.6×10^{5}	8.7×10^{4}	2.7×10^{4}	1.15×10^{4}
$\frac{1}{2}$ $IrO_2 + H_2 = \frac{1}{2}$ $Ir + H_2O$	2.2×10^{18}	1.3×10^{15}	8.5×10^{12}	2.2×10^{11}
$H_2O + CO = H_2 + CO_2$[c]	1542	137.1	28.31	9.42

[a] Change of structure of NiO in this interval.
[b] Zinc metal melts in this interval.
[c] $K = p_{H_2}p_{CO_2}/p_{H_2O}p_{CO}$.

Table 6
Reduction of Sulfides

Reaction	*Equilibrium constant* $K = p_{H_2S}/p_{H_2}$, *at*			
	400 K	*500 K*	*600 K*	*700 K*
$FeS + H_2 = Fe + H_2S$	5.2×10^{-9}[a]	4.3×10^{-7}[a]	7.0×10^{-6}	4.7×10^{-5}
$1.124\ CoS_{0.89} + H_2 = 1.124\ Co + H_2S$	2.4×10^{-9}	3.6×10^{-7}	9.6×10^{-6}	9.7×10^{-5}
$NiS + H_2 = Ni + H_2S$	1.2×10^{-7}	8.6×10^{-6}	1.4×10^{-4}	—
$Cu_2S + H_2 = 2\ Cu + H_2S$	1.9×10^{-7}	3.9×10^{-6}	2.5×10^{-5}[b]	8.6×10^{-5}
$ZnS + H_2 = Zn + H_2S$	2.4×10^{-21}	1.1×10^{-16}	1.4×10^{-13}[c]	2.15×10^{-11}
$\frac{1}{2}$ $MoS_2 + H_2 = \frac{1}{2}$ $Mo + H_2S$	5.1×10^{-13}	5.0×10^{-10}	4.8×10^{-8}	1.2×10^{-6}
$\frac{1}{2}$ $RuS_2 + H_2 = \frac{1}{2}$ $Ru + H_2S$	2.8×10^{-8}	3.5×10^{-6}	8.3×10^{-5}	7.6×10^{-4}
$\frac{1}{2}$ $WS_2 + H_2 = \frac{1}{2}$ $W + H_2S$	6.2×10^{-12}	3.8×10^{-9}	2.6×10^{-7}	5.1×10^{-6}
$\frac{1}{2}$ $ReS_2 + H_2 = \frac{1}{2}$ $Re + H_2S$	1.9×10^{-6}	1.0×10^{-4}	1.4×10^{-3}	8.5×10^{-3}
$\frac{1}{3}$ $Ir_2S_3 + H_2 = \frac{2}{3}$ $Ir + H_2S$	3.0×10^{-5}	1.0×10^{-3}	0.0102	0.0497

[a] Structure of FeS changes in this interval.
[b] Structure of Cu_2S changes in this interval.
[c] Zinc metal melts in this interval.

Table 7
Relative Stability of Oxides and Sulfides

Reaction	*Equilibrium constant* $K = p_{H_2S}/p_{H_2O}$, *at*			
	400 K	*500 K*	*600 K*	*700 K*
$NiS + H_2O = NiO + H_2S$	2.1×10^{-10}	2.0×10^{-8} [a]	4.2×10^{-7}	—
$Cu_2S + H_2O = Cu_2O + H_2S$	2.1×10^{-18}	3.4×10^{-15}	4.5×10^{-13} [b]	1.4×10^{-11}
$ZnS + H_2O = ZnO + H_2S$	2.25×10^{-10}	2.0×10^{-8}	3.9×10^{-7}	3.2×10^{-6}
$\frac{1}{2} MoS_2 + H_2O = \frac{1}{2} MoO_2 + H_2S$	1.1×10^{-8}	5.5×10^{-7}	7.4×10^{-6}	4.7×10^{-5}
$\frac{1}{2} WS_2 + H_2O = \frac{1}{2} WO_2 + H_2S$	1.5×10^{-7}	4.3×10^{-6}	4.0×10^{-5}	2.8×10^{-4}
$\frac{1}{2} ReS_2 + H_2O = \frac{1}{2} ReO_2 + H_2S$	4.1×10^{-12}	1.2×10^{-9}	5.0×10^{-8}	7.3×10^{-7}

[a] Structure of NiO changes in this interval.
[b] Structure of Cu_2S changes in this interval.

and sulfides where $K = p_{H_2O}/p_{H_2}$ and $K = p_{H_2S}/p_{H_2}$, respectively. Usually only data for the lower oxide or sulfide are given. Oxides of Co, Ni, Cu, Ru, Rh, and Ir have a large thermodynamic tendency to reduce. Oxides of Fe, Mo, and W can be reduced in H_2, provided that the partial pressure of water vapor is maintained low; ZnO is "nonreducible." Data for the water–gas shift are also given in Table 5, to permit calculating the K's for reduction by CO. Carbon monoxide is a stronger reducing agent than H_2 in this temperature range, but in some instances CO may also carburize the sample or form carbonyls.

Sulfides are inactive in the FTS and related reactions, and under most conditions can be produced from trace amounts of H_2S or other sulfur compounds in the synthesis gas, as shown in Tables 6 and 7. Iridium sulfide is the least stable, and the sulfur can be removed by H_2 at 600 K and above. The ratio of H_2/H_2S of 10^4–10^7 required for hydrogenating bulk sulfides of Fe, Co, Ni, Cu, and Ru could be achieved, but the process would be slow. As shown by the H_2S/H_2O ratios in Table 7, sulfides are more stable than oxides; sulfides cannot be converted to oxides by reactions with water vapor.

Table 8 presents equilibrium constants for the reduction of metal chlorides by H_2. These data are pertinent to the choice of chemicals for preparing catalysts, because chlorine is usually a severe poison for hydrogenation reactions. The chlorides of Ru, Rh, W, Re, and Ir are easily reduced to metal, and residual chlorine remaining from the preparation should be easily removed by treatment with H_2. At higher temperatures, $NiCl_2$ and CuCl may behave similarly, but Co and particularly Fe cannot

Table 8
Reduction of Chlorides

	Equilibrium constant $K = p_{HCl}^2/p_{H_2}$, *atm, at*			
Reaction	*400 K*	*500 K*	*600 K*	*700 K*
$FeCl_2 + H_2 = Fe + 2\ HCl$	1.6×10^{-13}	1.7×10^{-9}	7.8×10^{-7}	5.6×10^{-5}
$CoCl_2 + H_2 = Co + 2\ HCl$	6.2×10^{-9}	1.1×10^{-5}	0.00143	0.0433
$NiCl_2 + H_2 = Ni + 2\ HCl$	2.2×10^{-7}	2.7×10^{-4}	0.02851	0.762
$2\ CuCl + H_2 = 2\ Cu + 2\ HCl$	1.2×10^{-5}	0.00232	0.0709	0.7328
$\frac{2}{3}\ RuCl_3 + H_2 = \frac{2}{3}\ Ru + 2\ HCl$	1.3×10^{11}	4.1×10^{10}	1.6×10^{10}	7.8×10^{9}
$\frac{2}{3}\ RhCl_3 + H_2 = \frac{2}{3}\ Rh + 2\ HCl$	3.0×10^{9}	2.3×10^{9}	1.8×10^{9}	1.3×10^{9}
$WCl_2 + H_2 = W + 2\ HCl$	0.0109	0.6839	9.73	59.43
$\frac{2}{3}\ ReCl_3 + H_2 = \frac{2}{3}\ Re + 2\ HCl$	4.4×10^{10}	2.3×10^{10}	1.3×10^{10}	7.9×10^{9}
$\frac{2}{3}\ IrCl_3 + H_2 = \frac{2}{3}\ Ir + 2\ HCl$	7.4×10^{11}	2.7×10^{11}	1.3×10^{11}	9.7×10^{10}

be readily freed from chlorine by H_2 treatment. The data for bulk compounds, oxides, sulfides, and chlorides may be expected to be only a rough guide to processes occurring at a catalyst surface. Species chemisorbed on surfaces may be more tightly bound to the surface than the components of similar bulk compounds. Chlorides and sulfides are usually more stable than oxides, and they cannot be destroyed by treatment with water vapor.

2.3 Thermodynamics of Carbon, Carbides, Nitrides, and Carbonyls

The thermodynamics of the formation of elemental carbon and carbides is presented in Table 9. Iron, cobalt, and nickel form interstitial carbides, which are unstable with respect to the metal and carbon (as shown by a comparison of the values of the equilibrium constants for producing elemental carbon and the carbide). Copper, zinc, and ruthenium do not form carbides, whereas molybdenum and tungsten produce very stable carbides having an ionic structure.

The production of higher hydrocarbons by the hydrogenation of graphite is impossible except for lower paraffins at temperatures less than 150°C. Although the carbides of iron, cobalt, and nickel are less stable thermodynamically than the metal plus graphite, it is possible at 227°C to

Table 9
Formation of Elemental Carbon and Carbides

Reaction	*Equilibrium constant $K = p_{CO_2}/p^2_{CO}$, atm^{-1}, at*			
	400 K	*500 K*	*600 K*	*700 K*
$2\ CO = C + CO_2$ [a]	2.1×10^{13}	6.1×10^{8}	5.9×10^{5}	4102.
$3\ Fe + 2\ CO = Fe_3C + CO_2$	1.3×10^{11}	1.9×10^{7}	5.5×10^{4}	859.
	1.3×10^{11}	1.9×10^{7}	5.5×10^{4}	859. [b]
$2\ Fe + 2\ CO = Fe_2C(\text{Hägg}) + CO_2$		1.2×10^{7}	3.0×10^{4} [b]	
$2\ Co + 2\ CO = Co_2C + CO_2$		5.5×10^{7} [c]		
$3\ Ni + 2\ CO = Ni_3C + CO_2$	1.0×10^{9}	3.2×10^{5}	1550.	33.81
		4.4×10^{5} [c]		
$2\ Mo + 2\ CO = Mo_2C + CO_2$	3.01×10^{19}	5.7×10^{13}	8.8×10^{9}	1.7×10^{7}
$2\ W + 2\ CO = W_2C + CO_2$	1.9×10^{17}	1.4×10^{12}	5.7×10^{8}	2.4×10^{6}

[a] Carbon as graphite.
[b] Data from Browning *et al.* (*6*).
[c] Data from Browning and Emmett (*7*).

produce from these carbides only paraffins up to C_4 and no olefins (*1*). However, it is energetically possible to incorporate a limited fraction of carbon atoms in the organic molecule from elemental carbon or iron-group carbides into synthesis reactions of $H_2 + CO$. Reactions of the iron carbides with water to produce magnetite plus graphite and magnetite plus paraffins are possible thermodynamically.

Table 10 presents data for the preparation of nitrides and for the synthesis of ammonia. Iron-group nitrides are relatively unstable and cannot be prepared by reaction of molecular nitrogen with the metal; NH_3 must be used. Thermodynamic data are not available for carbonitrides.

Data on metal carbonyls, given in Table 11, are important because they define the temperature–pressure conditions under which catalysts can be

Table 10
Preparation of Nitrides

Reaction	*Equilibrium constant $K = p^{1.5}_{H_2}/p_{NH_3}$, atm$^{1/2}$, at*			
	400 K	*500 K*	*600 K*	*700 K*
$NH_3 + 4\ Fe = Fe_4N + \frac{3}{2}\ H_2$	0.01266	0.1263	0.6166	1.957
$NH_3 + 2\ Fe = Fe_2N + \frac{3}{2}\ H_2$	0.00142	0.0222	0.1483	0.6033
$NH_3 + 3\ Co = Co_3N + \frac{3}{2}\ H_2$	3.8×10^{-7}	1.2×10^{-5}	1.35×10^{-4}	—
$NH_3 + 2\ Mo = Mo_2N + \frac{3}{2}\ H_2$	4.6×10^{4}	5.4×10^{4}	6.5×10^{4}	7.8×10^{3}
$\frac{1}{2}\ N_2 + \frac{3}{2}\ H_2 = NH_3$ [a]	6.116	0.3225	0.0424	9.5×10^{-3}

[a] For this reaction $K = p_{NH_3}/p^{1/2}_{N_2}p^{3/2}_{H_2}$.

Table 11
Formation of Carbonyls

Reaction	Equilibrium constant K, at				
	300 K	*400 K*	*500 K*	*600 K*	*700 K*
$Fe + 5\,CO = Fe(CO)_5$ [b]	25.47[a]	3.85×10^{-8}	1.41×10^{-12}	1.71×10^{-15}	1.51×10^{-17}
$Ni + 4\,CO = Ni(CO)_4$ (gas)[c]	1.36×10^{6}	0.1396	9.48×10^{-6}	1.69×10^{-8}	1.96×10^{-10}
$Mo + 6\,CO = Mo(CO)_6$ (solid)[d]	4.14×10^{9}				
$W + 6\,CO = W(CO)_6$ (solid)[d]	1227				

[a] At 300 K, $Fe(CO)_5$ is a liquid with a vapor pressure of 0.0424 atm. $K = 1/p_{CO}^5$, atm^{-5}.
[b] $K = p_{Fe(CO)_5}/p_{CO}^5$, atm^{-4}.
[c] $K = p_{Ni(CO)_4}/p_{CO}^4$, atm^{-3}.
[d] $K = 1/p_{CO}^6$, atm^{-6}.

used without severe metal losses. In addition, metal carbonyls are toxic, $Ni(CO)_4$ being extraordinarily poisonous. At room temperature, $Ni(CO)_4$ is a gas, $Fe(CO)_5$ a liquid, and $Mo(CO)_6$ and $W(CO)_6$ are solids. The thermodynamic tendency to produce carbonyls increases sharply with increasing partial pressures of CO, but decreases sharply with increasing temperature. The operating pressure and temperature and the catalytic activity must be balanced to yield a suitable operating region. As an example, we may ask the maximum pressure at which the outlet mole fraction of $Ni(CO)_4$, $y_{Ni(CO)_4}$, does not exceed 10^{-8} when the mole fraction of CO, y_{CO}, is 0.1 at 500 K. The equilibrium constant K is given as

$$K = p_{Ni(CO)_4}/p_{CO}^4 = y_{Ni(CO)_4}/y_{CO}^4 P^3,$$

where P is the total pressure in atmospheres and $K = 9.48 \times 10^{-6}$. Solving for P^3,

$$P^3 = 10^{-8}/(0.1^4 \times 9.48 \times 10^{-6}) = 10.55,$$

and $P = 2.2$ atm. At 600 K, $P = 18.1$ atm. For iron at 500 and 600 K and other conditions the same, the maximum pressures are 163 and 874 atm, respectively.

2.4 Summary

The thermodynamic data presented in this chapter may be summarized as follows:

1. It is energetically possible to produce a vast array of hydrocarbons and organic molecules by the hydrogenation of CO and related processes such as the 3 H_2 + CO_2 and the H_2O + 3 CO reactions. Most of these reactions are highly exothermic.
2. A variety of reactions of synthesis products are thermodynamically possible, such as dehydration of alcohols, hydrogenation of olefins, hydrogenolysis of paraffins, and a number of isomerization reactions.
3. On an energetic basis, olefins and alcohols can be "incorporated" into synthesis reactions to produce higher molecular weight hydrocarbons in any amount. Paraffins, element carbon, or carbidic carbon can be "incorporated" into synthesis reactions to only a limited extent, that is, part of the carbon of the hydrocarbon must come from H_2 + CO.

For the active FTS metals, for Ni, Co, Fe, and Ru, the following statements can be made:

4. Oxides of Ni, Co, and Ru are easily reducible, and it seems unlikely that bulk oxides can be produced under FTS conditions. Iron can be present as metal or magnetite depending on the ratios H_2O/H_2 and CO_2/CO in the reactor.
5. Sulfides of Fe, Ni, Co, and Ru can be produced with trace amounts of H_2S in hydrogen or synthesis gas, and conversely these sulfides cannot be removed by treatment with hydrogen.
6. The formation of elemental carbon and carbides of Fe-group metals is thermodynamically possible under usual synthesis conditions. Carbides of Ni, Co, and Fe are energetically unstable with respect to the metal plus carbon. Elemental or carbidic carbon cannot be hydrogenated to higher hydrocarbons at usual synthesis conditions; however, elemental and carbidic carbon can be incorporated to a limited degree into synthesis reactions of H_2 + CO.
7. Operating conditions for synthesis reactions must be chosen so that loss of the active metal by forming volatile carbonyls is not rapid. Carbonyl formation is a severe problem with Ni catalysts. Thermodynamic data are available for the carbonyls of Ni and Fe, but not for carbonyls of Co and Ru.

References

1. Anderson, R. B., *in* "Catalysis" (P. H. Emmett, ed.), Vol. 4, Chapter 1. Van Nostrand-Reinhold, Princeton, New Jersey, 1956.

2. Anderson, R. B., *in* "Catalysis" (P. H. Emmett, ed.), Vol. 4, pp. 335–367. Van Nostrand-Reinhold, Princeton, New Jersey, 1956.
3. Anderson, R. B., Lee, C. B., and Machiels, J. C., *Can. J. Chem. Eng.* **54,** 590 (1976).
4. Barin, I., and Knacke, O., "Thermochemical Properties of Inorganic Substances." Springer-Verlag, Berlin and New York, 1973.
5. Barin, I., Knacke, O., and Kubaschewski, O., "Thermochemical Properties of Inorganic Substances, Supplement." Springer-Verlag, Berlin and New York, 1977.
6. Browning, L. C., DeWitt, T. W., and Emmett, P. H., *J. Am. Chem. Soc.* **72,** 4211 (1950).
7. Browning, L. C., and Emmett, P. H., *J. Am. Chem. Soc.* **74,** 1680 (1952).
8. Pichler, H., and Schulz, H., *Chem.-Ing.-Tech.* **12,** 1160 (1970).
9. Pitzer, K. S., and Brewer, L., revision of Lewis, G. N., and Randall, M., "Thermodynamics," Appendix 7. McGraw-Hill, New York, 1961.
10. Storch, H. H., Golumbic, N., and Anderson, R. B., "The Fischer–Tropsch and Related Syntheses," Chapter 1. Wiley, New York, 1951.
11. Stull, D. R., Westrum, E. F. Jr., and Sinke, G. C., "The Chemical Thermodynamics of Organic Compounds." Wiley, New York, 1969.
12. Tillmetz, K. D., *Chem.-Ing.-Tech.* **48,** 1065 (1976).
13. Weitkamp, A. W., and Frye, C. G., *Ind. Eng. Chem.* **45,** 365 (1953).
14. White, W. B., Johnson, S. M., and Dantzig, G. B., *J. Chem. Phys.* **28,** 751 (1958). This method is described in Balzhiser, R. E., Samuels, M. R., and Eliassen, J. D., "Chemical Engineering Thermodynamics." Prentice-Hall, Englewood Cliffs, New Jersey, 1972, p. 514.

CHAPTER 3

Carbides, Nitrides, and Carbonitrides of Iron as Catalysts in the Fischer–Tropsch Synthesis

3.1 Introduction

Iron-group metals can be converted to interstitial compounds, carbides, nitrides and borides, and mixed compounds of two or more of the interstitial elements, such as carbonitrides. Borides can be produced in high-surface-area forms suitable for catalyst tests, but few data are available on synthesis tests of more or less authentic samples. Nitrides of cobalt and nickel apparently cannot be produced by treating catalytically active metals with NH_3. The carbides of cobalt (Co_2C) and nickel (Ni_3C) have been prepared and tested in synthesis reactions many years ago. The carbides had very low activity compared with the reduced metal at usual synthesis temperatures, 180–200°C. Treatment of the carbides at ~200°C with H_2 rather than H_2 + CO removed the carbidic carbon, and the activity of the metal was regained. Thus, carbides of cobalt and nickel are inactive in the Fischer–Tropsch synthesis (FTS), and these carbides apparently are not produced during synthesis. These early tests have been summarized in references *4, 64,* and *65*.

All of the carbides, nitrides, and carbonitrides of iron have been found to be active catalysts in the FTS, and the present chapter summarizes United States Bureau of Mines studies of the preparation and testing of these phases from 1948 to 1963; the present chapter is adapted from U.S. Bureau of Mines Bulletin 612 (*60*).

German work from 1940 to 1950 showed that pretreatments of iron catalysts involving carbonization had desirable effects, and Pichler and

Table 1
Carbides and Nitrides of Iron[a]

Name	*Designation in this chapter*	*Atom ratios*		*Arrangement of iron atoms*
		C:Fe	*N:Fe*	
Carbides				
Cementite	C	0.33		Orthorhombic
Hägg carbide	χ	~0.45		Orthorhombic
Hexagonal carbide	ϵ	~0.50		Hexagonal close-packed
Eckstrom and Adcock carbide		~1.0(?)		—
Nitrides				
	γ'		0.241–0.259	Face-centered cubic
	ϵ		0.314–0.498	Hexagonal close-packed
	ζ		0.498–0.508	Orthorhombic

[a] Phases present in samples reported in Chapter 3 were identified from powder X-ray diffraction (XRD) patterns. Since these experiments were completed, Mössbauer spectroscopy (MS) has emerged as an ideal research tool for the analysis of iron catalyst, particularly because MS is relatively little affected by the presence of small crystallites compared with XRD and thermomagnetic analysis. MS has identified three carbide phases with more-or-less hcp arrangement of iron atoms, ϵ, ϵ', and Fe_xC, and three variations of χ carbides. XRD is inadequate for identifying in catalysts these variants of the ϵ and χ phases. The reader is referred to reference *47a* for a critical review of iron carbides. In the more recent literature, the designation for cementite is θ. The Eckstrom and Adcock carbide (*31*) was not found in the work described in Chapter 3; its formula is now given as Fe_7C_3.

Merkel (*53*) made chemical and thermomagnetic studies of carbides in iron catalysts. Hofer reviewed his own and other work on iron carbides, as well as other interstitial compounds of iron-group metals (*39, 40*). Nitrides, carbides, and carbonitrides of iron were described by Jack (*45*). Carbides and nitrides of iron are described in Table 1. Reactions in the Fe–C–N system at 250–400°C may be summarized as follows:

$$\gamma'\text{-}Fe_4N \longrightarrow \quad \epsilon\text{- or }\zeta\text{-}Fe_2N \longrightarrow \quad \epsilon\text{- or }\zeta\text{-}Fe_2N_xC_{1-x} \quad \longleftarrow Fe_3C,\ \epsilon\text{-}Fe_2C,\ \chi\text{-}Fe_2C$$

$$CO \text{ or } H_2 + CO \longrightarrow \qquad\qquad \longleftarrow NH_3$$

Reactions with CO or synthesis gas proceed to the right; those with NH_3 to the left. Reactions of ϵ- or ζ-Fe_2N to ϵ- or ζ-carbonitride to ϵ-Fe_2C are often reversible; other reactions are not reversible.

Table 2
Catalyst Composition

Catalyst type and number	*K_2O*	*Cu*	*MgO*	*Al_2O_3*	*SiO_2*	*Cr_2O_3*	*ZrO_2*
Fused							
D3001	0.85		6.8		0.9	1.21	
D3008	1.39			2.83	0.28		
L3028	0.51			0.50	0.42		3.9
Precipitated							
P3003.24	0.5	10.					
Sintered							
A2106.11	0.94		0.20	0.60	0.7		

3.2 Preparation and Composition of Catalysts

Compositions of catalysts described in this chapter are given in Table 2. Procedures for preparing these catalysts are given in this section.

3.2.1 Fused Iron Oxide

Catalysts D3001, D3008, and L3028 were prepared by electrical fusion of iron oxide plus promoters. D3001 and D3008 (*20*) were commercial preparations, and L3028 was prepared by the procedure described in reference *60*.

3.2.2 Precipitated Iron Oxide

Catalyst P3003.24 was precipitated from a hot aqueous solution of ferric nitrate plus copper nitrate by adding, with stirring, a hot aqueous solution of sodium carbonate.

3.2.3 Sintered Iron Oxide

Agglomerates of sintered, fine particles of magnetite ore were crushed and sieved to 6 to 8 mesh, impregnated with an aqueous solution of potassium nitrate, and dried and heated overnight at 500°C.

3.3 Preparation of Carbides, Nitrides, and Carbonitrides of Iron in Catalytic Form

Experimental procedures for converting the iron in catalysts to relatively pure carbides, nitrides, or carbonitrides have been adapted from procedures evolved in small-scale glass equipment for use in metallic systems of adequate capacity for catalyst testing. All preparative steps were done at atmospheric pressure, unless specific pressures are mentioned.

3.3.1 Reduction in Hydrogen

The catalysts were reduced in hydrogen in a special metal-block reactor (*16, 59*) that was rotated about a transverse axis to permit easy introduction and discharge of the catalyst. Precautions were taken to avoid oxidation of the catalyst during cooling and subsequent handling. Finally, the catalyst was transferred through rubber connections in a stream of CO_2 to a special weighing bottle (*16*).

3.3.2 Nitriding in Ammonia

The metal-block reactor was used, and the same precautions were observed as in reductions. In addition, when the catalyst was removed from the reactor, the flow of ammonia was continued at the rate used during nitriding until the catalyst temperature was less than 50°C. Before transferring the catalyst to the weighing bottle, the system was flushed first with prepurified N_2 and then with CO_2 (*16*).

3.3.3 Preparation of Hägg Carbide

Conversion of the reduced catalyst to Hägg carbide was accomplished in the same reactor as described in Section 3.3.1. Two procedures were used for reduced fused catalysts:

1. With pure CO at an hourly space velocity of 100, the carbiding temperature was increased progressively from 150 to 350°C as required to maintain the CO_2 content of the exit gas at ~20% (*38*). Approximately 20 h were required to deposit carbon corresponding to the formula Fe_2C, Hägg carbide being the predominant phase detected by X-ray diffraction or thermomagnetic analysis. This procedure is particularly advantageous

for preparing Hägg carbide. Because the rate of carbiding is maintained low and approximately constant, the possibility of the following two side reactions occurring is minimized: (a) deposition of elemental carbon because of overheating and (b) oxidation of elemental iron or carbide by excessively large concentrations of CO_2, resulting from high conversion of CO. Carbiding freshly reduced catalysts at 325–350°C would produce cementite, elemental carbon, and magnetite in addition to Hägg carbide. In the carbiding schedule the catalyst was largely converted to Hägg carbide by the time that these temperatures were reached, and apparently only Hägg carbide was formed under these conditions.

2. In carbiding the reduced catalyst with 1 H_2 + 4 CO gas, the temperature was increased by steps, 12 h at 200°C, 12 h at 250°C, and 12 h at 275°C. This procedure similarly minimized possibilities of overheating and oxidation owing to excessively high rates of carbiding, and usually satisfactory samples of Hägg carbide, without excessive amounts of elemental carbon or magnetite could be prepared this way.

Precipitated catalysts were converted to Hägg carbide by the following methods. The catalysts were pretreated, without prior reduction, using 2 H_2 + 1 CO gas at an hourly space velocity of 2500, 310°C, and atmospheric pressure for 6 h in the synthesis reactors. The catalysts were not removed from the reactor before synthesis, but analyses after synthesis indicated that this activation procedure had converted a substantial portion of the catalyst to Hägg carbide and that an appreciable quantity of elemental carbon was deposited.

Other pretreatments of precipitated catalysts were made in the reactor mentioned in Section 3.3.1, using CO according to procedure 1. A reduced catalyst was largely converted to Hägg carbide in 13 h, the temperature being increased from 185 to 300°C as required to maintain the CO_2 content of the exit gas at 20%. In another test the raw catalyst was treated with CO. Although the temperature required to produce 20% CO_2 in the exit gas did not exceed 250°C, an appreciable amount of carbides was not found in the catalyst.

3.3.4 Preparation of Cementite

Three methods, all at atmospheric pressure, were employed to prepare cementite:

1. Conversion of Hägg iron carbide to cementite. This method is based on the observation that Hägg carbide reacts with metallic iron to form cementite (*41, 55*). This reaction proceeds more rapidly than the decomposition of Hägg carbide into cementite and free carbon. Reduced catalysts were carburized with CO in a metal-block reactor to convert two-

thirds of the iron to Hägg carbide. Helium was then passed over the catalyst; the temperature was increased to 475°C and it was maintained for 2 h. The catalyst was dropped into the receiving vessel and was cooled in He.

2. Direct preparation with 2 H_2 + 1 CO gas. Catalysts largely converted to cementite were prepared by carburizing the reduced sample with 2 H_2 + 1 CO gas at a space velocity of 2500 h^{-1} and 310°C for 6 h in the reactor. The carburized sample was then dropped into the receiving vessel and cooled in 2 H_2 + 1 CO. This preparation may be regarded as an example of formation of cementite during synthesis. Small yields of higher hydrocarbons are produced; however, the activity is low and the selectivity poor under these conditions.

3. Direct preparation with methane. Pure methane was passed over the reduced catalyst at an hourly space velocity of 1000 and 500°C for 4 h in the reactor mentioned in Section 3.3.1. The catalyst was dropped into the receiver in methane and cooled to room temperature.

3.3.5 Preparation of ε-Iron Carbonitrides

Carbonitrides were produced from fused catalysts in four ways:

1. During synthesis, nitrided catalysts were treated with 1 H_2 + 1 CO gas at 7.8–21.4 atm.
2. The nitrided catalyst was treated with a stream of pure CO at atmospheric pressure, 350–450°C, and an hourly space velocity of 100. The higher temperatures and longer times were required to produce a carbonitride rich in carbon and low in nitrogen.
3. Hägg carbide was produced according to procedure 1 in Section 3.3.3. The carbided catalyst was then treated with ammonia at an hourly space velocity of 1000 and 350°C for 28 h.
4. The reduced catalyst was treated with monomethylamine at an hourly space velocity of 1000 and 350°C for 28 h.

The preparation and reactions of iron carbonitrides are described in reference *36*.

3.4 Synthesis Tests with Carbides of Iron

3.4.1 Fused–Iron Oxide Catalysts Converted to Hägg Carbide

To determine the effect of method of pretreatment, synthesis pressure, and conversion, synthesis tests were made with fused–iron oxide cata-

Table 3
Converting Catalysts to Hägg Carbide[a]

	Reduction in H_2[b,c]			*Carburization*[b,d]				*Phases from X-ray diffraction*[e]
Test, X-	*Temp., °C*	*Time, h*	*Extent of reduction, %*	*Gas*	*Temp. range, °C*	*Time, h*	*Atom ratio C:Fe*	
152	450	43	90.	None	—	—	0	α
289	400–525[f]	72	97.5	CO	227–305[g]	18	0.56	χ,M,α
294	500	24	97	CO	150–350[g]	18	0.58	χ,α
323	500	41	97	1 H_2 + 4 CO	200–275[h]	48	0.43	χ,α
325	500	43	97	1 H_2 + 4 CO	200–275[h]	48	0.42	χ,M
329	500	40	98	1 H_2 + 4 CO	200–275[h]	48	0.42	χ
342	500	24	97	CO	150–350[g]	18	0.43	χ,α
394	500	24	99.5	CO	150–350[g]	29	0.53	χ
399	500	40	98	CO	150–350[g]	18	0.46	χ,α
408	500	24	99	CO	150–350[g]	20	0.50	χ
428	500	26	98	CO	160–320[g]	19	0.52	χ
515	500	24	98	None	—	—	—	α
380[j]	400–525[f]	48	100	None	—	—	—	α
416	400–525[f]	56	100	CO	150–350[g]	13	0.42	χ,α

[a] Unless noted, the catalyst is D3001.
[b] All pretreatments at atmospheric pressure.
[c] Space velocity in excess of 1000 h^{-1}.
[d] Space velocity ~100 h^{-1}.
[e] Phases in decreasing order of intensity of patterns: χ, Hägg carbide; α, metallic iron; M, magnetite.
[f] 24 h at 400°C, 24 h at 450°C, 12 h at 500°C, and 12 h at 525°C.
[g] Temperature increased to maintain CO_2 in exit gas at ~20%.
[h] 12 h at 200°C, 12 h at 250°C, and 24 h at 275°C.
[j] Catalyst D3008 used in this test.

lysts converted to Hägg carbide, using 1 H_2 + 1 CO gas at pressures of 7.8, 11.1, 14.5, and 21.4 atm (*58*). Conditions for reduction and carburization and the principal phases in pretreated catalysts are given in Table 3. Procedures for these pretreatments have been given. Catalysts used in these experiments were in the form of 6- to 8-mesh particles.

Data for the activity of reduced and reduced and carbided fused-iron catalyst as a function of time are presented in Figs. 1–4. Figures 1 and 2 compare activities of catalysts at operating pressures of 7.8 and 21.4 atm (absolute), respectively. Figure 3 shows the activities as a function of operating pressure, and Fig. 4 compares activities of catalysts under different conditions of operation. In tests at 7.8 atm, the CO_2-free contraction was maintained constant at 65 and 80%, and in tests at 21.4 atm the contraction was held at 22 and 63%. The CO_2-free contraction is defined

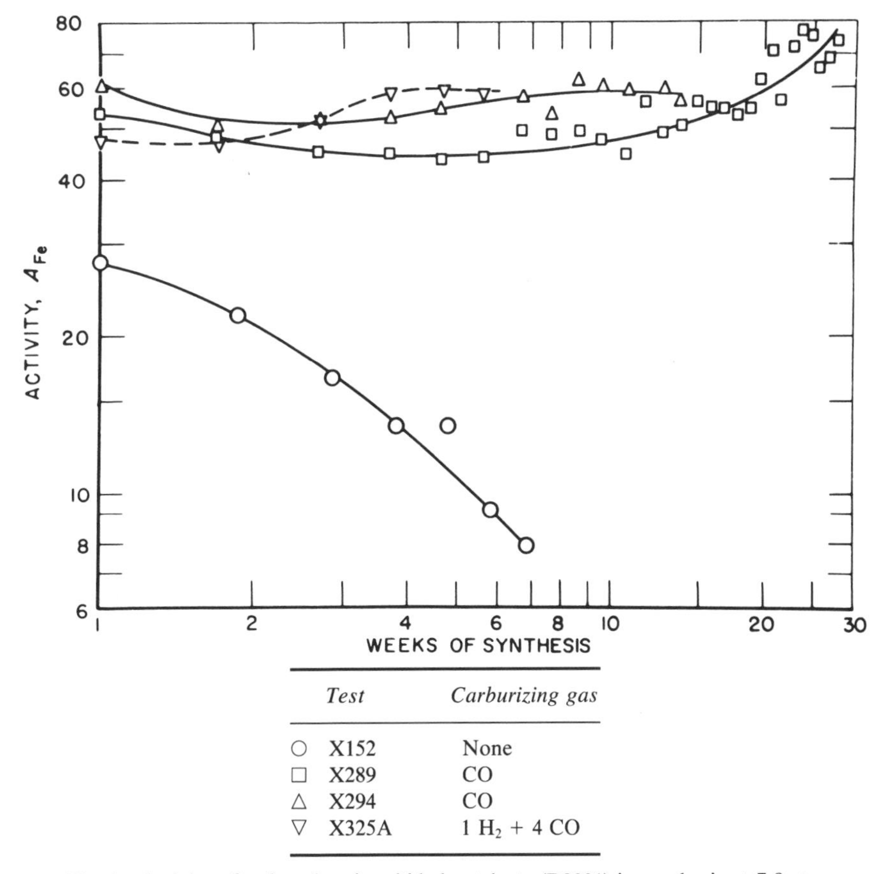

	Test	Carburizing gas
○	X152	None
□	X289	CO
△	X294	CO
▽	X325A	$1\ H_2 + 4\ CO$

Fig. 1 Activity of reduced and carbided catalysts (D3001) in synthesis at 7.8 atm.

as $1 - E/I$, where I is the volume (S.T.P.) of feed gas per hour and E the volume (S.T.P.) of exit gas per hour after removal of CO_2. The CO_2-free contraction is an easily obtained measure of conversion of $H_2 + CO$ and is usually only a few percent lower than the value for conversion. In these figures the activity was corrected to a standard synthesis condition (240°C and 65% conversion) by use of an empirical rate equation (*13*). Activities were computed from average weekly synthesis conditions. The activity designated A_{Fe} is defined as cubic centimeters (S.T.P.) of $H_2 + CO$ reacted per hour per gram of iron at 240°C when the conversion of $H_2 + CO$ is 65%. The logarithmic time scale was employed to include data of experiments of long duration without excessively compressing initial periods of synthesis.

Figure 5 compares product-distribution data from tests of both reduced

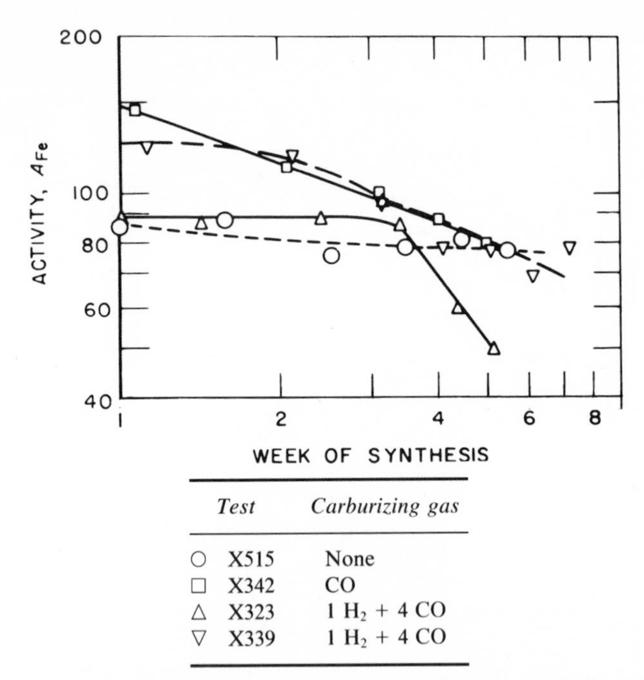

	Test	Carburizing gas
○	X515	None
□	X342	CO
△	X323	1 H_2 + 4 CO
▽	X339	1 H_2 + 4 CO

Fig. 2 Activity of reduced and carbided catalysts (D3001) in synthesis at 21.4 atm.

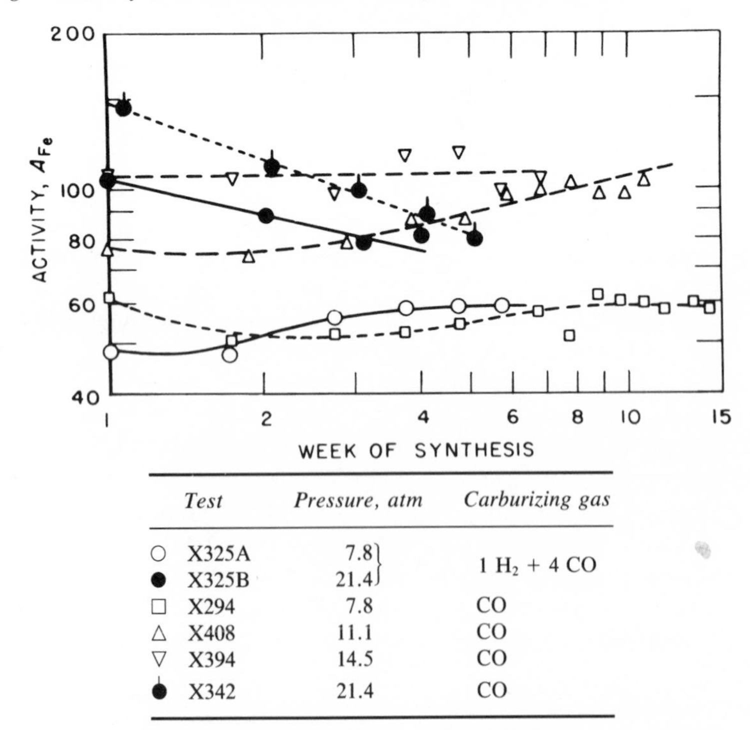

	Test	Pressure, atm	Carburizing gas
○	X325A	7.8	1 H_2 + 4 CO
●	X325B	21.4	1 H_2 + 4 CO
□	X294	7.8	CO
△	X408	11.1	CO
▽	X394	14.5	CO
● (with tick)	X342	21.4	CO

Fig. 3 Activity-time plots for carbided catalyst D3001 at several pressures.

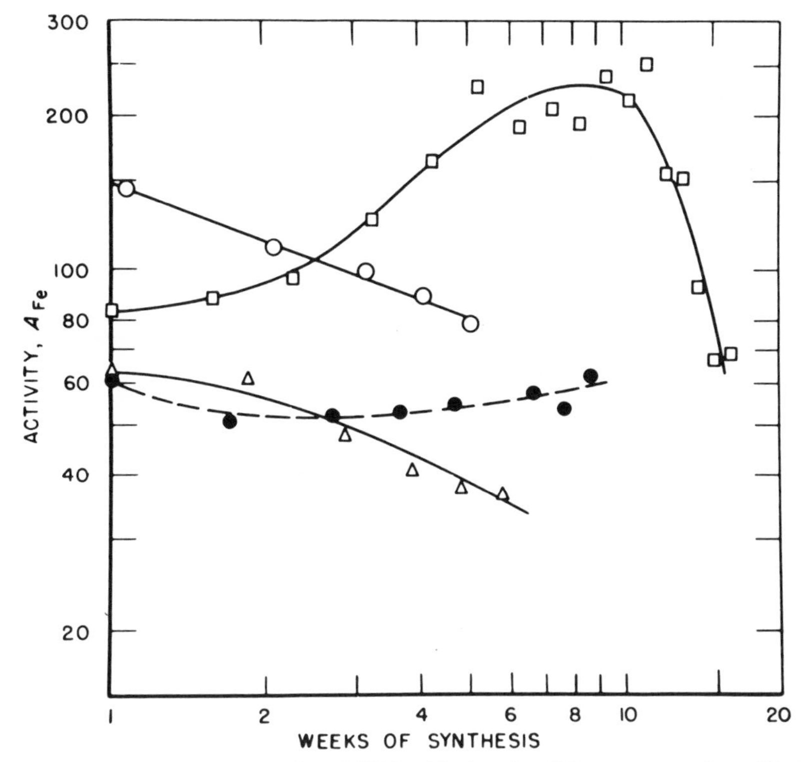

Fig. 4 Activity changes (catalyst D3001) with time for different contractions (%): □, X128, 22%; ○, X342, 63% (both at 21.4 atm); ●, X294, 65%; △, X399, 80% (both at 7.8 atm).

and reduced and carbided catalysts at 7.8 and 21.4 atm corresponding to activity data in Figs. 1 and 2. Figure 6 demonstrates the effect of operating pressure on selectivity of carbided catalysts (activities in Fig. 3).

The selectivity plots present distribution of total hydrocarbons in weight percentage, including oxygenated molecules dissolved in the condensed hydrocarbon phases, water-soluble oxygenated molecules being excluded. Gaseous hydrocarbons and distillation fractions of condensed hydrocarbons are shown. Numbers designated with (=) in gaseous hydrocarbon blocks indicate percentages of olefins. For the distillation fractions, Br is bromine number, OH the weight percent hydroxyl group, and CO the weight percent carbonyl group as aldehyde, ketones, and acids. The relative usage of H_2 and CO is given in Table 4. Selectivity data are averages of the second through fifth weeks of synthesis.

Catalysts containing principally Hägg carbide have greater initial activities than do reduced catalysts. At 7.8 atm the activity of the carbides remained essentially constant or increased slowly, whereas the activity of

	7.8 atm				21.4 atm		
Test number	X152	X289	X294	X325A	X339	X342	X515
Carbiding gas	None	CO	CO	1H$_2$+4CO	1H$_2$+4CO	CO	None
Phases present	α	χ,M,α	χ,α	χ	χ	χ,α	α
Temperature, °C	270	229	229	230	250	243	248
Space velocity, h^{-1}	97	103	106	101	292	334	303
Contraction, %	65.2	64.3	63.7	65.3	64.4	63.2	64.6
Activity, A_{Fe}	14	46	53	55	78	104	80

DISTRIBUTION OF TOTAL HYDROCARBONS, %
0 10 20 30 40 50 60 70 80 90 100
X152: C1; C2 = 22; C3+C4 = 64; <185° C Br 66 OH 0.2 CO 0.7; 185-352° C Br 33 OH 0.0 CO 0.4; 352-464° C; >464° C
X289: C1; C2 = 47; C3+C4 = 87; <185° C Br 77 OH 1.2 CO 1.1; 185-352° C Br 31 OH 0.2 CO 0.4; 352-464° C; >464° C
X294: C1; C2 = 33; C3+C4 = 76; <185° C Br 81 OH 2.1 CO 1.0; 185-352° C Br 5 OH 0.4 CO 0.7; 352-464° C; >464° C
X325A: C1; C2 = 27; C3+C4 = 75; <185° C Br 63 OH 1.6 CO 1.3; 185-352° C Br 27 OH 0.3 CO 0.5; 352-464° C; >464° C
X339: C1; C2 = 62; C3+C4 = 81; <185° C Br 86 OH 2.3 CO 1.7; 185-352° C Br 43 OH 0.9 CO 2.7; 352-464° C; >464° C
X342: C1; C2 = 52; C3+C4 = 80; <185° C Br 62 OH 2.4 CO 1.9; 185-352° C Br 35 OH 0.8 CO 2.0; 352-464° C; >464° C
X515: C1; C2 = 39; C3+C4 = 79; <185° C Br 96 OH 2.7 CO 2.4; 185-352° C Br 51 OH 1.1 CO 1.7; 352-464° C; >464° C

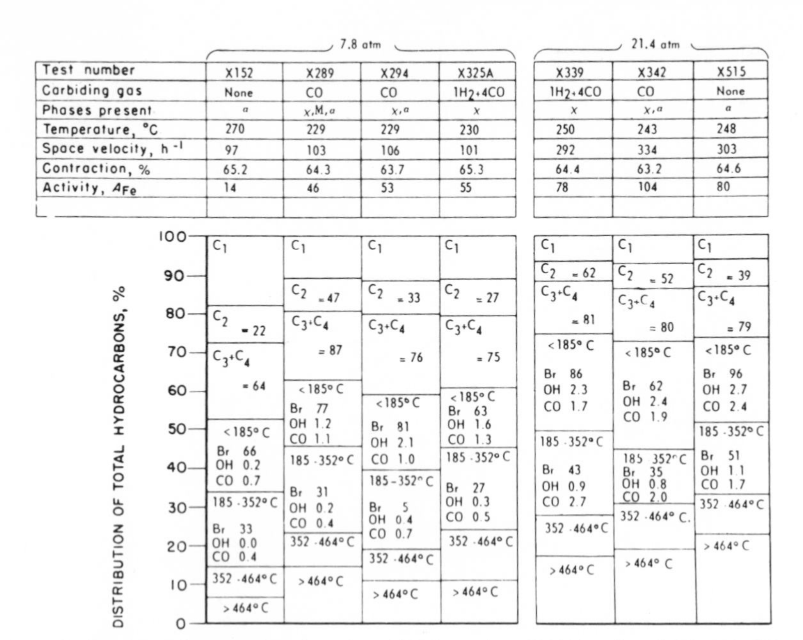

Fig. 5 Products from reduced and carbided catalyst D3001 using 1 H$_2$ + 1 CO gas. Symbols denoting phases are defined in Table 3. Total hydrocarbons include oxygenates dissolved in liquid phase.

Test number	X294	X408	X394	X342
Phases present	χ,α	χ	χ	χ,α
Pressure, atm	7.8	11.1	14.5	21.4
Temperature, °C	229	239	235	243
Space velocity, h^{-1}	106	245	247	334
Contraction, %	63.7	63.6	64.3	63.2
Activity, A_{Fe}	53	87	106	104

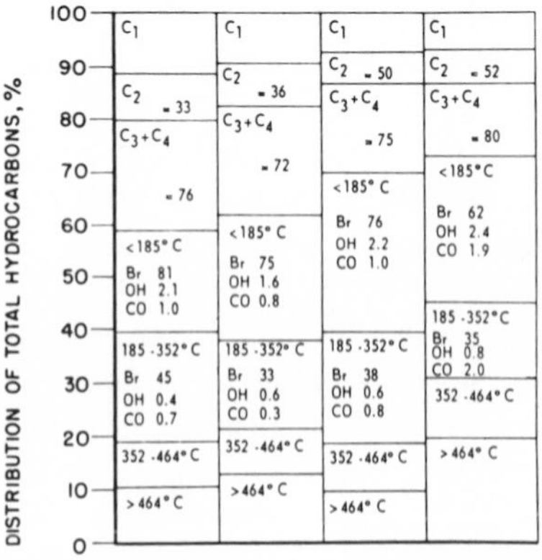

DISTRIBUTION OF TOTAL HYDROCARBONS, %
0 10 20 30 40 50 60 70 80 90 100
X294: C1; C2 = 33; C3+C4 = 76; <185° C Br 81 OH 2.1 CO 1.0; 185-352° C Br 45 OH 0.4 CO 0.7; 352-464° C; >464° C
X408: C1; C2 = 36; C3+C4 = 72; <185° C Br 75 OH 1.6 CO 0.8; 185-352° C Br 33 OH 0.6 CO 0.3; 352-464° C; >464° C
X394: C1; C2 = 50; C3+C4 = 75; <185° C Br 76 OH 2.2 CO 1.0; 185-352° C Br 38 OH 0.6 CO 0.8; 352-464° C; >464° C
X342: C1; C2 = 52; C3+C4 = 80; <185° C Br 62 OH 2.4 CO 1.9; 185-352° C Br 35 OH 0.8 CO 2.0; 352-464° C; >464° C

Fig. 6 Products from carbided catalyst D3001 at several pressures. Symbols denoting phases are defined in Table 3. Total hydrocarbons include oxygenates dissolved in liquid phase.

Table 4
Usage Ratios for Tests of Fused Catalysts with 1 H_2 + 1 CO Feed

Test, X-[a]	*Phased in pretreated catalyst*[b]	*Pressure, atm*	*Average*	
			Contraction, %	*Usage ratio $H_2:CO$*
152	α	7.8	65	0.72
289	χ,M,α	7.8	64	0.81
294	χ,α	7.8	64	0.75
399	χ,α	7.8	80	0.75
325A	χ,M	7.8	65	0.87
325B	—	21.4	64	0.88
408	χ	11.1	63	0.83
394	χ	14.5	64	0.84
342	χ,α	21.4	63	0.79
323	χ,α	21.4	64	0.85
339	χ	21.4	64	0.83
428	χ	21.4	22	1.46
515	α	21.4	65	0.76
380	α	21.4	65	0.67
416	χ,α	21.4	62	0.80

[a] All tests were made with D3001, except for X380 and X416 for which D3008 was used.

[b] χ, Hägg carbide; α, metallic iron; M, magnetite.

the reduced catalysts decreased to one-third of its initial value in 7 weeks of synthesis (Fig. 1). In test X194, described in references *10* and *66*, the activity remained essentially constant in the synthesis at 7.8 atm. However, in a similar test (X168 in reference *65*) the activity decreased in the same manner as shown in Fig. 1. In both tests, Hägg carbide was produced to some extent in the synthesis. The higher content of Hägg carbide in X194, resulting from a variation in the induction procedure, may possibly explain the constant activity observed in this test. At 21.4 atm the activity of carbides decreased rapidly with time, but the activity of reduced catalysts remained fairly constant (Fig. 2). At both synthesis pressures, activity was essentially the same for samples prepared by carbiding with either CO or 1 H_2 + 4 CO gas. The great differences in activity-time trends for reduced and carbided catalysts at 7.8 and 21.4 atm led to tests of carbides at intermediate pressures. At 7.8, 11.1, and 14.5 atm the activity of carbides was constant or increased slowly with time (Fig. 3). Initial activities increased approximately linearly with operating pressure. In test X325 the catalyst was operated at 7.8 atm for 6 weeks (Figs. 1 and 3), and the activity increased slightly with time. In the seventh week the operating pressure was increased to 21.4 atm. The activity was greater at the higher pressure, but it later decreased in a manner typical of catalysts tested entirely at 21.4 atm.

At a given pressure the distribution of products from catalysts converted to Hägg carbide by treatment with either CO or 1 H_2 + 4 CO gas was about the same as shown in Fig. 5. The average molecular weight of products from carbided catalysts was slightly higher than the average molecular weight from corresponding reduced catalysts; however, the difference in products can be largely attributed to the differences in operating temperature. The average molecular weight increased with increasing synthesis pressure in the range 7.8–21.4 atm, shown in Figs. 5 and 6. In all cases the amount of oxygenated molecules in the oil phase from carbided catalysts was low and was not significantly different from products from reduced catalysts. The carbided catalysts had somewhat higher usage ratios (H_2 : CO) than corresponding reduced preparations (Table 4). The high usage ratio in test X428 resulted chiefly from operation at low conversion. The selectivity of carbided catalysts with respect to the production of hydrocarbons and oxygenated molecules was about the same as the selectivity of reduced catalysts.

In tests at 7.8, 11.1, and 14.5 atm the activity remained essentially constant, and tests continuing as long as 25 weeks were possible, because catalyst disintegration was not severe. At 21.4 atm, however, the activity of carbided catalysts decreased rapidly, and catalyst deterioration caused plugging of the reactor and related difficulties after only a few weeks of synthesis.

Figure 4 demonstrates the influence of conversion on activity of carbides at 21.4 atm. In contrast to the rapid decline of activity and catalyst disintegration at 65% contraction, the activity of a carbide at low conversions, 22% (test X428), increased more than threefold in the first 12 weeks and then decreased to about the initial value by the sixteenth week. Apparently the improved operability in test X428 resulted from the lower concentration of reaction products, especially water, in the gas stream. Figure 4 also compares tests at 7.8 atm in which the apparent contraction was maintained at 65% (test X294) and 80% (X399). With time, the activity of the catalyst operated at the higher conversion decreased, whereas at the lower conversion the activity increased.

Another fused-iron catalyst (D3008, promoted with alumina and alkali) was tested with 1 H_2 + 1 CO gas at 21.4 atm in both the reduced and carbided states. The initial activity of the carbide was higher than that of the reduced catalyst, but the activity of carbided catalyst decreased rapidly with time, as observed for catalyst D3001. Activities of the reduced and carbided samples were those listed in the following tabulation:

	Activity A_{Fe}	
Duration of synthesis, weeks	*Reduced* (*X380*)	*Hägg carbide* (*X416*)
1	108	157
2	95	110
3	85	—
4	73	—
5	67	—
6	59	—

The particles of the sample converted to Hägg carbide deteriorated so rapidly that the converter tube plugged during the third week of synthesis.

Changes in composition of catalysts in the tests described in this section were followed as a function of time. Figures 7–11 present phases identified by X-ray diffraction and changes in carbon and oxygen contents from chemical analyses for tests at 7.8 to 21.4 atm. Data from chemical analyses are presented as atom ratios C : Fe and O : Fe. Because X-ray analyses indicated that α-iron was present in only minor amounts, an estimate of the content of Hägg carbide was made assuming that the iron of the catalyst was present as either Fe_3O_4 or Fe_2C. This value, expressed as atom ratio C : Fe, represents the maximum amount of carbide. A further complication of this simple approximation is the presence of carbonates in the samples. Analyses for CO_2 and iron were made on the final samples of each series, as shown in Table 5. Corrections were made to the atom ratios of total carbon, carbidic carbon, and oxygen to iron assuming that the CO_2 was present as $MgCO_3$. However, in most figures, uncorrected data were used, because these show trends of composition changes satisfactorily.

The data of Figs. 7–11 may be summarized as follows:

1. The atom ratio O : Fe increased during synthesis, approaching the value for magnetite, namely, 1.33 in tests at 14.5 and 21.4 atm. The rate of oxidation increased with operating pressure.

2. The atom ratio total C : Fe increased rapidly in the first 5 days of synthesis and then either remained constant or decreased. At 7.8 atm the total C : Fe remained essentially constant after the initial increase, but from 11.1 to 21.4 atm the atom ratio total C : Fe decreased with time. In the initial period the total C : Fe increased from about 0.5–0.6 to 0.7.

3. Although only approximations, the values for maximum carbidic carbon showed the same trends as the phases from X-ray diffraction. As the synthesis proceeded, X-ray diffraction lines of magnetite appeared.

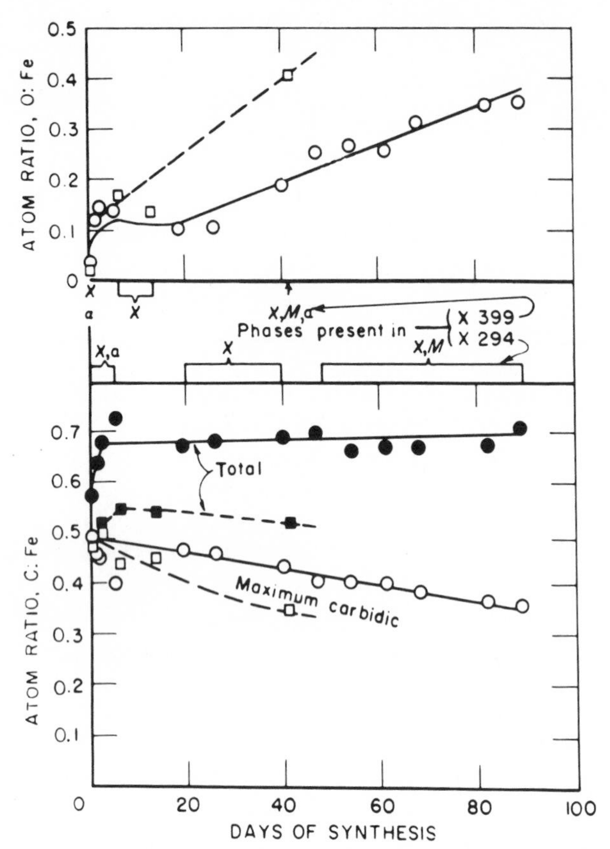

Fig. 7 Composition changes (top, oxygen; bottom, carbon) of carbided D3001 in synthesis at 7.8 atm. ○, X294, 65% contraction; □, X399, 80% contraction. Symbols denoting phases are defined in Table 3.

The lines of magnetite increased in intensity, and those of Hägg carbide decreased. In tests at 14.5 and 21.4 atm, the magnetite lines eventually became the prominent pattern, and in some experiments at 21.4 atm the lines of Hägg carbide disappeared. In tests at 21.4 atm, lines of metallic iron were found in the used catalysts. Magnetic analyses of two of these carbided catalysts agreed reasonably well with the estimates from chemical analyses.

4. At 21.4 atm the rate of oxidation and the rate of removal of carbon were greatly decreased by operation at low conversions, ~25% in test X428, compared with 65% in X320 and X342. It should be noted that the amount of synthesis gas converted per unit time was greater in test X428

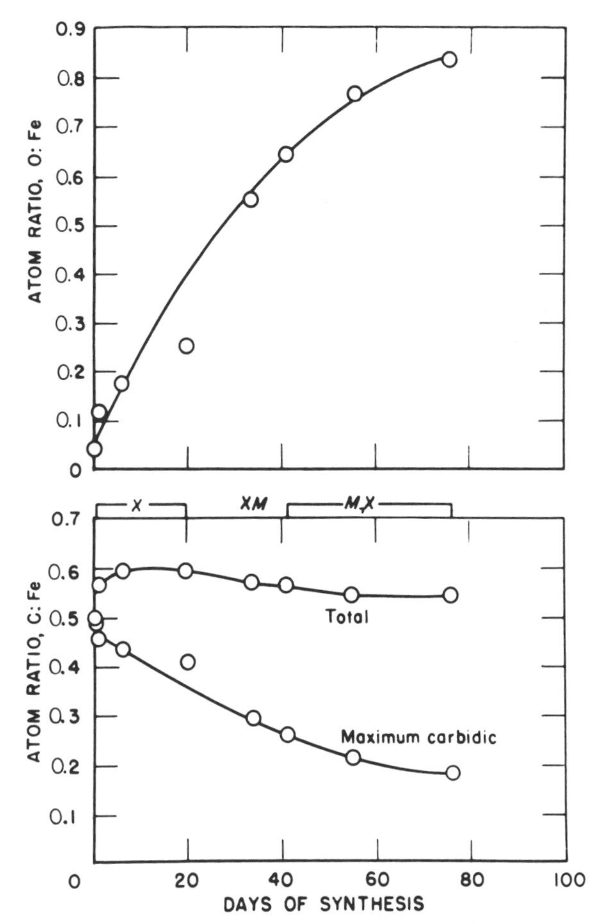

Fig. 8 Composition changes (top, oxygen; bottom, carbon) of carbided D3001 in synthesis at 11.1 atm. Symbols denoting phases are defined in Table 3.

at the lower conversion. Similarly, in tests at 7.8 atm, a catalyst operated at ~80% conversion (X399) oxidized more rapidly than the sample tested at 65% conversion (X294).

5. In tests at 7.8–14.5 atm, the catalyst retained sufficient mechanical stability to permit long periods of synthesis without plugging the reactor tube, even in experiments in which the catalyst was repeatedly removed from the reactor for sampling. The stability of the sample in the test at 14.5 atm was very good in view of the high degree of oxidation in this test. With Hägg carbide at 21.4 atm and high conversions, particle deterioration and plugging of the reactor terminated most of the tests in 2–4 weeks.

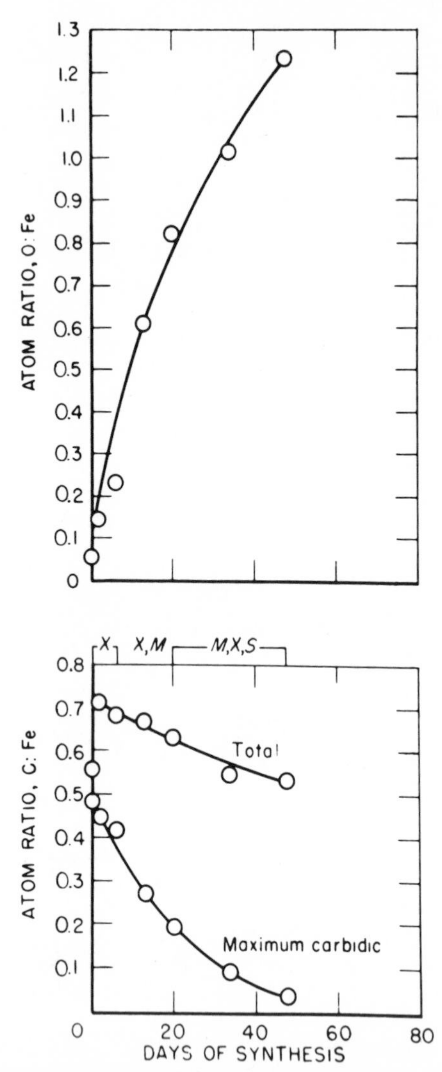

Fig. 9 Composition changes (top, oxygen; bottom, carbon) of carbided D3001 in synthesis at 14.5 atm. Symbols denoting phases are defined in Table 3.

Sampling was difficult because the catalyst was caked in the reactor; therefore, the analytical data shown in Fig. 10 are from two separate tests that were terminated after 3 and 6 weeks of synthesis.

6. With catalyst D3001 the CO_2 content of the used catalyst increased with operating pressure. In a few experiments at 14.5 and 21.4 atm, diffraction lines corresponding to $MgCO_3$ or $FeCO_3$ were found, isomorphous phases that cannot be distinguished by X-ray diffraction. However,

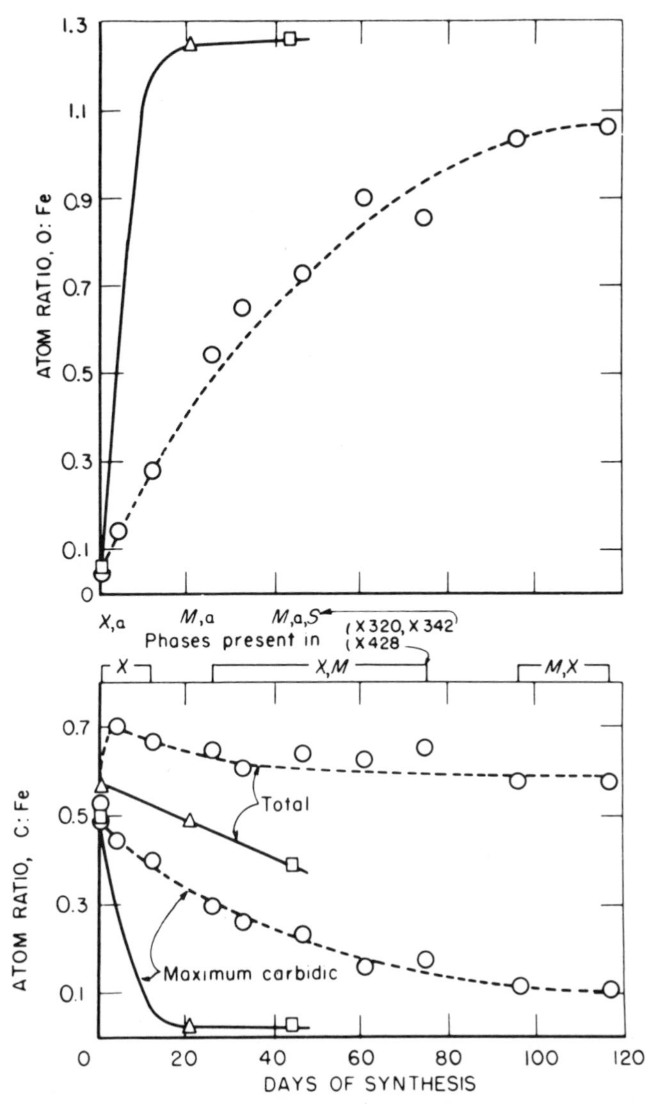

Fig. 10 Composition changes (top, oxygen; bottom, carbon) of carbided D3001 in synthesis at 21.4 atm. △, X320; □, X342 (both at 65% contraction); ○, X428 (22% contraction). Symbols denoting phases are defined in Table 3.

it seems likely that most of the CO_2 is present as magnesium carbonate, because sizable amounts of CO_2 were found in catalysts containing magnesia as promoter, and only small amounts were found in preparations promoted with alumina or zirconia (*61*). Furthermore, the ratio $CO_2:Fe$ exceeded the value 0.10, corresponding to complete conversion of magne-

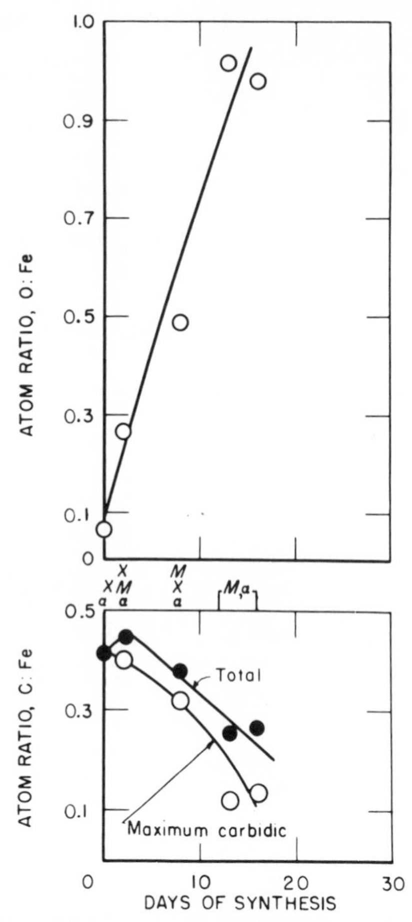

Fig. 11 Composition changes (top, oxygen; bottom, carbon) of carbided D3001 in synthesis at 21.4 atm. Symbols denoting phases are defined in Table 5.

sia and potassium oxide to carbonates by significant amounts in only two samples.

7. In a 21.4-atm test of fused catalyst D3008 (Fe_3O_4–Al_2O_3–K_2O) converted to Hägg carbide (Fig. 11), both oxidation and particle deterioration proceeded rapidly, and the reactor plugged during the third week of synthesis. The X-ray diffraction pattern of metallic iron was found in all samples.

8. The carbide of Eckstrom and Adcock (*29*), the hexagonal carbide, and cementite were not found in any of the catalysts after pretreatment or use in synthesis.

Table 5
Composition of Catalyst after Synthesis

Test, X-	*Operating pressure, atm*	*Duration, days*	*CO_2, wt %*	*Atom ratio, total C:Fe*	*Phases from X-ray diffraction*[a]
289	7.8	194	3.52	0.673	χ,M
408	11.1	76	4.15	0.546	M,χ
394	14.5	68	5.27	0.536	M,χ,S
323	21.4	39	5.45	0.504	M,α
339	21.4	57	7.70	0.347	M,S
416	21.4	16	0.46	0.265	M,α

[a] Phases presented in order of decreasing intensity of diffraction pattern: χ, Hägg carbide; M, magnetite; α, metallic iron; S, carbonate as $MgCO_3$ or $FeCO_3$.

Fused catalysts converted to Hägg carbide have about the same selectivity as the corresponding reduced samples. The only significant difference was the higher usage ratios, H_2:CO, observed for the carbided samples. Available data (*13*) indicate that the principal primary synthesis reaction has a usage ratio of ~2, and this value is decreased by the water–gas shift reaction. Thus, the magnitude of usage ratio may largely depend on the rate of the water–gas reaction, and if this reaction scheme is correct, it may be inferred that carbides are somewhat less active catalysts for this step. Data of Kölbel (*46*) showing that the water–gas shift reaction (1 H_2O + 1 CO at 240°C and 1 atm) proceeds more slowly on carbided than on reduced catalysts are consistent with these results.

In the synthesis with catalysts converted to Hägg carbide, the total carbon content increased in the first few days of testing. The initial increase of carbon results from two processes: (a) the formation of magnesium carbonate in catalyst D3001 and (b) conversion of residual metallic iron in the fresh catalyst to Hägg carbide. In experiments at 7.8 atm, diffraction lines of metallic iron in the fresh catalyst disappeared during this period of synthesis. Usually the initial increase in carbon exceeded by only a small amount the value corresponding to complete conversion of promoters to carbonates. After the initial increase, the carbon content either remained constant or decreased slowly. The estimated values for maximum carbidic carbon decreased more rapidly than total carbon owing to the formation of magnetite. The oxygen content of the catalyst increased continuously during the synthesis, and catalyst deterioration in the moderate temperature range 200–260°C was probably the result of structural changes accompanying oxidation rather than carbon deposition.

Table 6
Composition of Exit Gas in Typical Tests

Test, X-	Operating pressure, atm	Apparent contraction, %	Partial pressures, atm[a]					Pertinent ratios[b]		
			H_2	CO	H_2O	CO_2	CH_4	$H_2O:H_2$	$CO_2:CO$	$CH_4:(H_2)^2$
294	7.8	65	2.94	1.66	0.93	1.97	0.29	0.316	1.19	0.033
399	7.8	80	2.39	0.33	1.90	2.61	0.56	0.795	7.91	0.098
342	21.4	65	7.98	5.24	2.10	5.42	0.65	0.263	1.03	0.010
428	21.4	22	9.44	9.96	1.08	0.75	0.17	0.115	0.07	.0019

[a] Other gaseous components are present in minor amounts. Water is determined by difference.
[b] Thermodynamic data: Oxidation of iron to magnetite at 250°C requires $H_2O:H_2$ and $CO_2:CO$ in excess of 0.025 and 2.2, respectively (*6*). Hydrogenation of graphite to CH_4 requires $CH_4:(H_2)^2$ less than 1200 atm^{-1}. The free energy of formation of Hägg carbide at 250°C is +3.7 kcal/mol (*21*).

The observed composition changes must result from several competitive reactions, some limited by kinetic, and others by thermodynamic factors. Under most synthesis conditions the ratio $CO_2:CO$ is sufficiently low to reduce magnetite and to produce carbide or elemental carbon. Furthermore, carbides or elemental carbon can be produced by reactions of CO and H_2. On the other hand, the ratio $H_2O:H_2$ is usually sufficiently large to oxidize iron and carbides; the carbides may be oxidized according to the equations

$$3\ Fe_2C + 8\ H_2O \longrightarrow 2\ Fe_3O_4 + 3\ C + 8\ H_2 \tag{1}$$

and

$$3\ Fe_2C + 8\ H_2O \longrightarrow 2\ Fe_3O_4 + 3\ CH_4 + 2\ H_2 \tag{2}$$

Also, the ratio $CH_4:(H_2)^2$ is low enough to permit the hydrogenation of carbides or elemental carbon to methane. Removing elemental carbon as CO by water or CO_2 is not thermodynamically possible. Thus, the component responsible for oxidizing and possibly removing carbon appears to be water vapor reacting according to Eqs. (1) and (2). Thermodynamic tendency for Eq. (2) as well as for hydrogenation of elemental and carbidic carbon increases with operating pressure, which may account for the progressive decrease of total carbon and the presence of metallic iron in samples from tests at the higher operating pressures. Typical data for partial pressures of H_2, CO, H_2O, CO_2, and CH_4 and significant ratios of these concentrations in the exit gas are given in Table 6. These data represent the maximum values for products and minimum values for reactants, as observed in the main gas stream in synthesis tests. In all experiments the ratio $H_2O:H_2$ is large enough for oxidizing iron or carbide. Oxidation by water is possible at all conversions more than ~5%. The ratio $CO_2:CO$ exceeds that required for oxidation only in the test at conversions of ~80%. In all cases the ratio $CH_4:(H_2)^2$ is sufficiently low to permit the hydrogenation of carbides and elemental carbon.

Thus, the increased rate of oxidizing and removing carbon at the higher operating pressures and larger conversions must be related to the influence of the higher partial pressures of products, principally water vapor.

In the FTS using porous catalysts, the partial pressures of oxidizing components in the gas stream may represent only minimum values of these molecules in contact with the catalyst surface, for the gas in the pores very likely contains considerably greater concentrations of product molecules. Each pore may act like a long reactor, and possibly the gas entering a pore continues to react until hydrocarbons and iron oxide are formed. Within the pores these processes and the synthesis are impeded by the large amount of wax (liquid at synthesis temperature) adsorbed on the catalyst.

In the initial period, catalysts converted to Hägg carbide are more active than reduced catalysts at all synthesis pressures. Thus, the presence of Hägg carbide as well as other interstitial phases of iron, nitrides, and carbonitrides increases the catalytic activity (*16*). Because the present carbided catalysts have essentially the same selectivity as reduced preparations, the relative accessibility of the catalyst surface should be approximately equal for the reduced and the carbided catalyst. Surface areas of reduced, carbided, or nitrided catalysts before synthesis are approximately equal (*38*).

At pressures of 7.8–14.5 atm, the activity of carbides remained constant or increased with time, but at 21.4 atm the activity of carbides decreased rapidly. Although the rapid decrease in activity at 21.4 atm is probably related to the rapid oxidation and simultaneous destruction of the carbide and modification of the physical structure of the catalyst, the explanation is not simple. For example, the carbided catalyst used at 14.5 atm had a constant activity, even when more than 80% of the iron had been converted to magnetite. Similarly, although the reduced catalysts oxidize less rapidly at 7.8 atm than at 21.4 atm, the activities at the lower pressure usually decreased with time (X152, Fig. 1) but remained essentially constant at the higher pressure (X515, Fig. 2). Thus, other factors are important in determining catalytic activity. The activities of nitrided-iron catalysts, which oxidize at a slower rate than corresponding reduced catalysts, remained essentially constant for long periods of synthesis (*16*).

3.4.2 Synthesis with Fused–Iron Oxide Catalysts Converted to Cementite

Synthesis tests were made on 6- to 8-mesh samples of fused–iron oxide catalyst D3001 converted to cementite at 7.8 and 21.4 atm of 1 H_2 + 1 CO

Table 7
Converting Catalysts to Cementite and Hägg Carbide[a]

Test, X-	Reduction in H_2[b] Temp., °C	Time, h	Extent of reduction, %	Carbiding Gas	Temp., °C	Time, h	Atom ratio C:Fe	Phases from X-ray diffraction[c]
152	450	43	90	None	—	—	0	α
173	550	24	98	None	—	—	0	α
515	500	24	98	None	—	—	0	α
249	450	60	98	CO	240[d]	12	0.30	α,C
317	450	62	96	CO	200–276[d]	10	0.36	C
343	500	24	100	CO	245–252[d]	12	0.31	C,α
329	500	40	98	2 H_2 + 1 CO	310	6	0.42	C,α
294	500	24	97	CO	150–350	18	0.58	χ,α
342	500	24	97	CO	150–350	18	0.43	χ,α
268	550	20	96	CH_4	500	4	0.24	α,C
276	550	20	97	CH_4	500	6	0.40	C

[a] All steps at atmospheric pressure using catalyst D3001.
[b] Space velocity in excess of 1000 h^{-1}.
[c] Phases in order of decreasing intensity of patterns: C, cementite; χ, Hägg carbide; α, metallic iron.
[d] After carbiding, the catalyst was heated in helium for 2 h at 475°C.

gas to determine the influence of method of preparation and operating pressure on activity and selectivity (*57*).

The reducing and carbiding conditions and the principal phases, determined by X-ray diffraction, in the pretreated catalysts are given in Table 7. Three methods were used:

1. Reaction of Hägg iron carbide and iron at 475°C to form cementite
2. Direct preparation by carbiding the reduced catalyst with 2 H_2 + 1 CO gas at 310°C
3. Direct preparation by treating the reduced catalyst with methane at 500°C

Activities of reduced and carbided iron at 7.8 and 21.4 atm (absolute) are compared in Figs. 12 and 13, respectively. Activity, A_{Fe}, was corrected to standard synthesis conditions (240°C and 65% conversion) by an empirical rate equation and is expressed as cubic centimeters (S.T.P.) of synthesis gas consumed per gram of iron per hour. Although surface areas were determined on some of these samples, and could be estimated for the remainder, the activity per gram is usually independent of surface area. For catalyst D3001 the surface areas per gram of unreduced catalyst are principally a function of reduction temperature (*37, 38*) as follows: 450°C, 9.3 m^2; 500°C, 6.4 m^2; and 550°C, 3.5 m^2.

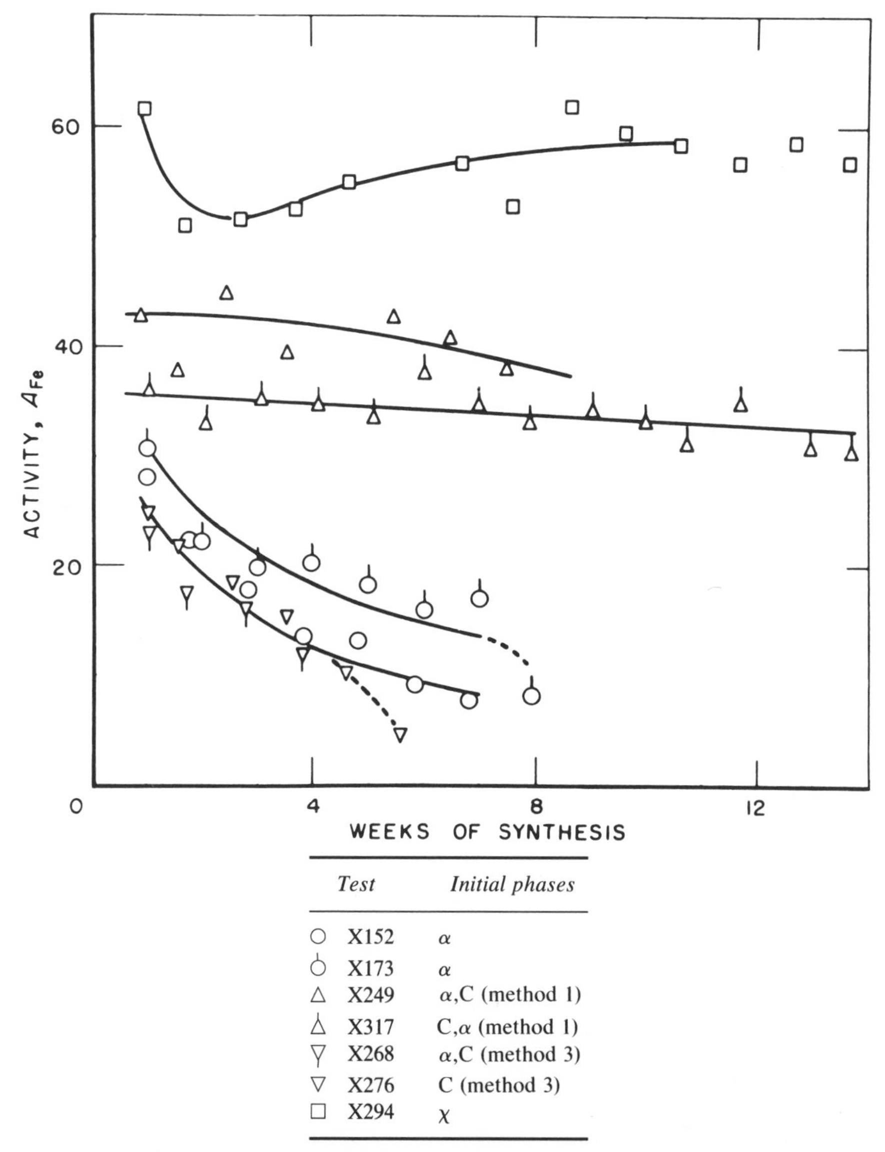

Fig. 12 Activity of reduced and carbided D3001 in synthesis at 7.8 atm. Symbols denoting phases are defined in Table 7.

Figure 14 compares product distributions of reduced and reduced and carbided catalysts at 7.8 and 21.4 atm, corresponding to data in Figs. 12 and 13. The histograms present distribution of total hydrocarbons, including oxygenated molecules dissolved in condensed hydrocarbons, water-

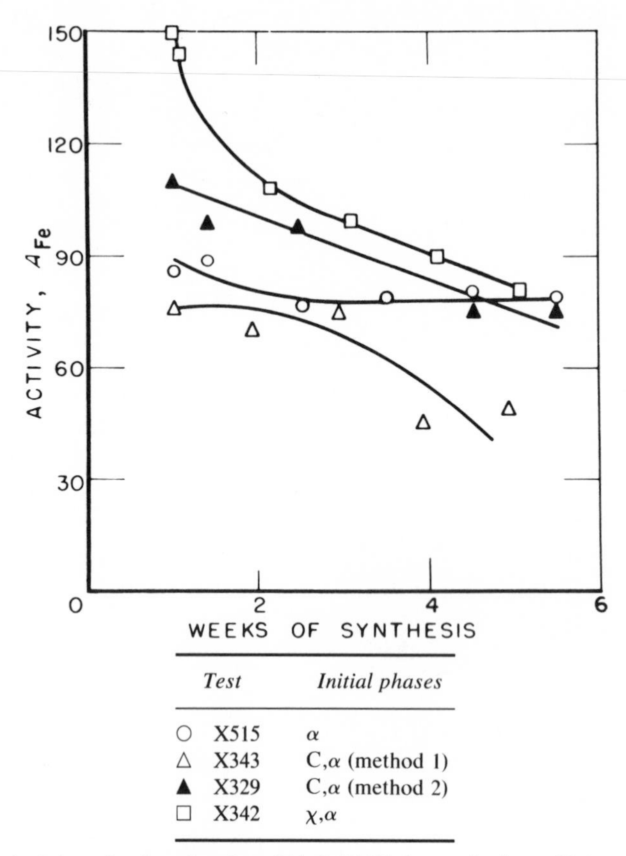

Fig. 13 Activity of reduced and carbided D3001 in synthesis at 21.4 atm. Symbols denoting phases are defined in Table 7.

soluble oxygenated molecules being excluded. Symbols in histograms were defined in Section 3.4.1. The relative usage of H_2 and CO is given in Table 8.

Reduction procedures for the catalysts described were not always the same (Table 7). These differences, however, do not seriously complicate the interpretation, because (a) synthesis behavior is relatively independent of reduction conditions in the temperature range 450–550°C, and (b) sufficient data are presented to permit cross comparisons.

At 7.8 atm the catalysts converted to cementite by thermal treatment of Hägg carbide and iron (X249 and X317) had high constant activities for as long as 14 weeks (Fig. 12). The average activity of these preparations was about twice that of the corresponding reduced sample (X152), and the activity of the reduced sample decreased sharply with time. The behavior

	7.8 atm					21.4 atm			
Test number	X152	X317	X294	X276	X173	X342	X343	X329	X515
Carbiding gas	None	CO	CO	CH_4	None	CO	CO	$2H_2+1CO$	None
Phases present	α	C	χ,α	C	α	χ,α	C,α	C,α	α
* Temperature, °C	270	235	229	270	259	243	257	247	248
* Space velocity, h^{-1}	97	83	106	100	97	334	287	322	303
* Contraction, %	65	63	64	63	63	63	62	64	65
* Activity, A_{Fe}	14	34	53	14	20	104	55	89	80

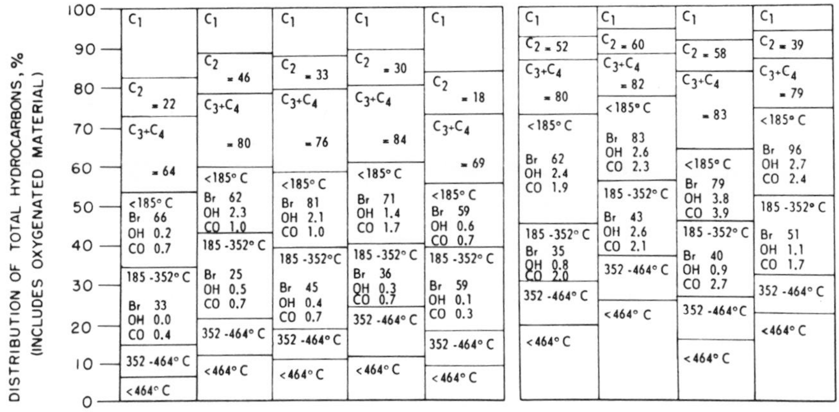

DISTRIBUTION OF TOTAL HYDROCARBONS, % (INCLUDES OXYGENATED MATERIAL) (scale 0–100)

	X152	X317	X294	X276	X173	X342	X343	X329	X515
C_1	C_1	C_1	C_1	C_1	C_1	C_1	C_1	C_1	C_1
C_2	C_2 = 22	C_2 = 46	C_2 = 33	C_2 = 30	C_2 = 18	C_2 = 52	C_2 = 60	C_2 = 58	C_2 = 39
C_3+C_4	C_3+C_4 = 64	C_3+C_4 = 80	C_3+C_4 = 76	C_3+C_4 = 84	C_3+C_4 = 69	C_3+C_4 = 80	C_3+C_4 = 82	C_3+C_4 = 83	C_3+C_4 = 79
<185° C	Br 66, OH 0.2, CO 0.7	Br 62, OH 2.3, CO 1.0	Br 81, OH 2.1, CO 1.0	Br 71, OH 1.4, CO 1.7	Br 59, OH 0.6, CO 0.7	Br 62, OH 2.4, CO 1.9	Br 83, OH 2.6, CO 2.3	Br 79, OH 3.8, CO 3.9	Br 96, OH 2.7, CO 2.4
185-352° C	Br 33, OH 0.0, CO 0.4	Br 25, OH 0.5, CO 0.7	Br 45, OH 0.4, CO 0.7	Br 36, OH 0.3, CO 0.7	Br 59, OH 0.1, CO 0.3	Br 35, OH 0.8, CO 2.0	Br 43, OH 2.6, CO 2.1	Br 40, OH 0.9, CO 2.7	Br 51, OH 1.1, CO 1.7
352-464° C	352-464° C	352-464° C	352-464° C	352-464° C	352-464° C	352-464° C	352-464° C	352-464° C	352-464° C
<464° C	<464° C	<464° C	<464° C	<464° C	<464° C	<464° C	<464° C	<464° C	<464° C

Fig. 14 Selectivity of reduced and carbided D3001. *, Average synthesis conditions. Symbols denoting initial phases are defined in Table 7.

Table 8
Usage Ratio for Tests of Reduced and/or Carbided Catalyst D3001 Using 1 H_2 + 1 CO Gas[a]

Test, X-	*Carbiding gas*	*Initial phases*[b]	*Usage ratio H_2:CO*
At 7.8 atm			
152	None	α	0.72
173	None	α	0.70
249	CO	α,C	0.81
268	CH_4	α,C	0.77
276	CH_4	C	0.79
294	CO	χ,α	0.75
317	CO	C	0.83
At 21.4 atm			
329	2 H_2 + 1 CO	C,α	0.85
242	CO	χ,α	0.79
343	CO	C	0.76
515	None	α	0.76

[a] Data are averages of weeks 2–6 with the contractions at 62–65%.
[b] Phases in order of decreasing intensity of X-ray diffraction pattern: C, cementite; χ, Hägg carbide; α, metallic iron.

of cementite was similar to that observed for Hägg carbide (X294), except that Hägg carbide was more active. The activity of cementite prepared from methane (X268 and X267) decreased rapidly with time, and the average activity was somewhat less than that observed for the corresponding reduced catalyst (X173).

In synthesis at 21.4 atm, however, the activity of the carbides decreased steadily with time, whereas the activity of the reduced catalyst remained essentially constant (Fig. 13). Hägg carbide (X342) was most active at this pressure. The activity of cementite prepared by method 2 (X329) was intermediate between the activities of Hägg carbide (X342) and reduced catalyst (X515). Cementite prepared from Hägg carbide and metallic iron (X343) had a lower activity than the reduced catalyst (X515).

At 7.8 atm the selectivity of catalyst converted to cementite (Fig. 14) appears to be essentially independent of operating temperature; however, the catalysts operating at the higher temperatures (X268 and X276) were those carbided with methane, the other two (X249 and X317) having been prepared by thermal reaction of Hägg carbide with iron. The carbided catalysts produced a material of higher average molecular weight than that from reduced catalyst (X152 and X173). Usually for a given catalyst with the same pretreatment, the average molecular weight decreases with increasing temperature, and the only positive statement that can be made from this group of data is that the catalysts carbided with methane yield a higher molecular weight product than reduced catalysts.

At 21.4 atm the average molecular weight of the product was greater than in corresponding tests at 7.8 atm. However, no simple explanation can be offered for these data based on initial phases or operating temperature.

At 7.8 atm the usage ratios ($H_2:CO$) for carbided catalysts (Table 8) were significantly higher than with reduced catalysts, but in most tests at 21.4 atm the usage ratios were about the same. In these experiments the usage ratio is principally determined by the relative production of water and CO_2 and probably reflects the relative rates of a primary synthesis reaction producing water and a subsequent water–gas shift reaction.

Analytical data for catalyst samples before and after synthesis are presented in Table 9, and Fig. 15 shows composition changes during the course of a 7.8-atm test of a catalyst converted to cementite by method 1. For comparison, typical data for oxygen contents of reduced catalysts and samples converted to cementite by two different methods are included. Samples that had been used in synthesis were extracted with boiling toluene before analyses. The phases present were determined by X-ray diffraction. Chemical analyses were made for total content of iron and carbon, and oxygen was estimated by difference. Analyses for carbonates were made on a few representative samples by measuring the

Table 9
Composition and Phase Changes of Reduced and Carbided Fused Catalyst D3001 during Synthesis

		Atom or mole ratios			
Test, X-	*Days of synthesis*	*O:Fe*	*CO_2:Fe*	*Total C:Fe*	*Phases from X-ray diffraction*[a]
At 7.8 atm					
249	0	—[b]	—[b]	0.30	α,C
	55	0.395	—[b]	0.35	C,M
268	0	0.03	—[b]	0.24	α,C
	26	0.415	0.03	0.32	C,α,M
276	0	0.02	—[b]	0.40	C
	40	0.63	0.04	0.56	C,M
294	0	0.02	—[b]	0.57	χ,α
	96	0.36	—[b]	0.71	χ,M
317	0	0.11	—[b]	0.36	C
	105	0.56	0.02	0.43	C,α,M
173	0	0.03	—[b]	0	α
	54	0.72	—[b]	0.061	M,α,χ
At 21.4 atm					
329	0	0.05	—[b]	0.42	C,α,M
	40	1.30	—[b]	0.35	M,α
342	0	0.06	—[b]	0.57	χ,α
	37	1.47	0.11	0.49	M,α,S
343	0	0.05	—[b]	0.31	C,α
	35	1.30	0.13	0.30	M,α,C
515	0	0.02	—[b]	0	α
	39	1.07	—[b]	0.26	M,α,χ

[a] Phases in order of decreasing intensity of diffraction pattern: α, metallic iron; M, magnetite; C, cementite; χ, Hägg carbide; S, magnesite or siderite.
[b] Not determined.

amounts of CO_2 evolved on acidifying the catalyst. The CO_2 evolved was 1–2 wt % at 7.8 atm, and 6.4 wt % at 21.4 atm. Previous work suggested that CO_2 was present principally as $MgCO_3$ (*16, 61*).

In used carbided catalysts, the concentration of metallic iron is usually very small, and an estimate could be made of the maximum amount of carbidic carbon by assuming that iron was present only as carbide or magnetite.

Pertinent results of Table 9 and Fig. 15 may be summarized:

1. Atom ratios O:Fe increased during synthesis, approaching the value of magnetite in tests at 21.4 atm. Catalysts converted to cementite by treatment with CO or 2 H_2 + 1 CO oxidized at about the

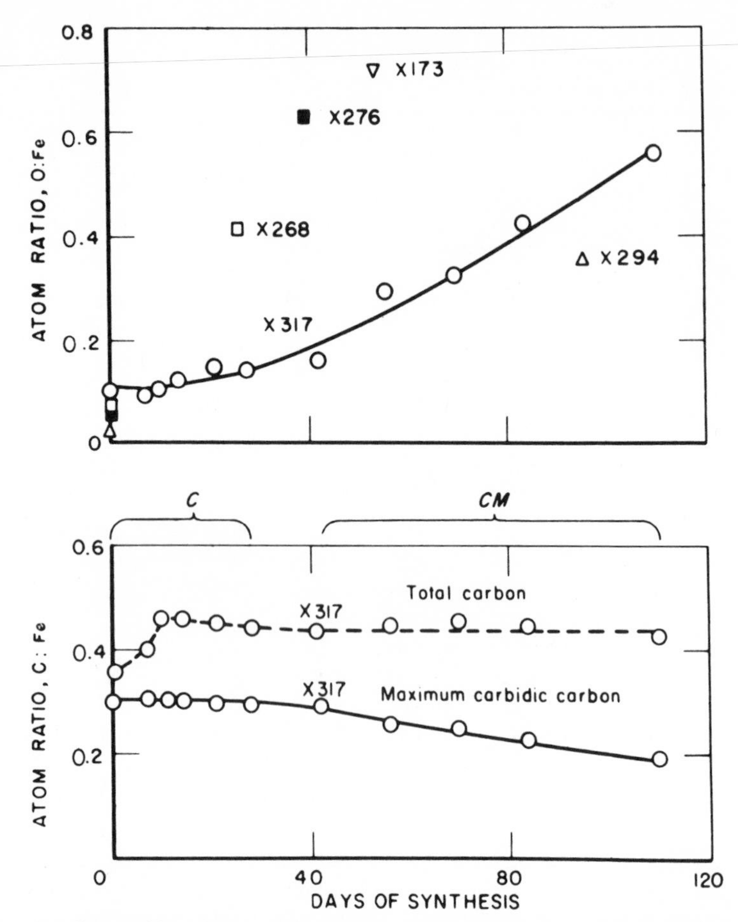

Fig. 15 Composition changes (top, oxygen; bottom, carbon) in carbided D3001 in synthesis at 7.8 atm. Symbols denoting phases are defined in Table 9.

same rate as those made from Hägg carbide and iron; however, cementite prepared from methane (X268 and X276) oxidized more rapidly. Its higher rate of oxidation may have resulted from operation at higher temperatures, required to maintain constant conversions. These preparations oxidized at about the same rate as the corresponding reduced catalyst.

2. At 7.8 atm (X317), atom ratios (total C : Fe) increased in the first 10 days of synthesis and then remained about constant. At 21.4 atm, the atom ratios (total C : Fe) for samples converted to cementite were lower after 35–40 days of testing than before synthesis. These results are similar to those obtained with Hägg carbide, mentioned previously.

3. Although values for maximum carbide carbon are only approximate, in most cases they show the same trends as phases found in X-ray and thermomagnetic analyses.
4. After synthesis with catalysts converted to cementite, this phase was the only carbide identified by X-ray diffraction.
5. Formation of carbonates during synthesis increased sharply with pressure, and the results were similar to those with catalysts converted to Hägg carbide (see Section 3.4.1), or iron nitride (*61*).
6. At the temperature studied, deposition of elemental carbon from synthesis gas does not appear to be an important factor in disintegration of catalysts during synthesis. Disintegration parallels oxidation and appears to be related to structural changes involved in forming magnetite.
7. Explanations advanced for composition changes of Hägg carbide during synthesis are adequate for cementite: (a) water is the principal oxidant, and (b) the decrease in carbon content of catalyst in synthesis at 21.4 atm may result from direct hydrogenation of carbide or reaction of water with carbide to produce methane and magnetite.

Fused-iron catalysts, converted to cementite, are active in the FTS, and their behavior is qualitatively similar to that of preparations containing Hägg iron carbide or metallic iron. Cementite prepared with CO was more active than cementite from methane; the reason for the difference is not known. In synthesis at 7.8 atm, catalysts converted to Hägg carbide or cementite by treatment with CO were more active and oxidized less rapidly than reduced catalysts. At 21.4 atm, carbided catalysts had little or no advantage over reduced catalysts with respect to either activity or rate of oxidation. The selectivity of fused catalysts containing metallic iron, Hägg carbide, or cementite as the initial phase was about the same, except that the carbides often had a slightly higher usage ratio ($H_2:CO$) than reduced catalysts.

These and older data demonstrate that cementite, Hägg iron carbide, iron nitride, and iron carbonitride are at least as active in the FTS as the metal. This behavior of interstitial compounds of iron is in sharp contrast to that of carbides of cobalt (*9, 67*) and nickel (*43, 52*), which have very low activities, compared with the metals.

3.4.3 Synthesis Tests of Carburized Precipitated-Iron Catalysts

The effect of a variety of pretreatments on 6- to 10-mesh precipitated–ferric oxide catalyst P3003.24 ($Fe:Cu:K_2CO_3 = 100:10:0.5$) was studied. Some of these tests, including nitriding experiments, were reported

Table 10
Pretreatment of Precipitated-Iron Catalyst P3003.24[a]

	Pretreatment				
Test, X-	*Gas*	*Hourly space velocity*	*Temp., °C*	*Duration, h*	*Phases from X-ray diffraction*[b]
149	1 H_2 + 1 CO	98	230	23	—[c]
101	1 H_2 + 1 CO	100	230	24	—[c]
304	2 H_2 + 1 CO	2500	310	6	—[c]
825	2 H_2 + 1 CO	2480	310	6	—[c]
341	CO	100	200–250[d]	16	M
245	H_2	1480	313	17	α,Cu
324	H_2	1940	296	25	χ
	CO	98	185–300[d]	13	
324C[e]	H_2	220	298	24	α,Cu
		190	391	24	

[a] All steps at atmospheric pressure.
[b] Phases in decreasing order of intensity of diffraction pattern: M, Fe_3O_4; α, metallic iron; Cu, copper; χ, Hägg carbide.
[c] Not determined, catalyst pretreated in synthesis reactor.
[d] Temperature increased as required to maintain CO_2 concentration of ~20% in exit gas.
[e] Catalyst from X324.

previously (*15, 62*). Pretreatment conditions are described in Table 10, and synthesis data and composition of used catalyst in Table 11. Selectivity data are given in Fig. 16. No correlation can be found between the carbide content of the used catalyst and activity or selectivity. Catalysts containing the largest quantities of elemental carbon had the highest activity, and those containing the larger amounts of oxide yielded products of higher molecular weight. The induction procedure described by Ackermann (*1*), used in X304 and X825, produced the most active conventional iron catalysts tested and produced high yields of hydrocarbons at temperatures less than 200°C. Apparently, pretreatment of the oxide under conditions that deposit substantial quantities of elemental carbon in the catalyst is required for preparing precipitated catalysts of very high activity. This carburization step would probably adversely affect the mechanical properties of the catalyst, but no difficulties were observed in the catalyst tests.

Catalysts that were reduced in hydrogen as a part of their pretreatment (X245, X324A, and X324C) yielded products containing lower amounts of wax and higher amounts of gaseous hydrocarbons than the catalysts treated directly with CO or synthesis gas. The catalysts that were pretreated only with CO or synthesis gas contained the largest amounts of

Table 11
Synthesis with Precipitated-Iron Catalyst P3003.24[a]

Test, X-	*Synthesis temperature, °C*		*Average*		*Phases in used catalyst*[c]	*Composition of used catalyst, % Fe as*		
	Second week	*Sixth week*	*Activity A_{Fe}*[b]	*Usage ratio $H_2:CO$*		*Metal*	*Fe_2C*	*Fe_3O_4*
149	229	219	136	0.57	M,ϵC,Cu	—	—	—
101	226	230	148	0.59	M,ϵC	0	31	69
304	203	195	479	0.76	χ,M	0	66	34
825	190	197	398	0.73	χ,M	0	69	31
341	224	214	217	0.62	M,Cu	0	10	90
245	228	229	145	0.66	ϵC,M,Cu,χ	4	82.5	13.5
324A	222	233	120	0.72	χ,S	0	81.5	18.5
324C	233	—	87	0.66	χ,M	0	84	16

[a] 1 H_2 + 1 CO gas at hourly space velocities of 100–139, 7.8 atm, and apparent CO_2-free contraction maintained at 62–65%.

[b] Activity expressed as cubic centimeters (S.T.P.) of H_2 + CO consumed per gram Fe at 240°C.

[c] Phases in decreasing order of intensity of X-ray diffraction pattern: M, magnetite; ϵC, hexagonal Fe_2C; Cu, metallic Cu; χ, Hägg carbide; S, siderite ($FeCO_3$).

oxygen after use in the synthesis, and these catalysts yielded the products of higher molecular weight, despite the presence or absence of substantial quantities of Fe_2C carbides.

3.5 Synthesis Tests with Nitrides and Carbonitrides of Iron

Studies of the preparation of iron nitrides and of their properties in the synthesis have been described in U.S. Bureau of Mines publications (*15, 59*) and elsewhere (*3, 14, 16, 56, 61, 62*). In the FTS, nitrided-iron catalysts show improved stability against the deterioration caused by oxidation and free carbon deposition, and they preferentially catalyze the synthesis of alcohols, other oxygenated products, and lower boiling liquid hydrocarbons. After a short period of operation, the nitride acquires appreciable amounts of carbon and becomes a carbonitride. Therefore, as the catalyst spends most of its lifetime in the carbonitride stage, this is the most important form to consider.

The usual form of freshly nitrided iron is ϵ-iron nitride. This phase is

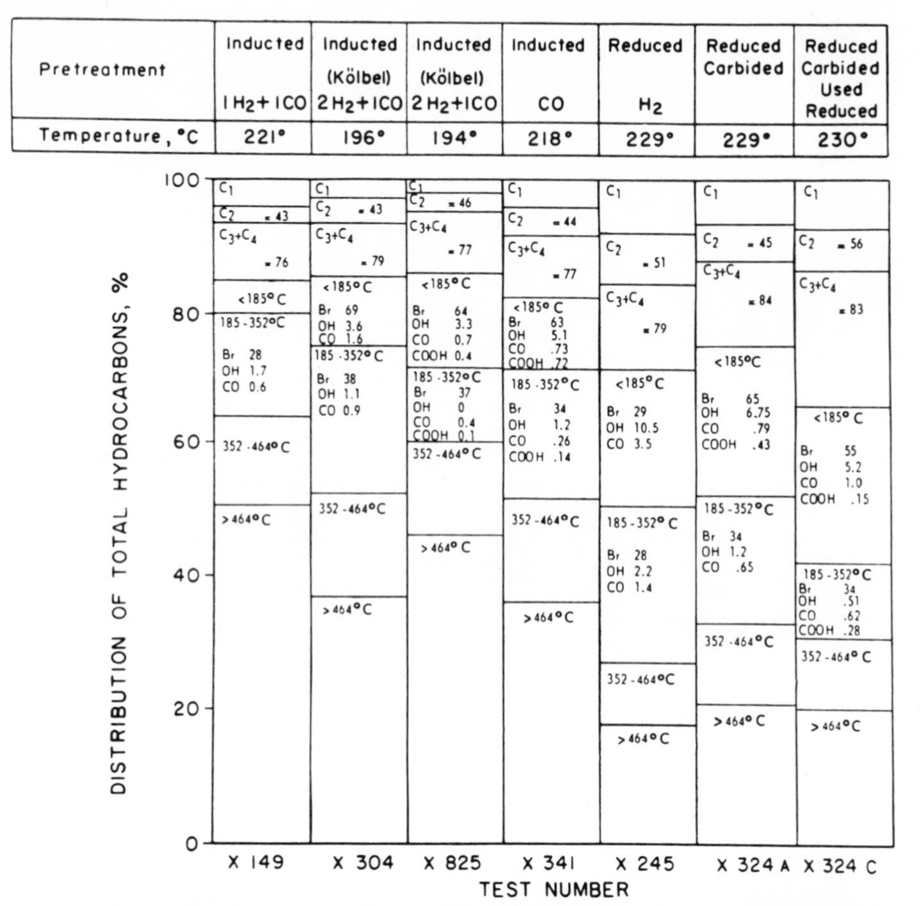

Fig. 16 Pretreatment of precipitated P3003.24 causes large changes in selectivity in synthesis at 7.8 atm. Total hydrocarbons includes oxygenates dissolved in liquid phase.

isomorphous with ϵ-iron carbide, and a continuous series of solid solutions in the ϵ phase, ranging from pure carbide to pure nitride, is theoretically possible. Although phases corresponding to the two terminal compositions have been realized, compositions of carbonitride have been prepared only in the range of C : N atomic ratios from zero to ~2 (*5, 45*). Not only are ϵ-carbonitrides with a wide range of C : N ratios experimentally realizable, but the ratio of iron to total interstitial elements (nitrogen plus carbon) can also vary between 2 to 1 and 3 to 1.

There is apparently a critical C : N ratio that, when exceeded, causes decomposition of ϵ-carbonitride to χ-iron carbide. In this transformation, the nitrogen from the ϵ-iron carbonitride (a) may go into the χ phase, which then would really be χ-iron carbonitride, (b) may go into a phase of such small crystallite diameter that the diffraction pattern is too diffuse to be detected, or (c) may be eliminated as gas. This critical ratio ranges from about 2 to 1 at 450°C to ∞ at 180°C. This conclusion is based on the following experimental facts.

Jack (*45*) and Hall *et al.* (*36*) have shown that ϵ-iron nitride is converted to carbonitride by substituting carbon atoms for nitrogen atoms when treated with CO. At 400°C, carburization proceeds in this way until the ratio of carbon to nitrogen is ~2. Further carburization leads to formation of χ-iron carbide.

Direct carbiding of iron catalysts at 190°C with CO leads to the formation of ϵ-iron carbide, isomorphous with ϵ-iron nitride (*42, 45*).

Thermal conversion of pure ϵ-iron carbide to χ-carbide proceeds at a measurable rate at 240°C and probably at lower temperatures. The critical C: Fe ratio is not affected by gas-phase composition, as the reactions involved can proceed entirely in the solid state.

This section describes, first, composition changes that occur when catalysts converted to ϵ-iron nitride are used in the synthesis, second, synthesis tests with catalysts nitrided only to α or γ' phase, and, third, tests of iron carbonitrides prepared in several ways.

3.5.1 Composition Changes of ϵ-Iron Nitrides during Synthesis

In the first series of experiments, catalysts of the fused, sintered, and precipitated types were analyzed after extended periods of synthesis (*61*). For chemical analyses, samples of the used catalysts were extracted with boiling toluene to remove adsorbed synthesis products, and the extracted material was analyzed for total iron, carbon, and nitrogen. Particular care was taken to remove the solvent and to prevent atmospheric oxidation of the catalyst. Oxygen values, determined by difference, are the most uncertain as they contain the combined errors of all of the other determinations. From weight percentages of iron, carbon, nitrogen, and oxygen, the atom ratios of carbon to iron, nitrogen to iron, and oxygen to iron were computed. The pretreatments of catalysts are described in Table 12, and the analyses of catalysts after synthesis are given in Tables 13 and 14.

The phases present in the pretreated and in used catalysts were identified by X-ray powder diffraction (*42, 45*), which showed that the used, nitrided catalysts usually contained ϵ-carbonitride and magnetite but no metallic iron. For quantitative estimation, the phases containing iron were assumed to be magnetite and ϵ-carbonitride, the latter being arbitrarily assumed to be a mixture of Fe_2C and Fe_2N. The possibility that used

Table 12
Pretreatment of Nitrided Catalysts[a]

			Reduction in hydrogen			*Nitriding in ammonia*		
Catalyst	*Type*	*Test, X-*	*Hourly space velocity*	*Temp., °C*	*Duration, h*	*Hourly space velocity*	*Temp., °C*	*Duration, h*
D3001	Fused	215	1000	550	20	5000	385	4
		218	1000	550	20	4000	380	8
		226	1000	550	20	1000	350	6
D3008	Fused	253	1000	550	40	1000	350	6
		337	2000	555	41	1000	350	6
L3028	Fused	355	2300	450	30	1000	350	6
A2106.11	Sintered	266	2000	400	24	1800	350	7
P3003.24	Precipitated	273	1000	300	25	1000	300	15
							325	7

[a] Both reduction and nitriding at atmospheric pressure.

catalysts may contain ferrous or promoter carbonates was examined by determining the CO_2 liberated when the sample was digested in hydrochloric acid. Less than 1% CO_2 by weight was found except for samples of tests X349 and X226 (Table 13), for which the CO_2 was 4.36 and 4.92%, respectively. Carbon dioxide values of ~1% produced no significant variations of the calculated composition data. However, in the calculations of X349 and X226, the CO_2 was assumed to be present as $MgCO_3$, because sizable CO_2 values were obtained only for catalyst containing magnesia. From the atom ratios of total carbon, of nitrogen, and of oxygen to iron, the percentages of iron as Fe_2C, Fe_2N, and Fe_3O_4 were calculated. When the sum of the percentages of iron in these phases exceeded 100, the apparent excess was attributed to the presence of elemental carbon, and an atom ratio of elemental carbon to iron was computed. This calculation as made only for samples containing essentially ϵ-carbonitride and magnetite.

Table 13 presents analytical data for tests of catalyst D3001 at 7.8 and 21.4 atm. In test X215 after 203 days of operation, the catalyst still showed only an X-ray diffraction pattern of ϵ-carbonitride and contained small amounts of elemental carbon and oxide. The catalytic activity remained essentially constant throughout this period. Nevertheless, the catalyst was renitrided after 203 days and again after 238 days with ammonia at 350°C and an hourly space velocity of 1000 for 6 h. After renitriding and using it in synthesis, the catalyst still produced an X-ray pattern of only ϵ-carbonitride. However, the fraction of nitrogen in the carbonitride phase

Table 13
Changes in Nitrided Catalyst D3001 during Synthesis

Test, X-	Previous treatment[a]	Total days of synthesis	Conditions of previous synthesis			Phases present[c]	Composition, atom ratios to iron		
			Temp., °C	Hourly space velocity	Contraction,[b] %		Total carbon	Nitrogen	Oxygen
Operating pressure, 7.8 atm									
215[d]	RN	0	—	—	—	ϵ	0	0.46	0
	RNU	203	227	99	65	ϵ	0.32	0.19	0.19
	RNUN[e]	238	217	100	64	ϵ	0.355	0.27	0.15
	RNUN[e] UN[e] U	252	223	101	63	ϵ	0.36	0.28	0.14
236[f]	RN	0	—	—	—	ϵ	0	0.46	0
	RNU	50	230	104	64	ϵ	0.26	0.30	0.12
Operating pressure, 21.4 atm									
349[d]	RN	0	—	—	—	ϵ	0	0.43	0.08
	RNU	37	241	309	64	ϵ,M,α	0.24	0.14	0.49
226[g]	RN	0	—	—	—	ϵ	0	0.45	0.07
	RNU	64	241	347	54	ϵ,M,α	0.22	0.14	0.33

[a] R, Reduced in H_2; N, nitrided in NH_3; U, used in synthesis.
[b] Apparent CO_2 equals free contraction.
[c] Phases in order of decreasing intensity of X-ray diffraction pattern: α, metallic iron; ϵ, nitride or carbonitride; M, magnetite.
[d] Synthesis gas, 1 H_2 + 1 CO.
[e] Renitrided in NH_3 at an hourly space velocity of 1000 and 350°C for 6 h.
[f] Synthesis gas, 1 H_2 + 1.5 CO.
[g] Synthesis gas, 2 H_2 + 1 CO.

Table 14
Changes in Nitrided Catalysts during Synthesis[a]

Test, X-	Catalyst	Previous treatment[b]	Total days of synthesis	Conditions of previous synthesis			Phases from X-ray diffraction[c]	Composition, atom ratios to iron		
				Pressure, atm	Temp., °C	Hourly space velocity		Total carbon	Nitrogen	Oxygen
355	L3028	RN	0	—	—	—	ϵ,(?)	0	0.45	0.08
		RNU	40	21.4	222	313	ϵ,(?)	0.19	0.29	0.22
253	D3008	RN	0	—	—	—	ϵ	0	0.43	0.09
		RNU	49	7.8	230	98	ϵ	0.18	0.33	0.13
337	D3008	RN	0	—	—	—	ϵ	0	0.44	0.10
		RNU	50	21.4	243	277	ϵ,M,α	0.25	0.26	0.18
		RNUR[d]	50	—	—	—	α,C,M	0.24	0.002	0.14
		RNURU	70	21.4	238	297	M,α,C	0.56	0.009	0.67
266	A2106.11	RN	0	—	—	—	ϵ,M	0	0.47	—
		RNU	60	7.8	216	102	ϵ	0.16	0.37	0.12
273	P3003.24	RN	0	—	—	—	ϵ,Cu,M	0	0.46	0.13
		RNU	39	7.8	229	138	ϵ,Cu	0.23	0.32	0.16
		RNUR[e]	39	—	—	—	ϵ,Cu,χ	0.24	0	0.12
		RNURU	67	7.8	230	137	M,χ	0.33	0	0.75

[a] All tests with 1 H_2 + 1 CO with CO_2-free contraction maintained at 63–66%.
[b] R, Reduced in H_2; N, nitrided with NH_3; U, used in synthesis.
[c] Phases in order of decreasing intensity of pattern: α, metallic iron; ϵ, ϵ nitride or carbonitride; χ, Hägg carbide; C, cementite; M, magnetite; Cu, metallic copper; (?), unidentified diffuse lines.
[d] Used catalyst reduced in H_2 at 385°C and an hourly space velocity of 1000 for 4 h.
[e] Used catalyst reduced in H_2 at 300°C and an hourly space velocity of 1000 for 9 h.

was increased, the ratio of oxygen to iron was decreased, and the ratio of elemental carbon to iron was increased.

In test X349 at 21.4 atm with 1 H_2 + 1 CO gas, the nitrogen content of the catalyst decreased more rapidly than at 7.8 atm, and oxidation and elemental carbon deposition were faster. After synthesis, the samples gave X-ray patterns of ϵ-carbonitride, magnetite, and metallic iron. About 36% of the iron was present as magnetite, and the ratio of elemental carbon to iron was 0.059.

In test X226 with 2 H_2 + 1 CO gas, the amounts of magnetite and elemental carbon found in the used catalyst were less than those observed in X349 with 1 H_2 + 1 CO gas. Although in 37 days of test X349 the nitrogen had decreased to about the same value as found after 64 days of test X226, removal of nitrogen was relatively independent of synthesis gas composition. In experiments at both 7.8 and 21.4 atm, the nitrogen content decreased rapidly in the first 20 days of synthesis, but it usually did not decrease appreciably after 30–40 days of synthesis. Thus, in test X225 (Table 15), which was a duplicate of X349 except for duration, the atom ratio of nitrogen to iron was the same (0.139) after 102 days as in X349; the nitrogen had decreased to about the same value as found after 64 days of test X226. Removal of nitrogen was relatively independent of synthesis gas composition.

Table 14 presents analytical data for nitrided fused catalysts containing alumina and zirconia as structural promoters and data for nitrided sintered and precipitated catalysts. Nitrided catalysts D3008 (Fe_3O_4–Al_2O_3–K_2O) differ from D3001 in being more resistant to oxidation during synthesis at 21.4 atm. Nitrogen was removed from D3008 at a greater rate at 21.4 atm (X337) than at 7.8 atm (X253). Analyses of nitrided sintered catalyst A2106.11 (Fe_3O_4–K_2O) and nitrided precipitated catalyst P3003.24 (Fe_2O_3–CuO–K_2CO_3) (Table 14) again demonstrated that nitrided-iron catalysts are resistant to oxidation in the synthesis at 7.8 atm.

The nitrided catalysts in tests X273 and X337 were reduced with hydrogen after 39 and 50 days of testing, respectively. This treatment removed virtually all of the nitrogen, whereas the carbon was essentially unchanged, and the oxygen decreased only slightly. The used fused catalysts in X337 and in X214 (*16*) contained metallic iron, cementite, and magnetite (in order of decreasing amounts) after reduction at 385°C. The used precipitated catalyst in X273 contained metallic iron, copper, and Hägg carbide after reduction at 300°C. After reduction, the products of the synthesis changed from the relatively low molecular weight, highly oxygenated products characteristic of nitrided catalyst to a high molecular weight, slightly oxygenated material characteristic of reduced or carbided catalysts. After several weeks of testing, the principal phase in the reduced nitrides was magnetite, more than 50% of the iron.

Table 15
Pretreatment and Tests of Carbonitrides Produced During Synthesis[a]

Test, X-	Reduction in H_2[b]		Nitriding in NH_3[b]		Composition of pretreated catalyst		Composition of used catalysts				Activity per gram Fe, A_{Fe}[e]
	Temp., °C	Time, h	Temp., °C	Time, h	Atom ratio, N:Fe	Phases[d]	C:Fe	N:Fe	O:Fe	Phases[d]	
517	550	24	—	—	0	M,α	0.28	0	1.05	ϵ',χ,S,M	69
274	550	20	500	1	0.07	α,γ'	0.23	0.02	0.99	M,α,χ	63
447	550	20	250	1	0.11	α,γ'	0.34	0.01	1.03	M,α,χ	90
766	450	40	350	0.5	0.16	α,γ'	0.17	0.09	1.11	M,α	75
445	550	20	450	1.5	0.18	γ',α	0.38	0.03	0.95	M,χ,α	99
219	550	20	385	4	0.26	γ',α	0.35	0.10	0.63	γ',ϵ	109
617	550	20	250	10	0.35	γ',ϵ,α	0.51	0.08	0.80	ϵ,M	88
711	450	40	350	8	0.38	ϵ,γ'	0.32	0.06	0.94	M,ϵ	76
225	550	20	350	6	0.41	ζ,ϵ	0.40	0.14	0.56	ϵ,M	115
612	550	20	350	6	0.45	ϵ	0.36	0.20	0.76	ϵ,M,S	81
620	550	20	350	12	0.46	ϵ,M	0.41	0.23	0.62	ϵ,M,S	126
518	550	24	350	6	0.46	ϵ	0.35	0.19	0.58	ϵ,M,S	95
670	550	20	350	22	0.47	ϵ	0.36	0.26	0.36	ϵ,M,S	111
251[f]	550	20	385	4	0.46	ϵ	0.36	0.31	0.23	ϵ	70
218[f]	550	20	385	8	0.43	ϵ	0.42	0.23	0.15	ϵ,M	68
361	500	24	—	—	0	α,M	0.18	0	1.14	M,α	54
194[f]	450	40	—	—	0	α	0.31	0	1.09	M,χ,α	27

[a] Catalyst D3001 used in all tests; pretreatments were at 1 atm and synthesis tests at 21.4 atm, except as noted.
[b] Hourly space velocity of H_2, 2300–2700, and reduction 96.4–100%.
[c] Hourly space velocity of NH_3, 950–1050.
[d] Phases in order of decreasing order of X-ray diffraction patterns: α, metallic iron; χ, Hägg carbide; M, magnetite; S, $MgCO_3$ or $FeCO_3$; γ', γ' nitride or carbonitride; ϵ, ϵ nitride or carbonitride; ζ, ζ nitride; ϵ', hexagonal Fe_2C.
[e] Activity is defined in Table 11.
[f] Synthesis at 7.8 atm.

When ϵ-iron nitrides were treated with CO at 450° and atmospheric pressure, ϵ- or ζ-iron carbonitrides were formed (*45*) in which carbon replaced nitrogen, with the ϵ and ζ structures remaining unchanged until more than 75% of the nitrogen was replaced by carbon. On further treatment with CO, the carbonitride changed to Hägg carbide. With massive iron nitride, no appreciable amount of oxide or elemental carbon was formed in the carburization (*45*). Apparently, the substitution reaction also occurs during the synthesis, but concomitantly with oxidation of the catalyst and deposition of elemental carbon. At synthesis temperatures of 210–250°C, the replacement of nitrogen by carbon proceeded somewhat more rapidly at 21.4 than at 7.8 atm, but in every case much slower than observed by Jack (*45*) at 450°C. The nitrogen content of the catalyst was never decreased below atom ratios of N : Fe of 0.14, nor was Hägg carbide observed in the used, nitrided catalysts. Although all of the nitrided catalysts were quite resistant to oxidation and deposition of elemental carbon at 7.8 atm, these reactions proceeded much faster in some catalysts at 21.4 atm. However, the stability and activity of nitrided catalysts were much greater at both pressures than those of reduced catalyst. The stability of fused catalysts at 21.4 atm varied considerably with the type of structural promoter; catalysts promoted with alumina or zirconia were much more resistant to oxidation than the magnesia-promoted catalyst.

The rate of removal of nitrogen from the catalyst was relatively independent of the composition of the synthesis gas, indicating that hydrogenation of nitrides was strongly inhibited by CO and/or synthesis products. In pure hydrogen, iron nitrides can be completely hydrogenated in a few minutes at synthesis temperatures.

When ϵ-carbonitrides were reduced in hydrogen, most of the nitrogen was removed; however, the carbon content changed very little, and a part of the carbon appeared as carbide. That cementite was formed in X337 and Hägg carbide in X273 may be attributed to the differences in temperatures of hydrogenation; however, the differences in catalyst type and composition, especially the presence or absence of copper, may be important in determining the nature of the carbide phase. In subsequent synthesis, the hydrogenated, used catalysts behaved like reduced catalysts in regard to their selectivity and tendency to oxidize.

In the second series, the changes in catalyst composition were determined as a function of time for fused catalyst D3001 with 1 H_2 + 1 CO gas at 7.8 and 21.4 atm (*56*). The nature and sequence of the pretreatment steps, chemical composition, and phases identified in the pretreated catalyst are given in Table 15.

Figure 17 shows the composition changes that occurred in a typical completely nitrided ϵ-iron nitride catalyst (test X670, Table 15) during a

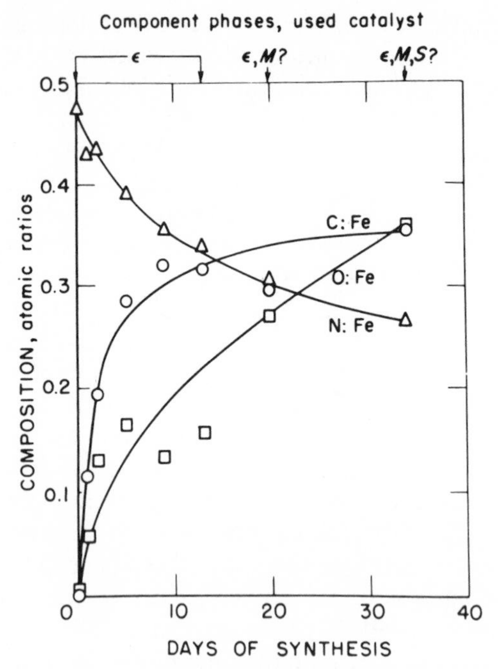

Fig. 17 Composition changes of a nitrided D3001 during synthesis at 21.4 atm. Symbols denoting phases are defined in Table 15.

synthesis experiment at 21.4 atm. Carbonitride was formed during synthesis by reaction of the nitride with CO in the synthesis gas. The comparatively rapid rate of nitrogen loss during the first few days of the experiment, and the rapid carbon uptake and the persistence of the ε phase as determined by X-ray diffraction, are evidence of ε-carbonitride formation. Carbon gain, however, was considerably greater than nitrogen depletion, which is interpreted to mean deposition of free carbon occurred during this period. With time, the rate of loss of nitrogen decreased substantially, so that at 35 days of operation the ratio N : Fe was still more than half the initial value.

As shown by Fig. 18, retention of nitrogen was not significantly affected by operating pressure during synthesis at 7.8 and 21.4 atm; refer to test X218 (*62*), Table 15.

Iron catalysts are gradually oxidized during the FTS. In Fig. 17 the rate of this reaction may be compared with rates of gain of carbon and loss of nitrogen. The slope of the O : Fe curve and the appearance of magnetite (M) and possibly siderite or magnesite (S) as major phases after 20–30 days of operation show that oxidation of iron is progressive in nitrided

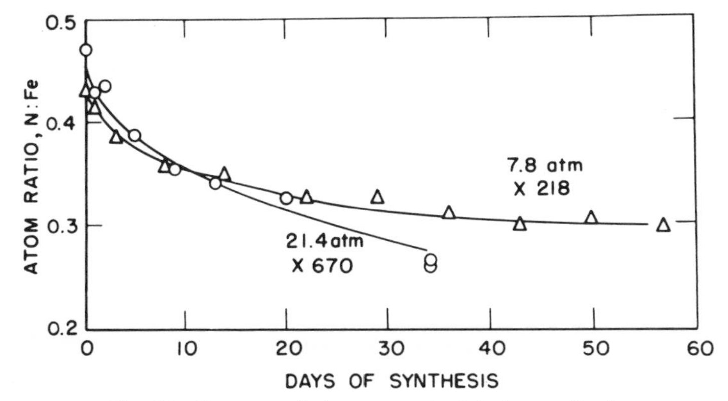

Fig. 18 Decrease of nitrogen content during synthesis.

catalysts, as it is in reduced ones. Both nitrided (tests X218 and X670, Table 15) and reduced catalysts (tests X194 and X361) showed increased rates of oxidation with increasing pressure (Fig. 19). Comparison of the performance of a nitrided with a reduced catalyst tested at the same operating pressure shows that nitriding substantially retarded oxidation (*10, 57, 58*).

As shown in Fig. 20, the nitrided catalyst at 21.4 atm gained more carbon during synthesis than the reduced catalysts, whereas in tests at 7.8 atm the carbon gain was slightly larger for the reduced catalyst. For the nitrided catalyst the gain in carbon was almost doubled by increasing the pressure from 7.8 to 21.4 atm; for the reduced catalyst the carbon content was almost the same at both pressures.

3.5.2 Activity, Selectivity, and Composition Changes of Partly Nitrided Catalysts

The influence of initial nitrogen content of catalysts on their performance in the synthesis was studied at 21.4 atm with 1 H_2 + 1 CO synthesis gas using fused catalyst D3001 (*56*). Time and temperature of nitriding were varied over wide limits to attain the desired range of nitrogen content (Table 15). This temperature variation probably does not influence the behavior of the catalysts significantly.

At the lower ratio of nitrogen to iron, to 0.26, nitriding produced γ'-iron nitride and metallic iron as the major components (Table 15, control test X517 and tests X274, X447, X766, X445, and X219). In the synthesis, the γ' phase proved unstable; in every experiment except X219 it disappeared completely and is assumed to have been transformed to magnetite, χ-iron carbide, and metallic iron.

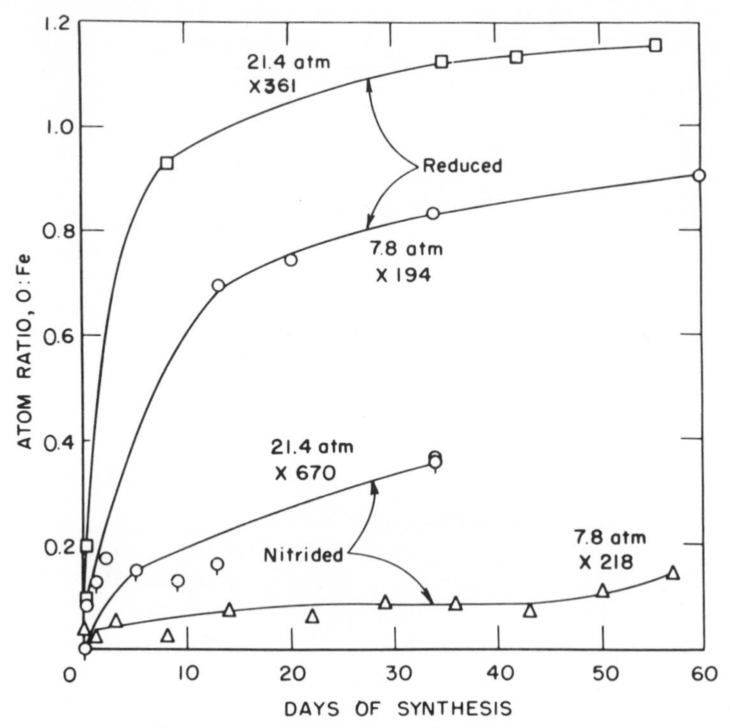

Fig. 19 Increase of oxygen in catalyst during synthesis.

In tests X617 and X711 with N : Fe of 0.35 and 0.38, respectively, the pretreated catalyst contained both the γ' and ϵ phases, but after synthesis the catalyst contained ϵ-carbonitride and magnetite.

On nitriding to N : Fe values above 0.40, ϵ-nitride was the principal phase formed (Table 15, tests X612, X620, X518 and X670). The ζ phase was also found in test X225. During synthesis, these phases were converted to ϵ-iron carbonitride, which persisted as a major component. Despite the nitrogen content of the catalysts decreasing with time, the ϵ-iron carbonitride structure was maintained, and the products of synthesis retained the characteristics associated with nitrided catalysts, even at N : Fe ratios as low as 0.14.

In Fig. 21, activity is plotted as a function of the initial ratio of nitrogen to iron. In a number of tests with nitrided-iron catalysts, more variability in activity was observed than in other synthesis tests. Examination of the catalyst after reducing and nitriding indicated that some particles had disintegrated to a fine powder, and because activity increases with decreasing particle size, marked deviations of activity from normal values were observed. Fines are also formed in the reduction of catalysts, but to

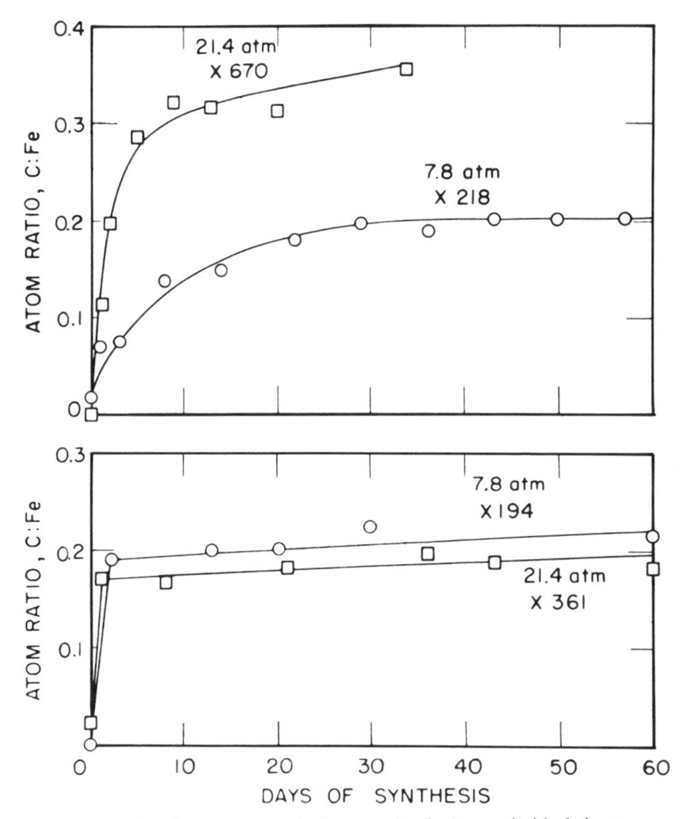

Fig. 20 Increase of carbon content during synthesis (top, nitrided; bottom, reduced).

a smaller extent, and reproducibility of activity is not seriously affected. Particle disintegration during pretreatment was fully appreciated only after the first part of the data for Fig. 21 had been obtained. Subsequently, several tests were made with catalysts that had been sieved under heptane to remove fines. In both sets of data, activity increased with nitrogen content to N : Fe values ranging from 0.20 to 0.25 and then remained substantially constant. Examination of catalysts after synthesis by X-ray diffraction showed that those in which the initial N : Fe was 0.25 or greater generally contained the ϵ-phase as the major component, whereas in those containing smaller amounts of nitrogen, χ-iron carbide and magnetite predominated.

Product distribution or selectivity as a function of initial N : Fe ratio is shown in Figs. 22 and 23. In these curves, the term "total hydrocarbons" includes oxygenated organic compounds dissolved in the oil phase. Figure 22 is a plot of synthesis products exclusive of CO_2, water, and water-

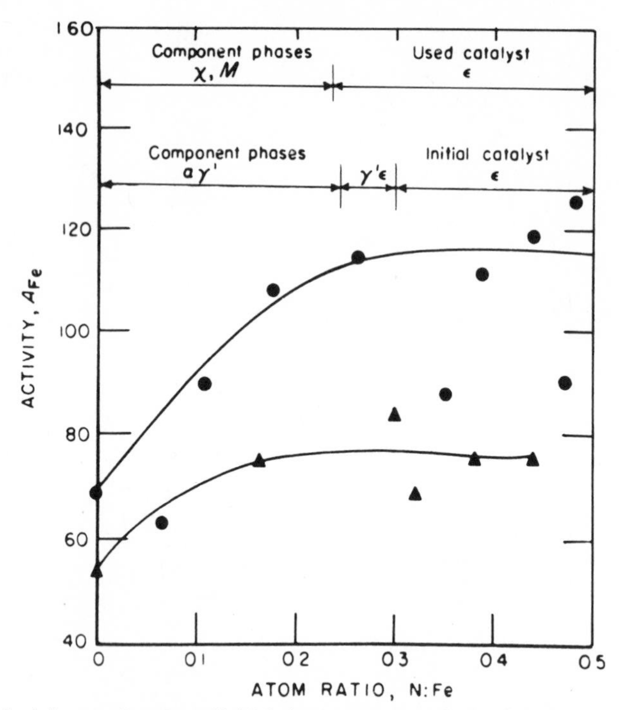

Fig. 21 Activity as a function of initial nitrogen content. ●, Unsieved catalyst; ▲, sieved catalyst. Symbols denoting phases are defined in Table 15.

soluble chemicals. The curves are based on mass spectrometric analysis of product gases and a simple distillation of liquid fractions. Figure 23 shows the weight percentage of C=C in the α and β positions and of total alcoholic hydroxyl in the fraction boiling from room temperature to 185°C, as determined by infrared spectrometry. Figure 24 shows the O : Fe atom ratio of the catalyst after 6 weeks of synthesis as a function of initial nitrogen content.

As the initial nitrogen content of the catalyst was increased, the activity increased to N : Fe = ~0.25 and then remained relatively constant. Also, at about this position a major change in selectivity occurred. Specifically, increased production of alcohols and products of lower molecular weight associated with nitrides and carbonitrides was observed. At N : Fe ratios less than 0.25, the products were characteristic of reduced or carbided catalysts.

Phase changes during synthesis (Table 15) indicate that the principal phases present in used catalysts with initial N : Fe less than 0.25 were magnetite and Hägg carbide, whereas for initial values greater than 0.25 ϵ-carbonitride was generally the principal phase. In previous experiments

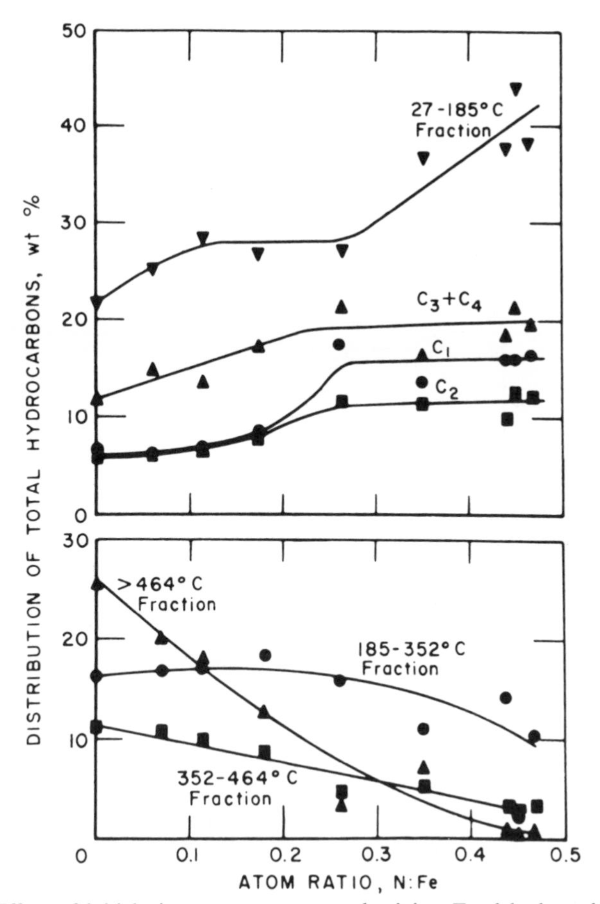

Fig. 22 Effect of initial nitrogen content on selectivity. Total hydrocarbons includes oxygenates dissolved in liquid phase.

(*36*) at atmospheric pressure with synthesis gas or CO, a sample with N : Fe = 0.08 was converted to Hägg carbide, another with N : Fe = 0.21 to ϵ-carbonitride. During synthesis, the carbon and oxygen contents of the partly nitrided samples increased more rapidly than for fully nitrided ϵ-iron nitrides under the same conditions.

Thus, the experiments on partly nitrided catalysts show that the ϵ-iron carbonitride does not form in the synthesis unless nitriding has already established the ϵ-iron nitride phase. Other forms of nitrides revert to metallic iron, iron carbides, magnetite, and perhaps other phases that are more readily oxidized and characteristic of the operation of reduced or carbided iron catalyst.

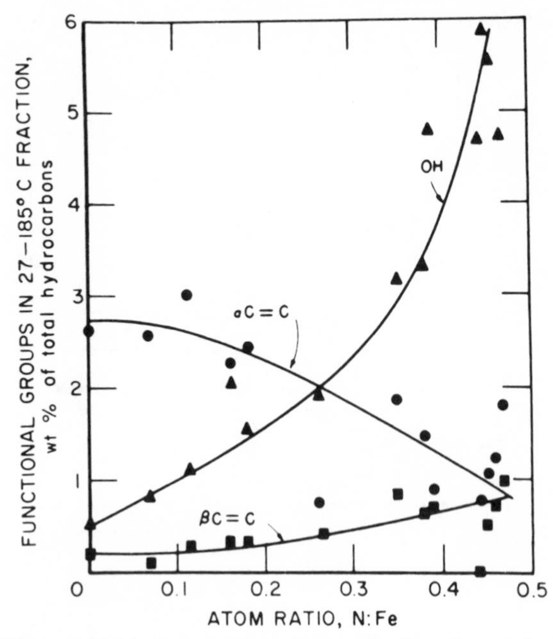

Fig. 23 Effect of initial nitrogen content on production of alcohols and olefins.

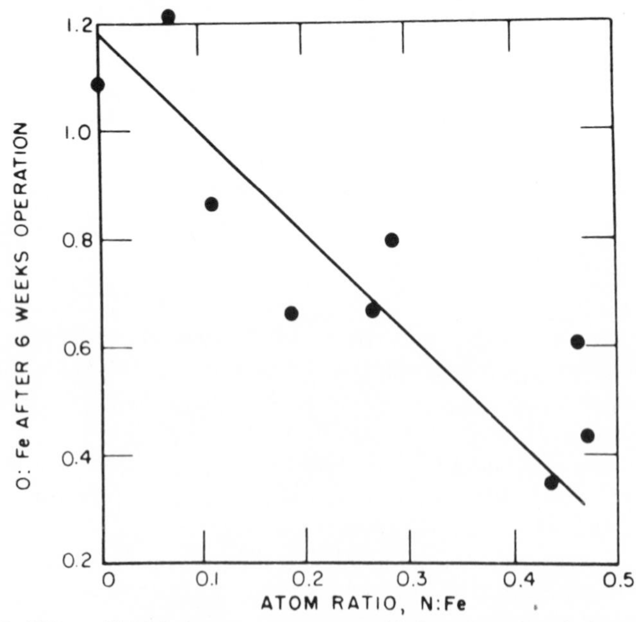

Fig. 24 Effect of initial nitrogen content on oxidation of catalyst during synthesis.

3.5.3 Synthesis Tests with Iron Catalysts Converted to ϵ-Iron Carbonitride

A series of ϵ-iron carbonitrides prepared prior to synthesis was also tested (*56*). All experiments were made with 6- to 8-mesh particles of catalyst D3001. Pretreated and used catalysts were analyzed by chemical methods for iron, carbon, and nitrogen, and oxygen was determined by difference. Phases present were identified by X-ray diffraction. The nature and sequence of pretreatment steps, chemical composition, and phases in pretreated and used catalysts are given in Table 16. The symbols, C : Fe, N : Fe, and O : Fe denote the atomic ratios of carbon, nitrogen, and oxygen to iron.

To study the effect of preparational variables (Table 16), carbonitrides were prepared by nitriding reduced catalysts in ammonia and then carbiding with CO (X287, X252, and X279), by carbiding with CO and then nitriding with ammonia (X407), or by simultaneous carbiding and nitriding with methylamine (X739).

In the synthesis tests, the ϵ-iron carbonitride persisted throughout the experiments. The catalyst of test X739 was not completely reduced before treatment with methylamine; however, based on the part of the catalyst that was reduced, N : Fe and C : Fe ratios were, respectively, 0.08 and 0.24. As only the outer layer of the catalyst is effective in the synthesis, this test is probably comparable, despite the lower extent of reduction.

In Figs. 25 and 26, the activities and selectivities of the carbonitrided catalysts of Table 16 are shown for experiments at 7.8 and 21.4 atm. As in previous presentations of selectivity, only hydrocarbons and oxygenated molecules dissolved in liquid hydrocarbons are considered. Data for carbided (tests X294 and X342) and nitrided (tests X215 and X225) catalysts are included for comparison. Increased production of alcohols and decreased production of high-boiling products were characteristic of the catalysts containing nitrogen, with the exception of test X287, wherein nitriding was insufficient to produce the ϵ phase when subsequently carbided with CO. As a result, only the χ phase was formed, and therefore this catalyst behaved more like the carbide in test X294.

Activity values for ϵ-nitrides and ϵ-carbonitrides prepared by carburizing nitrides were usually greater than for reduced or carbided catalysts, or for ϵ-carbonitrides prepared from carbide.

Both nitrided and reduced catalyst oxidized more rapidly at 21.4 than at 7.8 atm but the nitrides oxidized at lower rates at both pressures. Carburized catalysts oxidized more rapidly at the higher pressures (*16, 57*). The depletion in nitrogen was not noticeably affected by pressure. Specifically, the percentage of iron present as ϵ-carbonitride decreases more rapidly at the higher pressure, and the ratio of carbon to nitrogen in the

Table 16
Tests of Carbonitrides[a]

	Test, X-						
	342	*294*[b]	*287*	*252*[b]	*279*	*407*	*739*
Reduction in H_2[c]							
Temp., °C	500	500	550	550	550	550	450
Time, h	24	24	20	20	20	22	24
Second pretreatment[d]							
Reagent	CO	CO	NH_3	NH_3	NH_3	CO	CH_3NH_3
Temp., °C	150–350	150–350	350	400	400	150–250	250
Time, h	18	18	6	1	1	12	16
Third pretreatment[d]							
Reagent	—	—	CO	CO	CO	NH_3	—
Temp., °C	—	—	450	350	350	350	—
Time, h	—	—	10	6	6	28	—
Composition of pretreated catalyst, atom ratios							
C : Fe	0.57	0.58	0.62	0.19	0.30	0.21	0.15
N : Fe	0	0	0.02	0.29	0.23	0.29	0.05
O : Fe	0.07	0.03	0.16	0.03	0.08	0.13	0.85
Phases[e]	χ,α	χ,α	χ,M	ϵ	ϵ	ϵ	M,ϵ
Catalyst composition after synthesis, atom ratios							
C : Fe	0.49	0.71	0.63	—	0.42	0.44	0.24
N : Fe	0	0	0.2	—	0.15	0.08	0.02
O : Fe	1.48	0.37	0.32	—	0.60	0.81	1.09
Phases[e]	χ,α,S	χ,M	χ,M	ϵ	ϵ,M,χ	ϵ,M	M,ϵ
Activity[f]	95	52	65	69	108	93	114

[a] All tests with D3001 at 21.4 atm, except as noted; all pretreatments at 1 atm.
[b] Synthesis at 7.8 atm.
[c] Hourly space velocity of H_2, 2300–2700. Samples 96–98% reduced except for X739, for which reduction was 63%.
[d] Hourly space velocities: NH_3, 950–1050; CO, 95–105; methylamine, 1000.
[e] Phases from X-ray diffraction in order of decreasing intensity of pattern: α, metallic iron; χ, Hägg carbide; M, magnetite; S, $MgCO_3$ or $FeCO_3$; ϵ, ϵ nitride or carbonitrile.
[f] Activity per gram Fe at 240°C, see Table 11.

carbonitride increases more slowly at the higher pressure. Therefore, nitrogen is replaced by carbon more slowly, relative to the oxidation reaction at the higher pressure, and carbonitrides rich in carbon are more rapidly oxidized than those that are carbon poor.

ϵ-Carbonitrides prepared by treating nitrides with CO had about the same activity as carbonitrides produced during synthesis, as shown in all the experiments here reported. ϵ-Carbonitride prepared by nitriding

Test number	X294	X287	X252	X215
Initial N:Fe	0	0.02	0.29	0.46
Initial C:Fe	0.58	0.62	0.19	0
Temperature, °C	229	227	222	222
Space velocity, h^{-1}	106	100	94	99
Contraction, %	63.7	67.8	66.8	62.4
Activity, A_{Fe}	53	63	68	65
Initial component phases	χ, α	χ, M	ε	ε

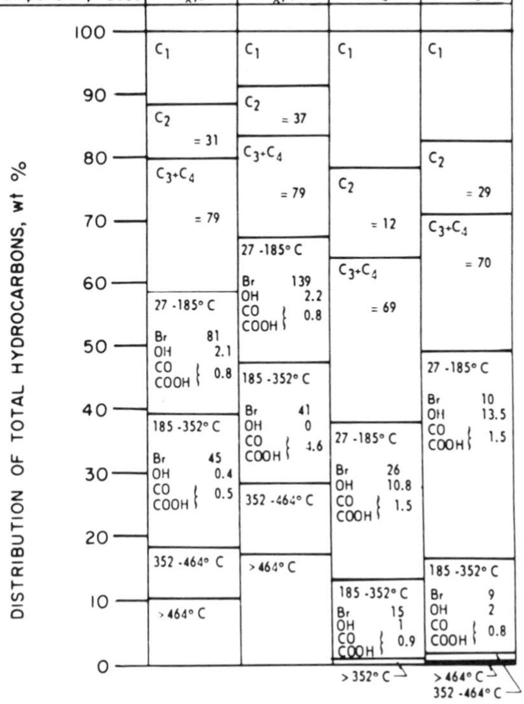

DISTRIBUTION OF TOTAL HYDROCARBONS, wt %

100 90 80 70 60 50 40 30 20 10 0

X294: C_1; C_2 = 31; C_3+C_4 = 79; 27-185° C: Br 81, OH 2.1, CO/COOH 0.8; 185-352° C: Br 45, OH 0.4, CO/COOH 0.5; 352-464° C; > 464° C

X287: C_1; C_2 = 37; C_3+C_4 = 79; 27-185° C: Br 139, OH 2.2, CO/COOH 0.8; 185-352° C: Br 41, OH 0, CO/COOH 4.6; 352-464° C; > 464° C

X252: C_1; C_2 = 12; C_3+C_4 = 69; 27-185° C: Br 26, OH 10.8, CO/COOH 1.5; 185-352° C: Br 15, OH 1, CO/COOH 0.9; > 352° C

X215: C_1; C_2 = 29; C_3+C_4 = 70; 27-185° C: Br 10, OH 13.5, CO/COOH 1.5; 185-352° C: Br 9, OH 2, CO/COOH 0.8; > 464° C; 352-464° C

Fig. 25 Activity and selectivity of carbides and carbonitrides in synthesis at 7.8 atm. Symbols denoting phases are defined in Table 16. Total hydrocarbons includes oxygenates dissolved in liquids.

χ-iron carbide with ammonia showed lower activity and formed products with a slightly higher molecular weight than carbonitrides produced in the reverse order. More extensive carburization of a nitride to remove the nitrogen fairly completely produced χ-iron carbide that had the same catalytic properties as the χ-iron carbide produced by direct carburization of a reduced catalyst. The ε-carbonitride from methylamine produced the low molecular weight product characteristic of ε-carbonitrides; however,

Test number	X342	X279	X407	X225	X739
Initial N:Fe	0	0.23	0.29	0.41	0.05
Initial C:Fe	0.57	0.30	0.21	0	0.15
Temperature, °C	243	238	245	238	239
Space velocity h^{-1}	334	294	313	297	300
Contraction, %	63.2	64.9	63.2	64.6	64.1
Activity, A_{Fe}	104	112	93	119	114
Initial component phases	χ, α	ϵ	ϵ	ζ, ϵ	ϵ, M

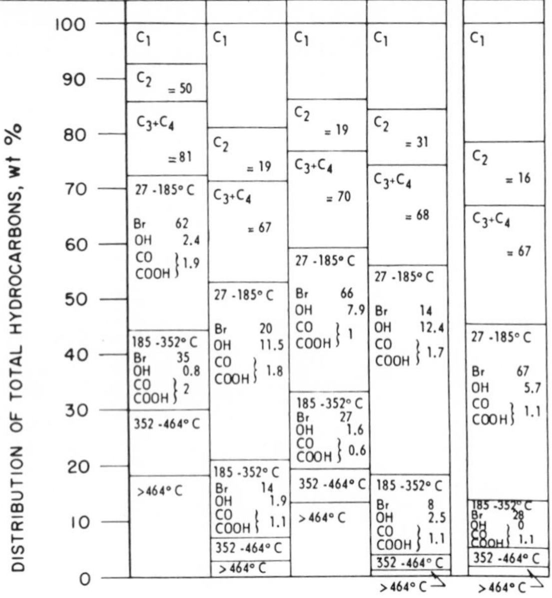

DISTRIBUTION OF TOTAL HYDROCARBONS, wt %

100, 90, 80, 70, 60, 50, 40, 30, 20, 10, 0

X342: C_1; C_2 = 50; C_3+C_4 = 81; 27-185° C: Br 62, OH 2.4, CO, COOH 1.9; 185-352° C: Br 35, OH 0.8, CO, COOH 2; 352-464° C; >464° C

X279: C_1; C_2 = 19; C_3+C_4 = 67; 27-185° C: Br 20, OH 11.5, CO, COOH 1.8; 185-352° C: Br 14, OH 1.9, CO, COOH 1.1; 352-464° C; >464° C

X407: C_1; C_2 = 19; C_3+C_4 = 70; 27-185° C: Br 66, OH 7.9, CO, COOH 1; 185-352° C: Br 27, OH 1.6, CO, COOH 0.6; 352-464° C; >464° C

X225: C_1; C_2 = 31; C_3+C_4 = 68; 27-185° C: Br 14, OH 12.4, CO, COOH 1.7; 185-352° C: Br 8, OH 2.5, CO, COOH 1.1; 352-464° C; >464° C

X739: C_1; C_2 = 16; C_3+C_4 = 67; 27-185° C: Br 67, OH 5.7, CO, COOH 1.1; 185-352° C: Br 28, OH 0, CO, COOH 1.1; 352-464° C; >464° C

Fig. 26 Activity and selectivity of carbides and carbonitrides in synthesis at 21.4 atm.

a smaller amount of oxygenated compounds and more olefins were formed.

In other studies, reduced catalyst D3001 was treated in methylamine at 200, 250, and 300°C in a glass apparatus; however, the principal phase was Hägg carbide in all of these experiments.

3.5.4 Other Studies of Nitrides and Carbonitrides of Iron

Tests were made of nitrided iron catalysts at the U.S. Bureau of Mines in small pilot-plant reactors of the fluidized, fixed-bed; slurry; and oil-circulation types. Table 17 compares products from these reactors (*6, 28*). Table 18 compares these data with products from the fixed-bed and entrained catalyst processes of SASOL (*44, 54*). The SASOL fixed-bed product is typical of reduced catalysts operated at 230–260°C, and the

Table 17
Selectivity of a Nitrided, Fused-Iron Catalyst in Three Types of Reactors[a]

	Reactor type		
Operating conditions and yields	*Fluidized*	*Slurry*	*Oil circulation, fixed bed*
Pressure, atm	21.4	21.4	21.4
Temperature, °C	251	247	230
Space velocity, h^{-1}	1000[b]	100[c]	200[d]
Conversion of H_2 + CO, %	68	69	69
Recycle ratio	9/1	2/1	1/1
Usage ratio, H_2:CO	1.2	0.95	0.75
Yield, g/m³, of H_2 + CO converted			
$C_1 + C_2$	72	60	58
C_3 + gases	54	59	58
Oxygenated compounds			
In oil phase	29	30	6
In aqueous phase	60	38	41
Total	89	68	47
Liquid hydrocarbons	14	21	36
Oxygenated compounds plus liquid hydrocarbons	103	89	82
Water	88	65	16
CO_2	329	422	535

[a] Adapted from Demeter and Schlesinger (*28*) and reproduced by courtesy of Marcel Dekker, Inc. from Anderson (*6*).
[b] Based on settled-bed volume.
[c] Based on slurry volume.
[d] Based on fixed-bed volume.

entrained product is typical of reduced iron at 325°C, including fluidized fixed-bed reactors.

The nitrides usually produced a larger fraction of $C_1 + C_2$ and larger yields of alcohols. The alcohols can be used in a variety of ways, but the $C_1 + C_2$ fraction will probably have value only as fuel or reformer feedstock. Nitrided iron in the hot-gas-recycle, entrained, or tube-wall reactors may be expected to yield products similar to those of the fluidized reactor (Tables 17 and 18). Nitrided catalysts in the fixed-bed reactors of SASOL would be expected to have selectivities similar to those of the fixed-bed tests (Table 18).

In the pilot-plant tests, nitrogen was lost only very slowly and was largely replaced with carbon. Oxidation of the iron was less rapid than with reduced catalysts, and the deposition of elemental carbon was not excessive. For example, the analysis of nitrided catalyst D3001 after 35

Table 18
Products from Nitrided Iron and from SASOL Reactors[a]

	Weight percentage					
			Oxygenates[b] *in*			
Reactor types	$C_1 + C_2$	$C_3 + C_4$	*Aqueous phase*	*Oil phase*	*Oil*	*Wax*
SASOL reduced-iron catalysts						
Fixed bed	11.9	10.7	3.4	[c]	47.6	26.4
Entrained	23.2	27.5	10.9	[c]	36.9	—
Nitrided-iron catalysts						
Fluidized fixed bed	30.9	23.2	25.7	13.6	5.8	—
Oil circulation	29.2	29.2	20.5	2.9	17.8	—
Slurry	28.8	28.4	18.2	14.6	9.9	—
Fixed bed, principal promoters						
Al–K	22.5	20.5	34.7		22.3	—
Mg–K	25.1	22.0	23.4		29.3	—

[a] Reproduced with permission from Anderson (*6*).
[b] Mostly alcohols.
[c] Probably included in "oil."

days of synthesis in a fluidized–fixed bed reactor was iron, 76.0; nitrogen, 4.0; carbon, 6.7; oxygen, 2.5 (by difference); hydrocarbons, 3.7; and promoters, 7.0 wt % (*28*). Sieve analysis suggested that the particles increased slightly in size during synthesis; the fraction of fines, less than 230 mesh, increased from 1.6 to 3.9 wt %.

Synthesis gas containing 1 H_2 + 1 CO has generally been used with nitrides, and the usage ratio can be readily adjusted to equal the feed ratio by changing the recycle ratio. The operation of nitrided catalysts was virtually trouble free, and nearly all of the U.S. Bureau of Mines pilot plant tests were terminated voluntarily with the activity and mechanical form of the catalyst still adequate for further use. The length of operation of nitrided catalysts exceeded substantially the lives of all other catalysts in these pilot-plant reactors. The length of operation for three types of reactors and the space–time yield in terms of volumes (S.T.P.) of H_2 + CO consumed per volume of catalyst space per hour are given in Table 19; the space–time yields of 200–600 h^{-1} are typical of the productivity of other iron catalysts in these reactors. Anderson (*6*) has considered the usefulness of nitrided iron catalysts in the production of gasoline.

Aristoff *et al.* (*17*) studied the synthesis with reduced and reduced-and-nitrided fused catalysts (Fe_3O_4–Al_2O_3–K_2O) at ~285°C in a fixed-bed reactor that simulated conditions in a fluidized system. Results of these

Table 19
Long Tests of Nitrided Catalysts[a]

	Type of reactor		
Operating conditions and yields	*Fluidized, fixed bed*	*Slurry*	*Oil circulation*
Length of test, days	35	133	154
Feed ratio, H_2/CO	1.0	1.0	1.0, 0.7
Hourly space velocity	500–1000	200–500	200–500
Recycle ratio	8–12	0–0.5	0.6–1.0
Maximum space–time yields, h^{-1}	610	170	210

[a] All tests were made at 21.4 atm and were terminated voluntarily. Space velocities are based on settled volume of catalyst. Reproduced with permission from Anderson (*6*).

tests are presented in Table 20. The nitrided catalyst was more active and produced less gaseous hydrocarbons than the corresponding reduced catalyst. At 285°C the nitrogen content of the nitrided catalyst decreased rapidly, 84.6% of the nitrogen being removed in 153 h. On this basis it would be expected that the nitride was largely converted to either Hägg carbide or magnetite, and the satisfactory selectivity of the nitride may possibly be related to the presence of these other phases during most of the test. Carbon deposition was 0.8 and 0.6 g C per cubic meter of synthe-

Table 20
Tests of Reduced and Nitrided Catalysts at High Space Velocity and Temperature[a]

Atom ratio, N : Fe	0	0.40
Gas composition, H_2 : CO	1.3	1.0
Space velocity, h^{-1}	2300	2820
Temperature, °C	316	288
Contraction, %	51	48
Yield, wt % of total "hydrocarbons"		
C_1	10.3	8.6
C_2	14.8	13.7
C_3	20.4	16.6
C_4	16.8	14.2
Condensed products		
Up to 204°C	26.2	36.3
Heavy oil	7.2	9.5
Wax	4.3	1.1

[a] Operating pressure, 25 atm. Reproduced with permission from Anderson (*17*).

sis gas for the reduced and the nitrided preparations, respectively. These data suggest that nitriding is advantageous even for operation at temperatures at which the nitrides are relatively unstable.

Basu and Murthy studied a nitrided precipitated iron catalyst in a fluidized reactor at 3.4–10 atm and 250–320°C (*18*). Yields of $C_1 + C_2$ and liquids were 65–80 and less than 60 g/m^3 of $H_2 + CO$ converted, respectively, and the liquids contained up to 60% oxygenated chemicals. The tests were short, of 8–10 h duration, and at the end of the series waxy deposits on the particles caused pressure drop. Used catalyst analyses were not given, but it may be inferred that the nitrogen had been largely eliminated by the end of the series.

Borghard and Bennett (*19*) studied a number of commercial catalysts in the FTS at 20 atm. Nitrided fused iron and cobalt on silica were the most active.

3.6 Structural and Chemical Changes in Catalysts during Pretreatment and Synthesis

This section describes investigations of the pore geometry of iron Fischer–Tropsch catalysts during various stages of pretreatment and during synthesis. These data provide a picture of the pore structure of the catalyst, which is required for a rational interpretation of the reaction.

3.6.1 Structural Changes in Reduced Iron Catalysts on Formation of Oxide and Interstitial Phases

Although catalysts may sinter or otherwise change their physical structure in many catalytic processes (*9, 22, 31, 32, 51*), their chemical composition is seldom appreciably altered. When chemical changes do occur, they usually have a marked effect on the process. The FTS over initially reduced iron is one instance of a catalyst undergoing chemical change in the course of operation, namely, reoxidation and carburization. Various authors (*16, 27, 47, 66*) have attributed deactivation and short catalyst life to some of these reactions, and attempts have been made to find promoters and iron phases (*16, 61, 62*) that catalyze the synthesis while resisting chemical changes. The present study deals with the physical changes accompanying the nitriding, carbiding, and reoxidation processes on a fused-iron catalyst (*37*).

Structural changes accompanying the initial reduction of this catalyst and its chemical changes during the synthesis have been described (*11*, *37*). Kölbel and Engelhardt (*47*) measured the rate of reoxidation of inducted iron catalysts with water vapor. The carbides and nitrides of iron have been extensively studied (*16, 25, 26, 30, 35, 48*). Podgurski *et al.* (*55*) showed that little or no change in surface area occurred on carbiding, but CO chemisorption decreased 10- to 30-fold as the reaction progressed. Our experiments confirmed this work. The densities of the iron carbides and nitrides are known to be lower than the density of iron, but nowhere have the changes in catalyst structure been described with formation of these phases or the oxide phase.

All experiments were carried out in glass tubes equipped with special four-way stopcocks (*2, 59*). These sample vessels were heated in a horizontal position in a small resistance furnace with automatic temperature control to ±3°C. Changes in the samples were determined by weighing the entire tube on an analytical balance, and the loss or gain of oxygen, carbon, or nitrogen was calculated. D3001 catalyst, 6- to 8-mesh had ~45% porosity and an average pore diameter of ~800 Å when completely reduced at 550°C (*37*). The oxidations were carried out with nitrogen, saturated with freshly boiled distilled water. Preparation of the nitrides has been described in the previous sections. The hourly space velocity of the gas was 1000 in reduction, oxidation, and nitriding.

Hägg carbide was prepared by treating the reduced catalyst with CO at an hourly space velocity of 100. Carbiding was started at 150°C and the temperature was increased, as required to maintain the concentration of CO_2 in the exit gas at ~20%, to a maximum of 350°C. The exit gas was passed through a thermal conductivity gas-analysis apparatus, which controlled the rate of temperature increase of the electric furnace. Nearly pure Hägg carbide was formed in this way.

The methods for making the adsorption and density measurements have been described (*7–9, 12, 37, 51*). No wetting of the catalysts by mercury was observed. Surface areas were calculated from the V_m values obtained from the simple BET equation (*24*), using 16.2 Å^2 as the cross-sectional area of the adsorbed nitrogen molecule.

Figure 27 shows the extent of reduction as a function of time during reduction and oxidation. After initial reduction with pure, dry H_2 at 550°C (curve O–A), the rate of oxidation of the catalyst at 250°C with water vapor (A′–B) was rapid at first and then decreased to a low value. Rereduction of the partly reoxidized catalyst at only 300°C (B′–C) was almost as rapid as the initial reduction at 550°C. The sample was again oxidized (C′–D), this time at 325°C, with similar results. At the end of ~100 h, the temperature was increased to 450°C in an attempt to increase the rate of oxidation. An increase was observed, but this may have been caused by

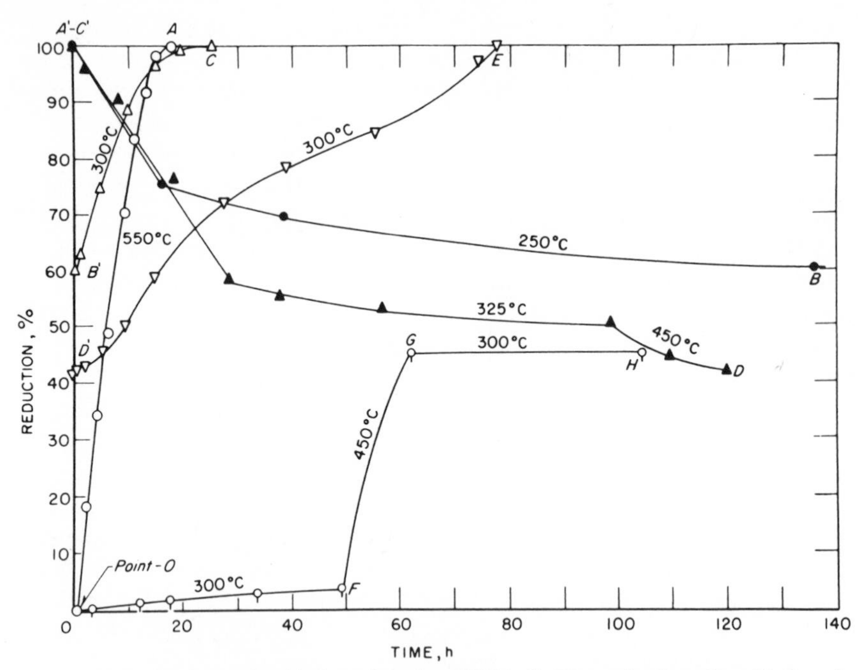

Fig. 27 Reduction and oxidation of catalyst D3001. ○, First reduction; ●, oxidation; △, second reduction; ▲, oxidation; ▽, third reduction; ♀, first reduction of duplicate sample.

surface oxidation to hematite rather than by bulk oxidation to magnetite, indicated by the reddish color of the catalyst. On rereduction at 300°C (D′–E), the rate of reduction was again much greater than that of the raw catalyst under the same conditions (O–F). After a raw sample had been reduced nearly 50% at 300 and 450°C (O–F–G), it was again treated with hydrogen for ~40 h more at 300°C (G–H). These data show clearly that the reoxidized catalyst was much easier to reduce than the raw or the partly reduced catalyst.

Adsorption measurements made at the end of the various steps shown in Fig. 27 are given in Table 21. In all cases the surface area decreased on reoxidation and increased to about its initial value when rereduced. The chemisorption of CO was sharply decreased by reoxidation; the ratios $V_{CO} : V_m$[1] indicate that the relative amount of free iron on the surface of the reoxidized catalyst was less than half the amount of the reduced catalyst. As the total surface area was decreased proportionally by reox-

[1] V_{CO} and V_m are the volumes of chemisorbed CO and the nitrogen monolayer capacity in cubic centimeters (S.T.P.), respectively. Carbon monoxide chemisorbs only on the part of the surface that is iron (*23*). The ratio $V_{CO} : V_m$, therefore, gives a measure of the fraction of the surface that is iron.

Table 21
Adsorption Measurements on Reduced and Reoxidized Catalyst D3001

Gas	*Treatment* Temp., °C	Time, h	Point on Fig. 27	*Extent of reduction, %*	*Surface area, m^2 per gram unreduced catalyst*	V_m[a] (S.T.P.)	V_{CO}[b] (S.T.P.)	V_{CO}/V_m	Area[c]/f	Area[c]/$f^{1/2}$
	—	—	0	0	0	0	0			
H_2[d]	550	18	A	99.7	4.7	1.07	(0.4)[e]	(0.4)[e]	4.7	4.7
$N_2 + H_2O$	250	136	B	60.4	2.6	0.59	0.1	0.2	4.3	3.3
H_2	300	27	C	100.0	4.5	1.04	0.5	0.5	4.5	4.5
$N_2 + H_2O$	{325 450	98 21.5}	D	42.1	1.7	0.39	0.04	0.1	4.1	2.6
H_2[d]	550	19	—	99.9	—	—			—	—
$N_2 + H_2O$	250	70	[f]	68.4	2.9	0.66			4.2	3.5
H_2[d]	550	18.5	[f]	99.8	—	—			—	—
$N_2 + H_2O$	250	66	[f]	68.2	2.8	0.64			4.1	3.4
H_2[d]	{450 500	56 5}	—	98.0	8.2	1.88			8.4	8.3
$N_2 + H_2O$	250	65	—	49.8	3.9	0.89			7.8	5.5

[a] Calculated by use of the simple BET equation (*24*); per gram of raw catalyst.
[b] Difference between the N_2 and CO isotherms at −195°C per gram of raw catalyst.
[c] f, Extent of reduction, expressed as a fraction.
[d] Original reduction of raw catalyst.
[e] V_{CO}/V_m from previous work (*37*); V_{CO} was calculated using V_m for this sample.
[f] Density measurements were made on this sample.

idation, the amount of iron (on the surface) available for oxidation was very small when the rate of reoxidation became low. The same behavior was observed for the sample initially reduced at 450 and 500°C (Table 21), except that the surface areas were about twice as large as those measured on the sample initially reduced at a higher temperature (*37*). The last two columns of Table 21 are discussed later.

Structural data for various solid phases prepared from this catalyst are summarized in Table 22, where the quantities used for comparing catalyst—volume of mercury displaced (V_{Hg}), pore volume, and surface area—are expressed per gram of unreduced catalyst. Thus, the reduced and reoxidized samples displaced the same volume of mercury as the raw sample, but the carbided and nitrided catalysts displaced 12–18% more (V_{Hg}, column 5). On reduction, the catalyst developed nearly all its pore volume (column 6), which decreased on reoxidation but increased enough on nitriding and carbiding to keep the percentage porosity approximately constant (column 8). The average pore diameters, calculated from the equation of Emmett and DeWitt (*33*), increased from that of the reduced catalyst, with both reoxidation and the formation of interstitial phases. The densities of the reoxidized catalysts agree almost exactly with those observed for the same catalyst at the same extent of reduction during its initial reduction (*37*).

The density of Fe_2N, calculated from the data of Eisenhut and Kaupp (*30*), is 7.02 g/cm^3, compared to 6.82 from our data; thus, by assuming that the volumes of the phases present were additive, the density of our nitrided catalyst calculated from their data should be 6.50 g/cm^3, compared to the observed 6.37. Similarly, according to Jack (*45*) the density of Fe_2C is 7.20 g/cm^3, compared to 6.77 calculated from our data.

Reactions of oxygen or water vapor with metals generally start with formation of a new phase at the surface. The oxide–metal interface moves into the metal either by diffusion of oxygen into the metal or by metal diffusing through the oxide layer. The work of Evans (*34*) indicates that after magnetite forms an interface on the iron surface, the oxide grows by migration of iron through the oxide layer. Thus, the rate of reoxidation decreases rapidly as the oxide layer is formed, and the rate eventually approaches the rate of diffusion of iron through the oxide layer. The reduction of this oxide layer is more rapid than the initial reduction, because the reoxidized catalyst has a larger surface area than the raw catalyst. By this mechanism oxide can form in the pores without expanding the pore system and increasing the volume of mercury displaced by the catalyst. All the results can thus be explained qualitatively. To account quantitatively for the observed changes in surface area (Table 21), however, it is necessary to consider the geometrical changes in the pores

Table 22
Structural Data as a Function of Solid Phase in Catalyst D3001[a]

Treatment of catalyst	Phases present, X-ray analysis	Density, g/cm^3		Volumes, cm^3 per gram unreduced catalyst			Porosity, %	Surface area, m^2 per gram unreduced catalyst	Average pore diameter,[d] Å
		Helium, ρ_{He}	Mercury, ρ_{Hg}	V_{Hg}	Pore volume[b] Measured	Pore volume[b] Calculated			
None, raw	Fe_3O_4	4.96	4.91	0.203	0.002	0	1	0	—
Reduced at 550°C	α-Fe	6.95	3.71	0.201	0.093	0.093	47	4.7	785
Reoxidized[e] $f = 0.684$	α-Fe, Fe_3O_4	5.94	4.13	0.200	0.061	0.064	30	2.9	820
Reoxidized[e] $f = 0.682$		5.93	4.13	0.200	0.061	0.063	30	2.8	860
Carbided[f]	χ-Fe_2C, α-Fe(?)	6.28	3.63	0.225	0.096		42	[g]	820[g]
Nitrided[h]	ϵ-Fe_2N	6.37	3.48	0.237	0.107		45	4.8	898

[a] Except for the raw catalyst, all samples were reduced for 19 h at 550°C at a space velocity of 1000 volumes H_2 per volume catalyst per hour.
[b] Pore volume is the difference of the volumes of mercury and helium displaced per gram of unreduced catalyst.
[c] Pore volume of partly reduced catalyst equals the pore volume of completely reduced sample times extent of reduction.
[d] Density $d = 4 \times$ pore volume per surface area (*33*).
[e] Density measurements were made on this sample; f, extent of reduction.
[f] Atom ratio, C : Fe = 0.494.
[g] Surface area not determined; several other samples showed no change in surface area with carbiding for this catalyst. The surface was assumed unchanged, and the pore diameter was calculated on this basis.
[h] Atom ratio, N : Fe = 0.482.

themselves. Two limiting cases are considered: (a) the pore volume decreases by decreasing the total length of pores of constant average diameter; (b) the pores maintain their length, but their diameters decrease as the reaction proceeds.

In the case (a) the area divided by the extent of reduction f should be constant, and in case (b) the area divided by $f^{1/2}$ should be constant. Both ratios in Table 21 are low for reoxidized catalysts but case (b) worse than case (a). Three plausible explanations for the observed surface areas being smaller than those predicted by the two simplified mechanisms may be suggested:

1. The smaller pores are preferentially and more or less completely filled with oxide.
2. Some of the pores develop bottlenecks, that is, become partly plugged with oxide, so that they are permeable to helium at 30°C, but impermeable to nitrogen at −195°C.
3. The pore walls, originally rough, become smoother as the reoxidation proceeds.

The first two ways seem improbable, because in a real system complete blocking of a fraction of the pores would be expected. This would be accompanied by an excessive decrease in the helium density, which is not observed. The last possibility has several points in its favor, although there is little direct evidence to support it. This catalyst developed average pore diameters ranging from 330 to 2400 Å, at temperatures of the reduction from 450 to 650°C (*37*). This effect was almost entirely the result of change in surface area, rather than in pore volume. Further, a catalyst reduced completely at 450°C, when heat treated in helium for 16 h at 550°C, had the same pore diameter as that of the 550°C reduction. McCartney and Anderson (*49*) made electron micrographs of replicas stripped from a polished surface of this catalyst after reduction at 450 and 550°C. Although these showed pore openings that compared reasonably well with the pore diameters previously calculated from the surface-area–pore-volume data (*51*), no definite differences were observed in the dimensions of the replica structure from samples reduced at the two temperatures.

The changes in structure that occur concurrently with formation of the interstitial carbides and nitrides are easier to interpret. As the metal expands, the pore volume increases, but the surface area changes only slightly if at all. Consequently, the average pore diameter increases, but the porosity remains constant, the behavior of an expanding matrix. If a solid containing a hole is expanded, the apparent volume, pore volume, and pore diameter increase, the surface area increases only slightly, and the percent porosity remains constant. The available data fit this simple picture.

3.6.2 Structural Changes of Iron Catalysts during Pretreatment and Synthesis

In this section structural changes occurring in iron catalysts during pretreatment and synthesis are described and related to the results reported in Section 3.6.1 (*63*).

For iron catalysts, structural changes during synthesis are more complex than for cobalt catalysts. Studies of a used cobalt catalyst (*9*) have shown that whereas the total volume of the catalyst does not change during synthesis, sizable decreases in the surface area and pore volume occur. These decreases are attributed to the accumulation of waxy products in the pores of the catalyst. On removing the wax by a simple hydrogen treatment, the original porous structure of the catalyst was regained. Carbiding or oxidation of the cobalt does not occur, nor are appreciable amounts of elemental carbon deposited. With iron catalysts, elemental carbon is deposited as well as wax, and most of the iron is converted to carbides and magnetite.

The present section reports structural data obtained from two types of iron catalysts after they had been used in the synthesis: precipitated iron oxide catalyst P3003.24 and fused–iron oxide catalyst D3001. Preparation, composition, and structural information pertaining to these catalysts are described in Section 3.6.1.

Catalyst P3003.24 was pretreated in a catalyst testing unit at atmospheric pressure for 24 h with 1 H_2 + 1 CO gas at 230°C and operated (*16, 35, 64*) for ~7 weeks at 7.8 atm and an average temperature of 221°C. A reduced sample of D3001 was operated for 2 weeks at 253°C and a pressure of 21.4 atm, and a nitrided sample of the same catalyst was operated for 6 weeks at 249°C and 21.4 atm. All of the tests were made with 1 H_2 + 1 CO synthesis gas and demonstrated typical activities (*13, 16, 62*) for these catalysts.

At the end of each synthesis experiment the catalyst was cooled to room temperature in the reactor by flowing nitrogen over the catalyst bed. The catalyst was then dropped into a glass jar, which was continuously flushed with CO_2. To prevent atmospheric oxidation, all subsequent handling and transfer of the catalyst was done in CO_2.

Surface-area and pore-volume measurements were made on three samples of each used catalyst. The first was a sample of the catalyst as removed from the reactor. The second was extracted with boiling toluene in a modified Soxhlet apparatus (*50*) for ~2 days. This treatment removed essentially all of the waxy reaction products from the catalyst. The third sample was extracted in the same manner as the second and then was reduced with hydrogen.

All adsorption and density determinations were carried out in Pyrex tubes equipped with special four-way stopcocks (*2, 38*). The method used for obtaining pore volumes from mercury and helium densities was that described previously (*12*). Surface areas were calculated from nitrogen adsorption data using the simple BET equation (*24*) with 16.2 $Å^2$ (*37*) taken as the cross-sectional area of the adsorbed nitrogen molecule. Chemical analyses were made on each catalyst sample by conventional methods. The amount of oxygen present was not determined directly, but was calculated by difference.

Comparison of the structural changes in the used catalyst may be conveniently made by studying the histograms shown in Figs. 28–30. The total volume of the catalyst (volume of solids plus pores), indicated by displacement of mercury, is shown by the total histogram. The hatched area represents the true volume of the catalyst (volume of solids only), indicated by displacement of helium. The open area is the pore volume, that is, the difference between the displacement of mercury and the displacement of helium. The numbers in the histogram following the letters $\bar{C}$, $\bar{O}$, or $\bar{N}$ represent the atom ratios in the catalysts of carbon to iron, oxygen to iron, and nitrogen to iron, respectively. At the top of the histograms the surface areas and the average pore diameters are tabulated; the average pore diameters are calculated from $d = 4V/A$, where V is the pore volume and A the surface area. All values of areas and volumes are per gram of unreduced catalyst.

Structural data for catalyst P3003.24 are presented in Fig. 28 and Table 23. The raw catalyst is principally a ferric oxide gel having a large surface area. Reduction to iron with dry hydrogen caused a drastic decrease in the surface area (*37*). The total volume of the catalyst was likewise decreased, the decrease being more pronounced on reduction to iron. After induction and use in the synthesis, the catalyst had virtually no surface area or pore volume, indicating that the pores had been almost completely filled with wax. After extracting the wax with toluene, the surface area and pore volume of the catalyst increased appreciably. Reduction of the extracted catalyst resulted in a moderate decrease in surface area and a slight increase in the pore volume, approaching the values of the completely reduced sample. The total volumes of used and extracted portions remained about the same as the volume of the sample reduced to magnetite. Reduction of the used catalyst caused a slight decrease in total volume.

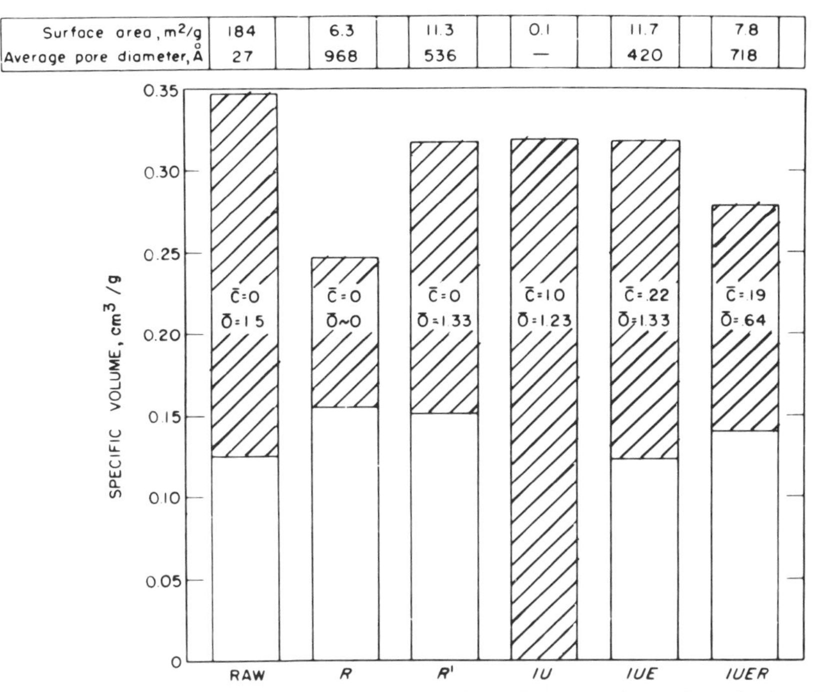

Fig. 28 Changes in pore geometry (□, pore volume; ▨, true volume of catalyst) and composition of precipitated P3003.24 in pretreatment and synthesis. R, Reduced in H_2; R′, reduced in H_2 + H_2O; I, inducted; U, used; E, extracted.

In Fig. 29 and Table 24, structural data for the fused catalyst D3001 are presented. The raw catalyst had no pore volume or surface area, but on reduction the porosity increased to ~40%, and the area increased to 9.4 m^2/g. On partial reoxidation the surface area (Table 25) and pore volume both decreased. However, the volume of mercury displaced by the catalyst was unchanged in the reduction or the subsequent oxidation. This behavior has been demonstrated in this laboratory (*37*), and Westrik and Zwietering (*68*) have also shown that a single crystal of magnetite does not change its shape or dimensions during reduction. When the reduced catalyst was converted to Hägg carbide by treatment with CO, both the pore volume and the volume of mercury displaced were increased, suggesting that the carbon had entered into the arrays of iron atoms, increasing the size of the matrix forming the pore system. This behavior is quite different from the oxidation of the catalyst, where the oxygen apparently entered and filled the pores that were available in the reduced catalyst. Reduction of the carbided catalyst in hydrogen at 450°C removed most of

Table 23

Structural Data for Precipitated Catalyst P3003.24 (Fe_2O_3–CuO–K_2CO_3)

Treatment[a]	Phases present, X-ray analysis	Densities, g/cm^3		Surface area, m^2 per gram unreduced catalyst	Pore volume, cm^3 per gram unreduced catalyst	Volume Hg displaced, cm^3 per gram unreduced catalyst	Average pore diameter, Å
		Helium, ρ_{He}	Mercury, ρ_{Hg}				
None, raw	Amorphous	4.43	2.83	184	0.125	0.347	27
R	α-Fe, Cu	7.60	2.73	6.3	0.155	0.247	968
R′, $H_2 + H_2O$	Fe_3O_4, Cu	5.25	2.75	11.3	0.151	0.317	536
IU	Fe_3O_4	3.06	3.10	0.1	0.000	0.320	—
IUE	Fe_3O_4	4.51	2.76	11.7	0.123	0.319	420
IUER″	α-Fe, Fe_3O_4, Cu	5.50	2.73	7.8	0.140	0.279	718

[a] R, Reduced with H_2 at 300°C; R′, reduced to Fe_3O_4 with $H_2 + H_2O$ at 250°C; I, inducted with 1 H_2 + 1 CO at 230°C; U, used; E, extracted with boiling toluene; R″, reduced with H_2 at 300°C.

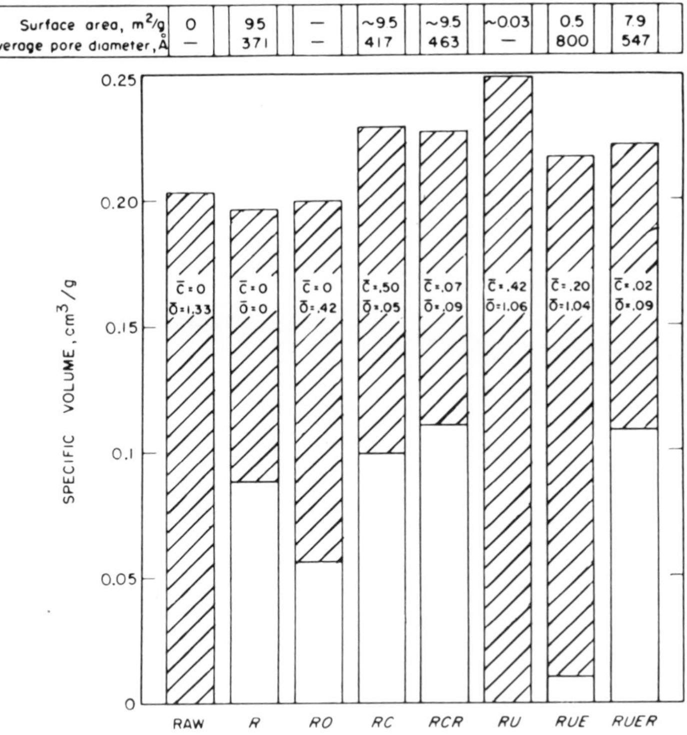

Fig. 29 Changes in pore geometry (□, pore volume; ▨, true volume of catalyst) and composition of catalyst D3001 in carbiding and synthesis. R, Reduced; O, oxidized; C, carbided; U, used; E, extracted.

the carbon. However, the total volume of the catalyst did not return to the original value of ~0.20 cm³/g.

After the catalyst had been used in the synthesis, the total volume had increased to ~0.25 cm³/g. The pore volume and surface area were essentially zero, because the pore volume was decreased by both wax in the pores and oxidation. Extraction of this catalyst with boiling toluene increased the pore volume and surface area slightly and decreased the total volume of the catalyst particles, which may be interpreted as removal of wax from the external surface of the particle as well as from the pores. The final histogram (Fig. 29) represents the pore geometry of the used catalyst after extraction and after reduction at 450°C. Here the pore volume and total volume of the particle were similar to those observed for the reduced, carbided, and reduced catalyst. The total volume of the reduced, used, extracted, and reduced catalyst was ~13% higher than that of the original reduced catalyst.

Table 24
Structural Data for Fused Catalyst D3001, Reduced

Treatment[a]	*Phases present, X-ray analysis*	*Densities, g/cm³*		*Surface area, m² per gram unreduced catalyst*	*Pore volume, cm³ per gram unreduced catalyst*	*Volume Hg displaced, cm³ per gram unreduced catalyst*	*Average pore diameter, Å*
		Helium, ρ_{He}	*Mercury,* ρ_{Hg}				
None, raw	Fe_3O_4	4.96	4.91	0	0.002	0.203	—
R	α-Fe	6.82	3.78	9.5	0.088	0.197	371
RO	α-Fe, Fe_3O_4	5.94	4.13	—[b]	0.061	0.200	—[b]
RC	χ-Fe_2C	6.38	3.60	9.5[c]	0.099	0.229	417
RCR	α-Fe	6.58	3.38	9.5[c]	0.110	0.227	463
RU	Fe_3O_4, α-Fe	4.06	4.06	~0.03	0.000	0.24	—
RUE	Fe_3O_4, α-Fe	4.69	4.48	0.5	0.010	0.217	800
RUER	α-Fe, Fe_3O_4	6.68	3.44	7.9	0.108	0.222	547

[a] R, Reduced with H_2 at 450°C; O, oxidized with $N_2 + H_2O$ at 250°C; C, carbided with CO at 180–340°C; U, used; E, extracted with toluene.
[b] Not determined.
[c] Not determined, but measurements on other samples have shown that surface area does not change appreciably during carbiding.

Table 25
Structural Data for Fused Catalyst D3001, Nitrided

Treatment[a]	*Phases present, X-ray analysis*	*Densities, g/cm³*		*Surface area, m² per gram unreduced catalyst*	*Pore volume, cm³ per gram unreduced catalyst*	*Volume Hg displaced, cm³ per gram unreduced catalyst*	*Average pore diameter, Å*
		Helium, ρ_{He}	*Mercury,* ρ_{Hg}				
None, raw	Fe_3O_4	4.96	4.91	0	0.002	0.203	—
R	α-Fe	6.95	3.71	4.1	0.094	0.201	917
RN	ϵ-Fe_2N, γ'-Fe_4N	6.39	3.49	4.0	0.104	0.231	1040
RNR′	α-Fe	7.08	3.75	4.1	0.092	0.196	898
RNO, $\bar{O} = 0.22$	ϵ-Fe_2N, Fe_3O_4	6.23	3.62	3.9	0.093	0.229	954
RNOR″	α-Fe	7.04	3.48	4.1	0.107	0.211	1040
RNO′, $\bar{O} = 0.81$	Fe_3O_4, ϵ-Fe_2N, γ'-Fe_4N	5.09	4.01	0.8	0.049	0.229	1090
RNO′R″	α-Fe	6.71	3.33	4.8	0.113	0.225	942
RNU	ϵ-Fe_2X,[b] Fe_3O_4, $FeCO_3$	3.98	3.91	0.4	0.005	0.249	500
RNUE	ϵ-Fe_2X, Fe_3O_4, $FeCO_3$	5.00	3.98	4.6	0.046	0.228	400
RNUER‴	α-Fe, Fe_3C	6.93	3.51	7.5	0.109	0.220	580

[a] R, Reduced with H_2 at 550°C; N, nitrided with NH_3 at 350°C; R′, reduced with H_2 at 300°C; O, oxidized with $N_2 + H_2O$ at 250°C; R″, reduced with H_2 at 450°C; O′, oxidized with $N_2 + H_2O$ at 350°C; U, used; E, extracted with toluene; R‴, reduced in H_2 at 400°C.

[b] Iron carbonitride.

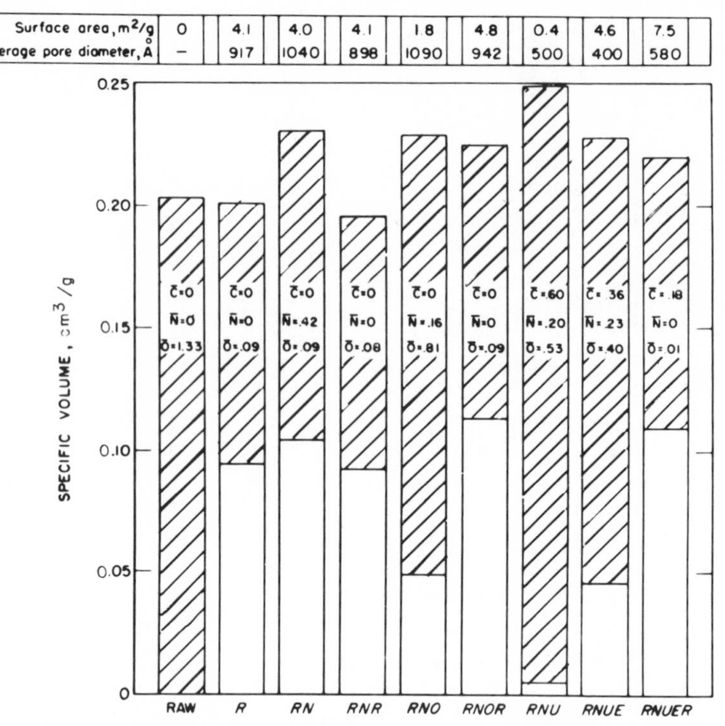

Fig. 30 Changes in pore geometry (□, pore volume; ▨, true volume of catalyst) and composition of catalyst D3001 in nitriding and synthesis. R, Reduced; O, oxidized; U, used; E, extracted; N, nitrided.

Figure 30 and Table 25 show a similar series on fused, nitrided iron catalysts. In this series the original reduction was done at 550 rather than 450°C. The higher reduction temperature resulted in a catalyst having a smaller surface area and a larger average pore diameter. After nitriding, the total volume and pore volume of the catalyst were appreciably increased, as shown in the third histogram of Fig. 30. When the nitrogen was removed by hydrogenation, however, the catalyst properties again were those of the reduced catalyst; the nitrogen apparently was removed completely and did not stabilize the expanded structure, as observed with carbided catalysts. Partial oxidation of the nitrided catalyst resulted in a considerable decrease in the surface area and pore volume, shown in the fifth histogram. Total volume remained about the same. Oxidation of a nitrided catalyst was more difficult than oxidation of a reduced catalyst. The nitrides represented by the fifth and sixth histograms were oxidized at a temperature of 350°C, compared with 250°C for the reduced catalysts (*38*). After reduction of the nitrided, oxidized catalyst, the pore volumes were larger than the pore volume of the reduced, nitrided, and reduced catalyst. The total volume decreased slightly and was almost the same as that of the original nitride. Oxidation of the nitride, therefore, had a stabilizing effect on the expanded structure.

After 6 weeks of synthesis, the seventh histogram (Fig. 30), the nitrided, used catalyst, had only a very small pore volume and displaced a greater amount of mercury than the reduced or even the reduced, nitrided catalyst. Its surface area was relatively low but appreciable. After extraction in toluene the pore volume was increased, and the volume of mercury displaced by the catalyst was decreased. Accompanying this change was a large increase in surface area. Finally, after reduction in hydrogen at 400°C, the nitrogen was removed completely, but the carbon remained in moderate amounts. The surface area increased significantly and the volume of mercury displaced by the catalyst was larger than that of the original reduced catalyst. Here again the amount of carbon remaining, or the intermediate formation of oxide, apparently was sufficient to stabilize the expanded structure.

Changes in a precipitated-iron catalyst during synthesis are relatively minor. The induction procedure largely destroys the gel structure of the raw material and results in a catalyst that is essentially magnetite and iron carbide. This inducted catalyst, although of low mechanical strength, is apparently quite stable during synthesis, as shown by the fact that after extraction to remove waxes, the used catalyst is very similar to the starting material in both chemical composition and physical structure.

However, reduced– or nitrided–fused catalysts undergo major changes in surface area and pore volume during synthesis. Here the structure and composition are affected by several reactions that probably occur simultaneously. To summarize the structural changes of fused-iron catalysts that occur during pretreatment and synthesis, let us consider briefly what happens to the catalyst during each of the following procedures:

1. *Reduction.* During reduction no change takes place in the total volume of the catalyst, but a pore structure is generated by removing oxygen (*37*). This process occurs with pure magnetite, but the presence of structural promoters greatly assists in maintaining a pore structure of moderate surface area. It is probably not necessary to postulate a framework of structural promoter that is continuous throughout the structure. Thus, crystallites of structural promoter may be considered as bricks in a wall of reduced catalyst.

2. *Reoxidation.* Oxidation of a reduced catalyst is rapid initially, but the rate decreases very quickly with time or extent of oxidation (*38*). The reaction proceeds at the internal surface and the layer of oxide progressively thickens and quickly smooths out the surface, thus decreasing the

pore volume and surface area. This reoxidation process occurs on a solid framework of iron and promoter crystallites that maintain the total volume of the catalyst constant.

3. *Nitriding and carbiding*. During these procedures the metal crystallites expand, the observed increase in the total volume being ~13% for carbides and ~17% for nitrides. The pore volume and average pore diameter increase, but the surface area changes only slightly.

4. *Reduction of nitride*. When iron nitride is reduced with hydrogen, the metal contracts to its original volume. Although this process is largely reversible, it does lead to disintegration of some of the catalyst particles. Differential expansion and movement about the structural promoter aggregates may be responsible for the breakup.

5. *Oxidation of nitride*. During the oxidation of iron nitride with steam, nitrogen is displaced only from the part of the catalyst that is oxidized. Hydrogen formed during the steam oxidation is apparently ineffective for reducing the nitride in the presence of water vapor. Thus, the oxide is built on a substrate of nitride.

6. *Reduction of oxidized nitride*. In this step both the oxide and the nitride are reduced to metallic iron. The oxide portion will retain the bulk volume of the original nitride, and the nitride will attempt to revert to the volume of the original reduced catalyst. The relative amounts of oxide and nitride will then determine the final total volume of the catalyst. The oxide may be considered as a layer of cement over the nitride. If this oxide layer is sufficiently thick, it will prevent overall shrinkage of the catalyst particle during reduction.

7. *Reduction of carbide*. Apparently some free carbon is produced during carburization as well as in subsequent hydrogenation, and some oxide is formed in carburization. In addition, carbidic carbon is probably not completely removed during reduction of the carbide with hydrogen. The presence of carbon in the catalyst after reduction prevents the solid from regaining its original density; that is, carbon aggregates between the crystallites may prevent the return of the overall structure to its original state. Oxidation of iron carbide is probably similar to oxidation of the nitride; the oxide layer is probably built on a substrate of carbide.

Thus, physical and chemical changes in an iron catalyst during synthesis are large. Reduced catalysts oxidize extensively and carbides are formed. Nitrides oxidize to a smaller extent than reduced catalysts, but they are largely converted to carbonitrides. The active portion of the catalyst is probably confined to a thin zone or shell on the external surface of the particles (*13*). The low surface areas obtained for the used catalyst in Fig. 29 confirm the fact that only a small fraction of the total surface area of the catalyst is required for the synthesis. The inner part of the catalyst then may be considered to be inert, but it is important in that it should provide a mechanically stable substrate for the active zone.

References

1. Ackermann, P. [Latest German Developments on the Gas Phase. Gasoline Synthesis at Middle Pressures, 1942–1945.] FIAT Reel X-116, Frames 1717–1787.
2. Anderson, R. B., *Ind. Eng. Chem., Anal. Ed.* **18,** 156 (1946).
3. Anderson, R. B., *Adv. Catal.* **5,** 355 (1953).
4. Anderson, R. B., *in* "Catalysis" (P. H. Emmett, ed.), Vol. 4, pp. 86–89. Van Nostrand-Reinhold, Princeton, New Jersey, 1956.
5. Anderson, R. B., *in* "Catalysis" (P. H. Emmett, ed.), Vol. 4, Chapter 3. Van Nostrand-Reinhold, Princeton, New Jersey, 1956.
6. Anderson, R. B., *Catal. Rev.—Sci. Eng.* **21**(1), 53 (1980).
7. Anderson, R. B., Hall, W. K., Hewlett, H. D., and Seligman, B., *J. Am. Chem. Soc.* **69,** 3114 (1947).
8. Anderson, R. B., Hall, W. K., and Hofer, L. J. E., *J. Am. Chem. Soc.* **70,** 2465 (1948).
9. Anderson, R. B., Hall, W. K., Krieg, A., and Seligman, B., *J. Am. Chem. Soc.* **71,** 183 (1949).
10. Anderson, R. B., Hofer, L. J. E., Cohn, E. M., and Seligman, B., *J. Am. Chem. Soc.* **73,** 944 (1951).
11. Anderson, R. B., Krieg, A., Seligman, B., and O'Neill, W. E., *Ind. Eng. Chem.* **39,** 1548 (1947).
12. Anderson, R. B., McCartney, J. T., Hall, W. K., and Hofer, L. J. E., *Ind. Eng. Chem.* **39,** 1618 (1947).
13. Anderson, R. B., Seligman, B., Shultz, J. F., Kelly, R., and Elliott, M. A., *Ind. Eng. Chem.* **44,** 391 (1952).
14. Anderson, R. B., and Shultz, J. F., U.S. Patent 2,629,728 (1953).
15. Anderson, R. B., Shultz, J. F., Hofer, L. J. E., and Storch, H. H., *Bull.—U.S., Bur. Mines* **580,** 1–25 (1959).
16. Anderson, R. B., Shultz, J. F., Seligman, B., Hall, W. K., and Storch, H. H., *J. Am. Chem. Soc.* **72,** 3502 (1950).
17. Aristoff, E., Shipley, C. S., and Ciapetta, F. G., presentation at the Catalysis Club of Philadelphia, 1951, in R. B. Anderson, *Adv. Catal.* **5,** 382 (1953).
18. Basu, S., and Krishna Murthy, V. A., *Proc. Symp. Chem. Oil Coal, 1969* p. 183 (1972).
19. Borghard, W. G., and Bennett, C. O., *Ind. Eng. Chem. Prod. Res. Dev.* **18,** 18 (1979).
20. Bridger, G. L., Pole, G. R., Beinlich, W. A., and Thompson, H. L., *Chem. Eng. Prog.* **43,** 291 (1947).
21. Browning, L. C., DeWitt, T. W., and Emmett, P. H., *J. Am. Chem. Soc.* **72,** 4211 (1950).
22. Brunauer, S., and Emmett, P. H., *J. Am. Chem. Soc.* **59,** 2682 (1937).
23. Brunauer, S., and Emmett, P. H., *J. Am. Chem. Soc.* **62,** 1732 (1940).
24. Brunauer, S., Emmett, P. H., and Teller, E., *J. Am. Chem. Soc.* **60,** 309 (1938).
25. Brunauer, S., Jefferson, M. E., Emmett, P. H., and Hendricks, S. B., *J. Am. Chem. Soc.* **53,** 1778 (1931).
26. Cohn, E. M., and Hofer, L. J. E., *J. Am. Chem. Soc.* **72,** 4662 (1950).
27. Crowell, J. H., Benson, H. E., Field, J. H., and Storch, H. H., *Ind. Eng. Chem.* **42,** 2376 (1950).

28. Demeter, J. J., and Schlesinger, M. D., *Rep. Invest.—U.S., Bur. Mines* **5603,** 1–32 (1960).
29. Eckstrom, H. C., and Adcock, W. A., *J. Am. Chem. Soc.* **72,** 1042 (1950).
30. Eisenhut, O., and Kaupp, E., *Z. Elektrochem.* **36,** 392 (1930).
31. Emmett, P. H., and Brunauer, S., *J. Am. Chem. Soc.* **59,** 310 (1937).
32. Emmett, P. H., and Brunauer, S., *J. Am. Chem. Soc.* **59,** 1553 (1937).
33. Emmett, P. H., and DeWitt, T. W., *J Am. Chem. Soc.* **65,** 1253 (1943).
34. Evans, U. R., *Nature (London)* **164,** [illegible]09 (1949).
35. Hägg, G., *Nova Acta Regiae Soc. Sci. Ups.* **7,** 1 (1929).
36. Hall, W. K., Dieter, W. E., Hofer, L. J. E., and Anderson, R. B., *J. Am. Chem. Soc.* **75,** 1442 (1953).
37. Hall, W. K., Tarn, W. H., and Anderson, R. B., *J. Am. Chem. Soc.* **72,** 5436 (1950).
38. Hall, W. K., Tarn, W. H., and Anderson, R. B., *J. Phys. Chem.* **56,** 688 (1952).
39. Hofer, L. J. E., *in* "Catalysis" (P. H. Emmett, ed.), Vol. 4, Chapter 4. Van Nostrand-Reinhold, Princeton, New Jersey, 1956.
40. Hofer, L. J. E., *Bull.—U.S., Bur. Mines* **631,** 1–60 (1966).
41. Hofer, L. J. E., and Cohn, E. M., *J. Chem. Phys.* **18,** 766 (1950).
42. Hofer, L. J. E., Cohn, E. M., and Peebles, W. C., *J. Am. Chem. Soc.* **71,** 189 (1949).
43. Hofer, L. J. E., Cohn, E. M., and Peebles, W. C., *J. Phys. Chem.* **54,** 1161 (1950).
44. Hoogendoorn, J. C., *Clean Fuels Coal, Symp. Pap., 1973* p. 353 (1973).
45. Jack, K. H., *Proc. R. Soc. London, Ser. A* **195,** 34–40, 41–55, 56–61 (1948).
46. Kölbel, H., and Engelhardt, F., *Erdoel Kohle* **2,** 52 (1949).
47. Kölbel, H., and Engelhardt, F., *Erdoel Kohle* **3,** 529 (1950).
47a. Le Caër, G., Dubois, J. M., Pijolat, M., Perrichon, V., and Bussière, P., *J. Phys. Chem.* **86,** 4799 (1982).
48. Lehrer, E., *Z. Elektrochem.* **36,** 383, 460 (1930).
49. McCartney, J. T., and Anderson, R. B., *J. Appl. Phys.* **22,** 1441 (1951).
50. McCartney, J. T., Hofer, L. J. E., Seligman, B., Lecky, J. A., Peebles, W. C., and Anderson, R. B., *J. Phys. Chem.* **57,** 730 (1953).
51. McCartney, J. T., Seligman, B., Hall, W. K., and Anderson, R. B., *J. Phys. Chem.* **54,** 505 (1950).
52. Perrin, M., The synthesis of aliphatic hydrocarbons. Doctoral Dissertation, University of Lyons, Lyons, France (1948).
53. Pichler, H., and Merkel, H., *Tech. Pap.—U.S., Bur. Mines* **718** (transl. by Ruth Brinkley), 1–108 (1949).
54. Pichler, H., and Schulz, H., *Chem.-Ing.-Tech.* **42,** 1162 (1970).
55. Podgurski, H. H., Kummer, J. T., DeWitt, T. W., and Emmett, P. H., *J. Am. Chem. Soc.* **72,** 5382 (1950).
56. Shultz, J. F., Abelson, M., Shaw, L., and Anderson, R. B., *Ind. Eng. Chem.* **49,** 2055 (1957).
57. Shultz, J. F., Hall, W. K., Dubbs, T. A., and Anderson, R. B., *J. Am. Chem. Soc.* **78,** 282 (1956).
58. Shultz, J. F., Hall, W. K., Seligman, B., and Anderson, R. B., *J. Am. Chem. Soc.* **77,** 213 (1955).
59. Shultz, J. F., Hofer, L. J. E., Cohn, E. M., Stein, K. C., and Anderson, R. B., *Bull.—U.S., Bur. Mines* **578,** 1–139 (1959).
60. Shultz, J. F., Hofer, L. J. E., Stein, K. C., and Anderson, R. B., *Bull.—U.S., Bur. Mines* **612,** 1–70 (1963).
61. Shultz, J. F., Seligman, B., Lecky, J. A., and Anderson, R. B., *J. Am. Chem. Soc.* **74,** 637 (1952).

62. Shultz, J. F., Seligman, B., Shaw, L., and Anderson, R. B, *Ind. Eng. Chem.* **44,** 397 (1952).
63. Stein, K. C., Thompson, G. P., and Anderson, R. B., *J. Phys. Chem.* **61,** 928 (1957).
64. Storch, H. H., Anderson, R. B., Hofer, L. J. E., Hawk, C. O., Anderson, H. C., and Golumbic, N., *Tech. Pap.—U.S., Bur. Mines* **709,** 1–213 (1948).
65. Storch, H. H., Golumbic, N., and Anderson, R. B., "The Fischer–Tropsch and Related Syntheses," pp. 208–211, 465–477. Wiley, New York, 1951.
66. Storch, H. H., Golumbic, N., and Anderson, R. B., "The Fischer–Tropsch and Related Syntheses." Wiley, New York, 1951.
67. Weller, S. W., Hofer, L. J. E., and Anderson, R. B., *J. Am. Chem. Soc.* **70,** 799 (1948).
68. Westrik, R., and Zwietering, P., *Proc. Ned. Akad. Wet., Ser. B: Phys. Sci.* **56,** 492 (1953).

CHAPTER 4

Tests of Catalysts

4.1 Introduction

Our discussion begins with platinum-group metals because in the Fischer–Tropsch synthesis (FTS), these metals are the least complicated. As shown in Table 1, the FTS on the platinum metals are not complicated by the formation of carbides or oxides and by the water–gas shift. Platinum metals other than ruthenium generally have low activity. Carbonyls can be produced from ruthenium catalysts at high pressures and low temperatures. Cobalt and nickel may be converted to carbide and to volatile carbonyls under some synthesis conditions. Iron, molybdenum, and tungsten may form carbides, nitrides, and oxides and are active catalysts for the water-gas shift. In addition, catalysts of all of these metals deposit elemental carbon in the presence of CO and/or hydrocarbons under some conditions, and almost all useful FTS catalysts are susceptible to poisoning by sulfur compounds. The catalysts are discussed in the order just presented.

4.2 Platinum-Group Metal Catalysts

Shultz *et al.* (*130*) studied the H_2 + CO reaction on supported catalysts of the Pt group and on Mo, W, and Re, as shown in Table 2. Ruthenium was active and produced substantial amounts of higher hydrocarbons. Other Pt-group catalysts and those of W and Re had low activity; methane was the major product. Supported Mo had modest activity, especially in CO-rich gas, as well as a modest tendency to produce higher hydrocarbons.

McKee (*94*) showed that CO is very strongly adsorbed on Pt, Rh, and Ir, and CO displaces chemisorbed hydrogen. At 150 and 200°C, H_2 + CO react very slowly, if at all. Ruthenium and Ru alloys are active catalysts.

Table 1
Complicating Factors in the Fischer–Tropsch Synthesis

	Metal						
Factor	*Ru*	*Rh*	*Other Pt metals*	*Co*	*Ni*	*Fe*	*Mo, W*
Carbon deposition	+[a]	+	+	+	+	+	+
Carbide formation	0	0	0	+	+	+	+
Volatile carbonyl formation	I[b]	0	0	I	+	I−	0?
Oxide formation	0	0	0	0	0	+	+
Water–gas shift	0	0	0	0	0	+	+
Forms nitrides	0	0	0	0	0	+	+
Low activity	0	0?	+	0	0	0	+

[a] +, Denotes a phenomenon that occurs readily; 0, does not occur; I, may be a problem in some cases.
[b] Ru may also produce polynuclear carbonyls of low volatility.

Some have postulated that CO is chemisorbed too strongly on Pt, Rh, and Ir for these elements to be useful FTS catalysts; however, Ponec and Barneveld (*119*) showed that the heats of adsorption of CO are ~32 kcal/mol on the most active metals (Ru, Co) and ~36 kcal/mol on the relatively inactive metals (Pd).

Vannice (*140*) studied the FTS on group VIII metals supported on alumina at 1 atm, reporting activity in terms of turnover number, constants of a power rate law equation, and selectivity graphs. The data are summarized in Table 3. Here, Ru has the largest activity, and the other Pt-group metals the lowest. Later, alumina was shown to be a better support for Pt and Pd than silica (*141*).

4.2.1 Ruthenium

In the late 1930s, Pichler began studies of Ru catalysts that led to the synthesis of very high molecular weight waxes, the polymethylene synthesis. This early work has been reviewed (*6, 112*). Schulz summarized Pichler's work on this subject (*126*), and Section 4.2.1.1 is based in part on that review.

Table 2
U.S. Bureau of Mines Tests of Catalysts of the Platinum Group[a]

Metal and support	*Feed gas, H_2/CO*	*Conversion of $H_2 + CO$, %*	*Temp., °C*	*Usage ratio, H_2/CO*	*Hydrocarbons, wt %*			
					CH_4	*C_2*	*$C_3 + C_4$*	*C_{5+}*
41% Mo on alumina	3	56	420	1.8	89	10	1	0
28% Mo on	2	67	419	1.4	78	17	5	0
alumina	1	66	279	1.3	72	20	8	0
0.5% Ru on	3	80	222	2.4	40	3	10	47
alumina	2	66	225	2.2	13	2	6	79
	1	34	239	1.9	8	2	6	84
0.5% Rh on	3	87	441	2.4	94	5	1	0
alumina	2	71	551	1.8	94	1	5	0
	1	83	583	1.2	96	4	0	0
0.5% Pd on alumina	3	36	499	2.2	97	2	1	0
31% W on alumina	1	44	599	1.3	91	8	1	0
0.5% Re on alumina	3	86	451	2.4	96	3	1	0
0.5% Os on charcoal	3	70	600	2.3	93	6	1	0
0.5% Pt on	3	56	513	1.8	69	9	22	0
alumina	2	72	577	1.6	84	9	7	0

[a] Hourly space velocity, ~300; pressure, 21 atm. Adapted from Shultz *et al.* (*130*).

Ruthenium catalysts are active over a wide range of operating conditions: temperatures from 100 to 300°C and pressures from 1 to 2000 atm; selectivity varies from the production of all CH_4 to "polymethylene." Water is usually the principal oxygen-containing product, and the stoichiometry for producing olefins may be represented by

$$2\ H_2 + 1\ CO = (1/n)\ (CH_2)_n + H_2O \tag{1}$$

At low temperatures and high pressures, carbonyls are produced, and the mode of conducting the synthesis under these conditions must be devised so that the rate of formation of carbonyls is slow. These carbonyls are not catalysts for the FTS (*115*), but they are active in hydroformylation (*127*). The carbonyls and hydrocarbonyls are

$Ru(CO)_5$	colorless	$Ru_3(CO)_{12}$	red / orange	$[H_2Ru_3(CO)_{11}]_x$
$H_2Ru(CO)_4$	colorless	$H_4Ru_3(CO)_{10}$	yellow / orange	yellow-brown to blue-green
Soluble in organic solvents and water				Insoluble

Table 3
Kinetics of Synthesis on Metals Supported on Alumina[a]

Metal	CO reacted per metal site per second at 275°C (×10^{-3})	Activation energy, kcal/mol	Exponent of H_2	Exponent of CO	CH_4 yield, mol % of hydrocarbons: CH_4, %	CH_4 yield, mol % of hydrocarbons: T, °C
5% Ru	325	18.3	1.6	−0.6	62	210
15% Fe	160	25.9	1.14	−0.05	65	240
5% Ni	38	23.5	0.77	−0.31	90	242
2% Co	28	26.7	1.22	−0.48	81	240
1% Rh	17	24.2	1.04	−0.20	89	265
2% Pd	13	19.7	1.03	0.03	—	
1.75% Pt	3.4	15.3	0.83	0.04	93	280
2% Ir	2.6	11.7	0.96	0.10	83	276

[a] Adapted from Vannice (*140*).

and alkyl carbonyls are known (*115*). Thermodynamic data on Ru carbonyls are not available. Carbonyls are readily produced from CO at 120°C and 50–100 atm and higher pressures; however, the rate and the type of carbonyl depend on the catalyst and its pretreatment. For example, hydrated RuO_2 forms carbonyls very much faster than the dried and completely reduced oxide. Catalysts submerged in liquid media such as nonane form carbonyl at the slowest rate. Kuznetsov *et al.* (*84*), described infrared spectra of Ru carbonyls supported on silica and alumina.

4.2.1.1 The Polymethylene Synthesis (126) On Ru at low temperatures and high pressures, waxes having very high molecular weights are produced; this process is called the polymethylene synthesis (PMS). Although active and selective catalysts can be made in a variety of ways, the most active catalyst was a Ru powder obtained by a thorough reduction in hydrogen of a dried RuO_2 powder. The RuO_2 powder was made by reducing an aqueous solution of K_2RuO_4 with methanol. The reduced Ru consisted of a porous wad of fine particles with a crystallite size of 10–20 Å and a surface area of 10–20 m^2/g (*113*). In electron diffraction, no lines were observed; the metal was amorphous.

No promoters, carriers, or supports that improved activity were found. In early work, fine-mesh copper screen was used for continuous reactors to keep the Ru powder in the reaction vessel. A catalyst made by intensive mixing of an Ru(VI) oxide hydrate with alumina granules was active for more than a month in a continuous system in which a liquid was circulated over the catalyst to remove wax. Active catalysts may be prepared by thermal decomposition of Ru carbonyls on supports.

The polymethylene synthesis operates at 100–140°C and 1000 atm. Under these conditions the process is slow, 2.2–11.2 mmol CO converted per gram Ru · hour. These rates may be compared with ~190 mmol CO per gram Ru · hour for 0.5% Ru on alumina at 225°C and 20 atm (*73*). The rate of synthesis increases with increasing temperature, as shown in Table 4, and the yields of wax and aldehydes decrease and CH_4 formation increases with increasing temperature. The rate increases with pressure but eventually levels out. The average molecular weight increases with pressure, and this increase continues up to 2200 atm, as shown in Fig. 1. Starting from the stoichiometric feed ratio, 2 H_2 + 1 CO, increasing the H_2 increases CH_4 yields, and increasing the CO increases the production of CO_2 slightly. Thus, 2 H_2 + 1 CO feed seems the best choice. The PMS also proceeds with a 3 CO + 1 H_2O feed or with pure CO and the catalyst immersed in water.

Data for six tests (Table 5) illustrate a number of aspects of the PMS. Tests 1–3 were batch experiments with the fine catalyst powder suspended in hexadecane. As temperature is increased, particularly from 120 to 140°C, the selectivity changes; more CH_4 and saturated hydrocarbons are produced at the higher temperature, as well as smaller amounts of wax, olefins, and oxygenated molecules. Test 5 is also a batch operation at 200 atm, with the catalyst suspended in water, the Kölbel–Engelhardt synthesis (Chapter 7). Here the wax and aldehyde yields are large and the methane yield low. For this reaction (3 CO + 1 H_2O) the rates should be divided by 3 to be the comparable with those of the other tests of Table 5.

In a continuous operation with the catalyst submerged in nonane (test 4), the activity and the yields of wax and aldehydes were large and

Table 4
CO Conversion and Selectivity as a Function of Temperature at 1000 atm[a]

Rate and selectivity	*Temperature, °C* 100	120	140
Rate, ml (S.T.P.) CO converted per gram Ru per hour	30	50	108
Ratio of solid to gaseous and liquid products on a carbon basis	3.1	2.1	0.56
Selectivity to CH_4, % of carbon in products	10.1	13.8	24.3
Selectivity to aldehyde in low molecular weight products, %	6.3	4.6	0.02

[a] Catalyst suspended in hexadecane. Adapted from Schulz (*126*).

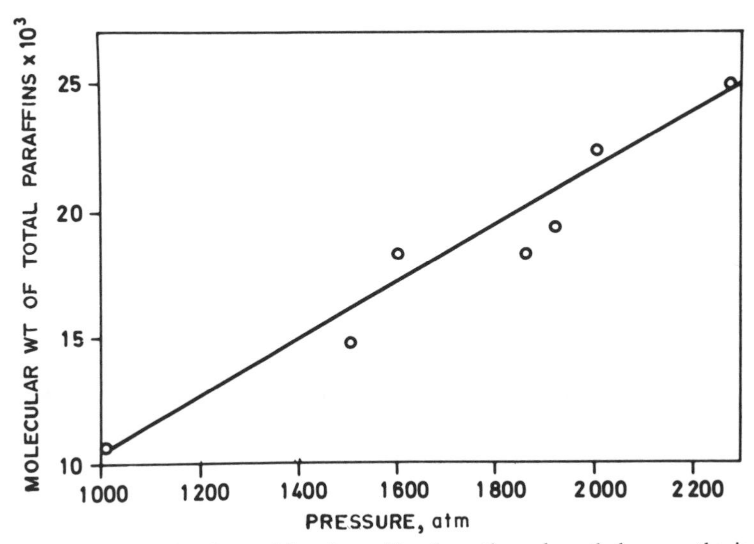

Fig. 1 Average molecular weight of paraffins from the polymethylene synthesis as a function of pressure. Reproduced with permission from Schulz (*126*).

Table 5
Tests of Polymethylene Synthesis[a]

Test	1	2	3	4	5	6
Pressure, atm	1000	1000	1000	1000	200	1000
Temperature, °C	100	120	140	120	120	180
Mode of operation	Batch	Batch	Batch	Continuous	Batch	Continuous
Liquid medium	Hexadecane	Hexadecane	Hexadecane	Nonane	Water	None[b]
Rate, conversion of CO, mmol per gram Ru per hour	1.3	2.2	4.8	7	8	25
Reaction products, wt % of products, excluding H_2O						
Wax	75.5	68	36	54	55	61
Liquids and gases						
Paraffins	5.4	11	39.3			
Olefins	0.3	0.5	0.1			
Aldehydes	6.3	4.4	0.02	39	~36	29
Alcohols	0.3	1	0.06			
Organic acids, ketones, esters	0.5	0.1				
CH_4	10.2	14	24.2	6	7–9	10
CO_2	1.5	1.0	0.3	1	~1	—

[a] Adapted from Schulz (*126*).
[b] Fixed bed.

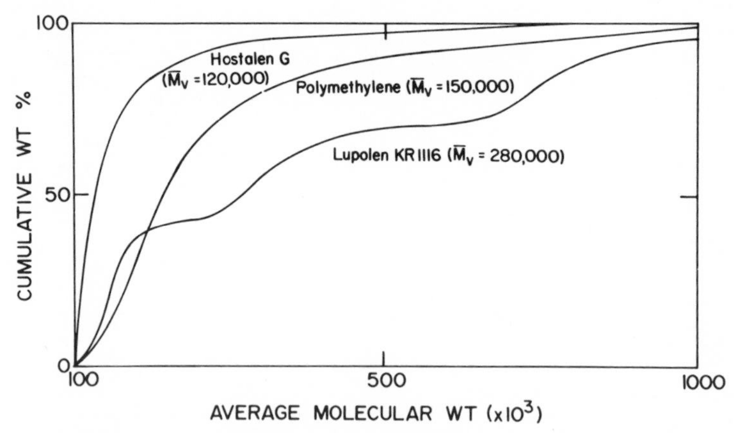

Fig. 2 Molecular weight distribution of "polymethylene" is compared with two polyethylenes. Reproduced with permission from Schulz (*126*).

methane production was small. The continuous fixed-bed operation, without liquid in test 6, produced large yields of wax and aldehyde despite the higher temperature. The rate was large.

Under usual polymethylene synthesis conditions, 1000 atm and 100–120°C, about 50–55% of the product was solid wax with an average molecular weight of 15,000–20,000, as determined by viscometry, and a melting point of 128–129°C. Half of the wax was soluble in toluene at 85°C; the insoluble portion had a molecular weight (M_v) of 30,000–40,000. The molecular weight distribution of a fraction separated from a PMS product and having an M_v of 150,000 is compared in Fig. 2 with two commercial polyethylenes, Hostalen G, a Ziegler high-density polyethylene (HDPE), and Lupolen, a gas-phase HDPE. Distribution curves were determined by sol–gel gradient elution and gel permeation chromatography. The infrared spectra of polymethylene and a Zeigler polyethylene were nearly identical. The bands at 7.25 μm corresponded to 8 CH_3 groups per 1000 C atoms for the polyethylene and a slightly smaller number of CH_3 groups for the polymethylene.

Polymethylenes with $M_v > 100{,}000$ have densities of 0.975 g/cm^3 and melting points in the range 132–138°C (*126*). When catalysts are used at temperatures below the melting point of the waxes, the high-melting components accumulate and block the active surface. Circulating a liquid hydrocarbon through the catalyst bed dissolves and removes these waxes and retards formation of carbonyls.

The products from C_1 to C_{15} from a batch test at 1000 atm with hexadecane as a suspension medium follow the pattern shown in Fig. 3. At the lower temperatures, aldehydes are the main products from C_2 to C_{10}, with

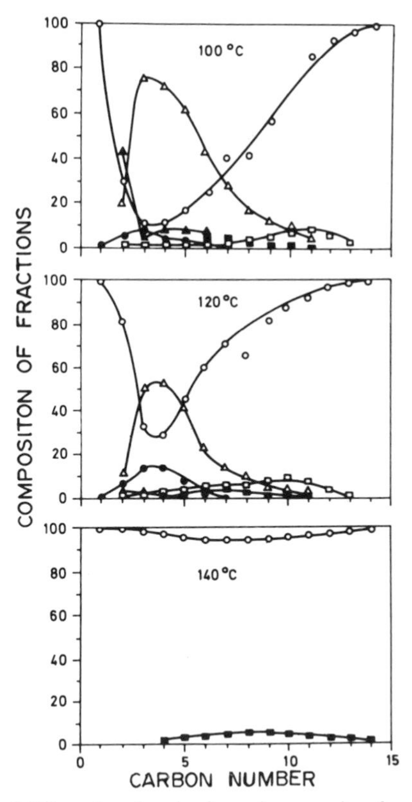

Fig. 3 Distribution of different molecules in carbon-number fractions for the lower molecular weight products from the polymethylene synthesis at three temperatures. ○, *n*-Paraffins; △, *n*-aldehydes; ●, *n*-primary alcohols; ▲, organic acids, □, olefins; ■, methyl-substituted paraffins. Reproduced with permission from Schulz (*126*).

small amounts of olefins, alcohols, and organic acids. Alcohols and olefins have maximum yields at 120°C, and at 140°C the product consists entirely of saturated hydrocarbons. The molecules have predominantly straight carbon chains, as shown in Fig. 3. At 140°C the maximum amount of isoparaffins in carbon-number fractions was ~5%. Carbon-number distributions were generated by summing the moles of each molecule at each carbon number. When plotted as percentage of reacted carbon, the same as on weight percentage for hydrocarbons, the plots were similar to those observed for distributions from Co and Fe catalysts. Methane values were high, C_2 was small, the yields increased to maximum values at C_4, and then the curves decreased monotonically. The distribution at 140°C seems significantly different from those at 100 and 120°C.

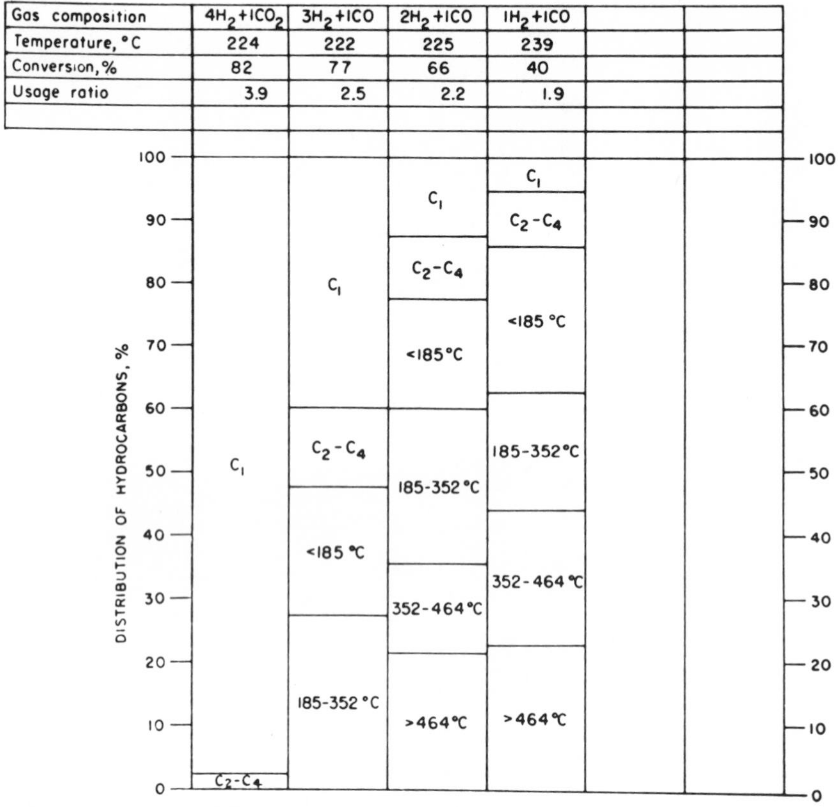

Fig. 4 Selectivity of Ru on alumina at 21.4 atm and an hourly space velocity of 300. Reproduced with permission from Karn *et al.* (*73*).

4.2.1.2 Tests at Moderate Pressure United States Bureau of Mines workers (*73*) studied 0.5% Ru on alumina in the form of $\frac{1}{8} \times \frac{1}{8}$ in. cylinders at 21.4 atm. This catalyst (produced by Engelhard) has essentially all of the Ru deposited near the periphery of the particles (*78*). Figure 4 presents the distribution of products as a function of gas composition, and Fig. 5 one of the conversion–reciprocal-space-velocity plots from a kinetic study. Figure 4 shows that the molecular weight of the hydrocarbons decreased sharply as the CO content of the gas decreased, as is shown by the CH_4 data in Fig. 5. The hydrocarbons contained moderate amounts of olefins but very little oxygenated material. Water was the principal oxygenated product, and the usage ratio $H_2:CO$ was usually more than 2. Only traces of C_{2+} were produced from 4 H_2 + 1 CO_2. The catalysts that make high molecular weight products *must be selectively poisoned by*

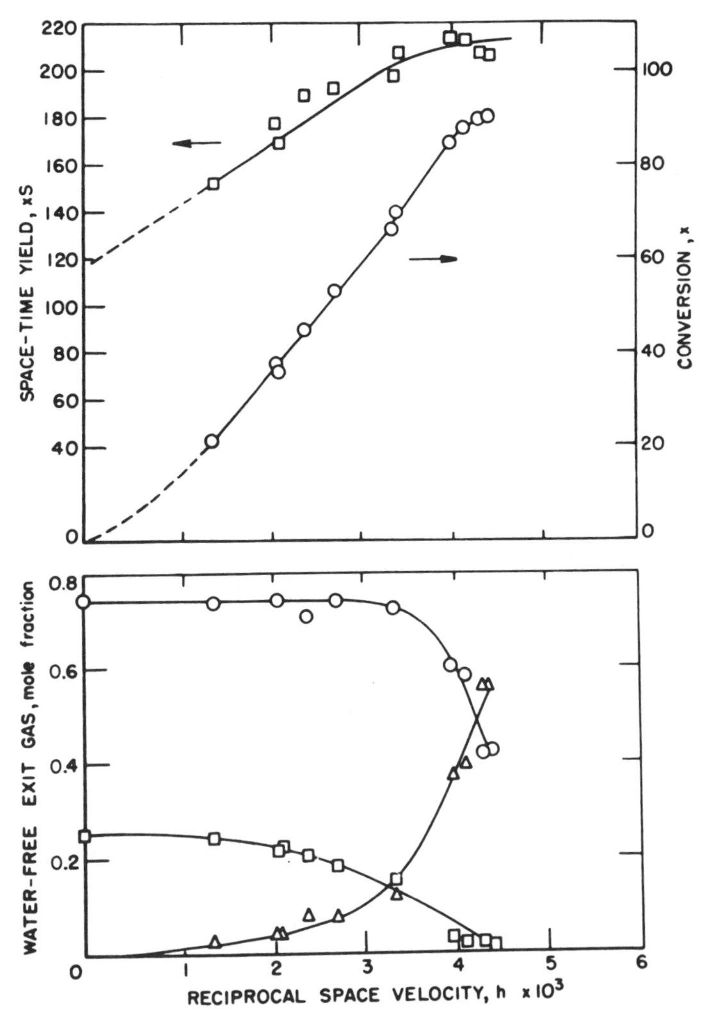

Fig. 5 Rate data for 3 H_2 + 1 CO feed over Ru catalyst at 222°C and 21.4 atm. For bottom graph: ○, H_2; □, CO; △, CH_4. Reproduced with permission from Karn *et al.* (*73*).

CO. We should also note that this same catalyst is excellent in the hydrogenolysis of hydrocarbons at 100°C (*79*). In the range 1–21 atm, the rate of the FTS increased with operating pressure to the 1.2 power, and the power rate law $r = kp_{H_2}^{1.33}p_{CO}^{-0.13}$ represented the rate data.

Lunde and Kester (*89*) reported on the kinetics of methanation of CO_2 on the Engelhard 0.5% Ru on alumina at 1 atm. The H_2 and CO_2 orders were 0.9 and 0.225, respectively, and the activation energy was 16.8 kcal/mol.

Table 6
Product Distributions from 2.5% $Ru-SiO_2$ Catalyst[a]

Product	*Product, mol %*		
	225°C	*250°C*	*275°C*
Methane	63	71	80.5
Ethane	4.8	7.6	6.6
Ethylene	1.3	0.3	0.1
Propane	3.4	6.6	5.1
Propylene	8.9	3.4	1.7
n-Butane	4.1	4.1	2.5
Isobutane	0.0	0.0	0.0
Butene-1	7.8	1.6	0.3
cis-Butene-2	0.9	1.0	0.8
trans-Butene-2	1.7	1.0	0.7
Isobutylene	0.0	0.0	0.3
1,3-Butadiene	0.0	0.0	0.0
n-Pentane	4.1	3.4	1.4
Isopentane	0.0	0.0	0.0
CO Conversion, %	3.1	12.9	17.6

[a] Reproduced with permission from King (*80*).

King tested a large group of supported Ru catalysts in 2 H_2 + 1 CO at 4 atm (*80*). Supports were alumina, silica, silica–alumina, chromia–alumina, thoria, thoria–alumina, and some zeolites; metal loading was 0.5–2.8%; and the dispersion 8–60%. Turnover numbers for CO and CH_4 varied inversely with dispersion. Products from 2.5% Ru on silica (Table 6) show substantial yields of straight-chain olefins and paraffins, but none with branched chains.

Table 7 compares olefin/paraffin rates and the ratio C_{2+}/C_1 for different supported catalysts. The olefin yields increased with increasing conversion of CO. The largest olefin yields were produced by a catalyst supported on $Cr_2O_3–Al_2O_3$ and by the unsupported Ru; lowest olefin yields were from catalysts supported on $SiO_2–Al_2O_3$ and on the ultrastable zeolite. The samples producing the least CH_4 were supported on $Cr_2O_3–Al_2O_3$ and thoria. Unsupported Ru produced the most CH_4. Yields of branched hydrocarbons were usually about zero, except for preparations supported on silica–alumina or some zeolites (Table 8). The author suggested that an isomerization subsequent to the synthesis occurs on acidic supports. Extensive conversion to branched molecules occurred on a mechanical mixture of 1.8% Ru on alumina and ultrastable zeolite. Catalysts supported on NaY produced large yields of branched isomers; however, NaY should

Table 7
Selectivity of Ruthenium Catalysts at 250°C[a]

Catalyst	*Ethylene*/*ethane*	*Propylene*/*propane*	*Butene-1*/*n-butane*	C_{2+}/C_1
0.5% on γ-alumina[b]	0.24	1.95	1.0	1.7
1.25% on γ-alumina[b]	0.08	1.12	0.60	2.1
1.8% on γ-alumina[b]	0.06	0.75	0.43	2.3
2.5% on γ-alumina[c]	0.02	0.33	0.21	1.4
2.5% on γ-alumina[d]	0.05	0.70	0.46	1.7
1.5% on silica	0.15	1.70	1.31	1.9
2.5% on silica	0.04	0.52	0.39	1.5
2% on SiO_2–90% Al_2O_3	0.02	0.19	0.21	1.3
2% on SiO_2–15% Al_2O_3	0.08	0.84	0.48	2.0
2% on Cr_2O_3–Al_2O_3[e]	0.05	0.65	0.33	2.4
2% on Cr_2O_3–Al_2O_3[f]	0.27	3.0	1.5	3.0
2.8% on ThO_2	0.13	1.2	0.62	2.5
2% on ThO_2–90% Al_2O_3	0.04	0.67	0.36	1.8
2.5% on NaX	0.04	0.69	0.44	1.9
2.5% on ultrastable zeolite	0.02	0.20	<0.02	1.8
Unsupported Ru	0.30	3.8	2.4	0.35

[a] Adapted from King (*80*).
[b] Pore diameter, 118 Å.
[c] 137 Å.
[d] 44 Å.
[e] Commercial preparation containing 15% Cr_2O_3.
[f] Laboratory preparation containing 15% Cr_2O_3.

Table 8
Production of Branched-Chain Hydrocarbons at 250°C[a]

Catalyst	$i\,C_4/n\,C_4$	$i\,C_5/n\,C_5$
2.5% Ru on NaX sieve	0.01	0.01
Ion-exchanged Ru on NaX sieve	0.01	0.01
2.5% Ru on NaY sieve	1.20	3.0
Ion-exchanged Ru on NaY sieve	0.30	0.80
2.5% Ru on Ultrastable sieve	3.5	13.8
1.8% Ru on Al_2O_3 + ultrastable sieve	4.0	11.9
2% Ru on SiO_2–Al_2O_3[b]	0.01	0.01
1.25% Ru on SiO_2–Al_2O_3[c]	0.06	0.34
2% Ru on SiO_2–Al_2O_3[d]	0.16	0.57
Thermodynamic equilibrium	0.75	3.3

[a] Reproduced with permission from King (*80*).
[b] 10% SiO_2.
[c] 74% SiO_2.
[d] 85% SiO_2.

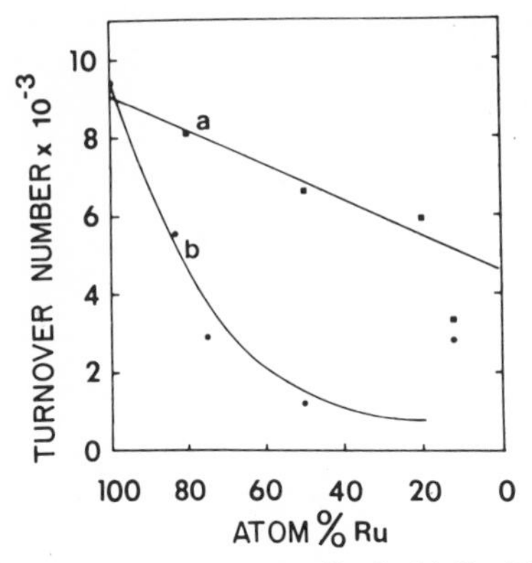

Fig. 6 Turnover number versus atom percent Ru in (a) Ru–Ni zeolites and (b) Ru–Cu zeolites. Reproduced with permission from Elliott and Lunsford (*54*).

not be acidic unless it contains substantially less than the stoichiometric amount of sodium.

Jacobs *et al.* (*67*) prepared and tested Ru on zeolite catalysts made by ion exchange using $Ru(NH_3)_6^{3+}$ complex. Reduction and chemisorption of H_2 and CO suggested that after ion exchange, part of the Ru was in sodalite cages and the rest in the supercages. As the reduction temperature was increased from 300 to 500°C, the Ru in the sodalite cages moved into the supercages. There was no tendency of Ru to aggregate into very large crystallites or to migrate to the periphery of the particles. Chemisorption ratios, H_2 : Ru, remained ~0.5 for samples reduced at all temperatures between 300 and 500°C, and the Ru could be reoxidized at 100°C. Turnover numbers (TONs) decreased with the Si/Al ratio of the zeolite, sharply for samples reduced at 300°C and slightly for those reduced at 500°C.

Elliott and Lunsford (*54*) prepared Ru, Ru–Ni, and Ru–Cu catalysts by ion exchange on NaY and CaY. In all cases the content of Ru was 0.5%. In methanation at 280°C, the catalysts aged rapidly. Ruthenium was deposited in large cavities of the zeolite, and the TON was about the same as Ru on alumina. Nickel was deposited near the exterior of the particle and had low TON with respect to Ni on alumina. Addition of Ni to the RuY decreased the activity linearly (Fig. 6), but Cu decreased the activity sharply.

Kellner and Bell (*76*) found that acetaldehyde was the major oxygenated product from the FTS on 1.2% Ru on silica, and methanol the principal oxygenate from 1.0% Ru on alumina at 1 and 10 atm. Tests were made with 3 D_2 + 1 CO as well as 3 H_2 + 1 CO feed (Figs. 7 and 8). For the oxygenated products on both catalysts and for CH_4 from Ru on alumina,

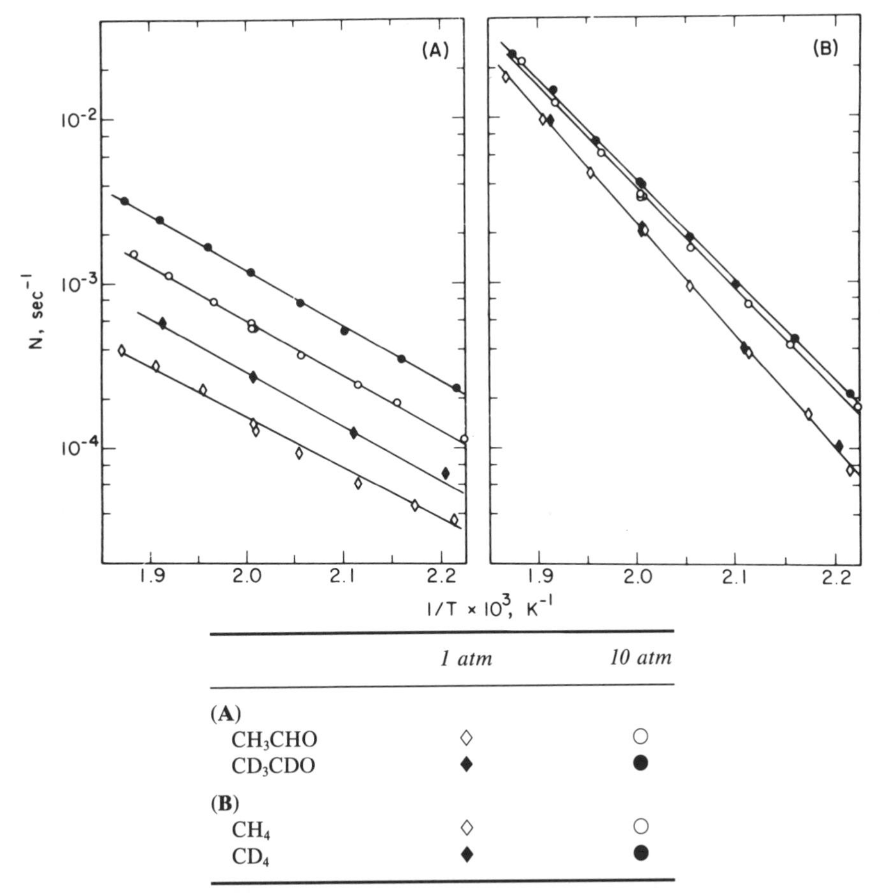

Fig. 7 Arrhenius plots for the synthesis of acetaldehyde (A) and methane (B) from H_2 (D_2) and CO [H_2 (D_2)/CO = 3] over a silica-supported Ru catalyst (1.2% Ru on SiO_2). Reproduced with permission from Kellner and Bell (*76*).

the rate was greater in D_2–CO than in H_2–CO, an inverse isotope effect. For Ru on silica, activation energies of 15.0 and 29.6 kcal/mol were obtained for acetaldehyde and CH_4, and with Ru on alumina, 21.6 and 28.0 kcal/mol for CH_3OH and CH_4.

Kellner and Bell (*75*) studied the kinetics of the FTS on well-dispersed Ru on alumina at 1 and 10 atm. Turnover numbers were reported for C_1–C_{10} straight-chain paraffins and C_2–C_{10} olefins. The rate equation for CH_4 was

$$N_{C_1} = 1.3 \times 10^9 \exp(-28{,}000/RT) p_{H_2}^{1.35} / p_{CO}^{0.99}, \qquad (2)$$

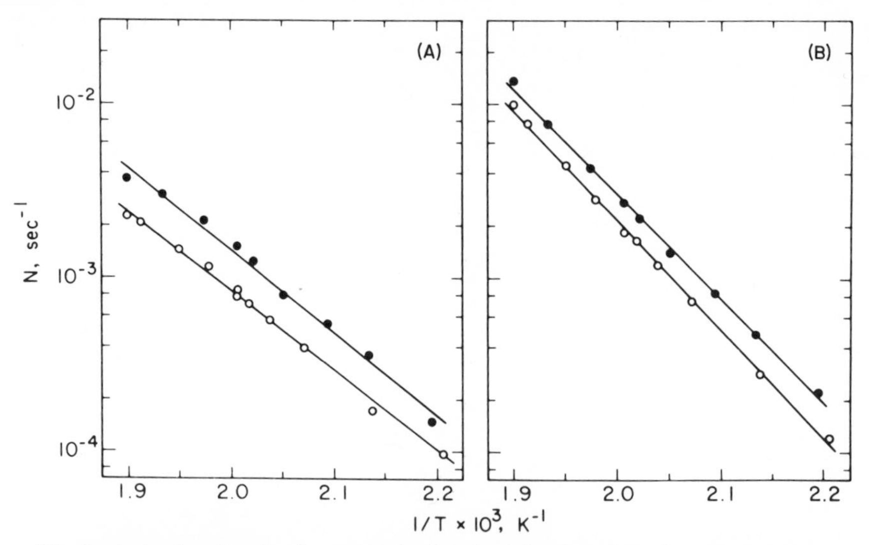

Fig. 8 Arrhenius plots for the synthesis of methanol and methane from H_2 (D_2) and CO over an alumina-supported Ru catalyst, [1.0% Ru on Al_2O_3; p = 10 atm; H_2 (D_2)/CO = 3]. (A) ○, CH_3OH; ●, CD_3OD. (B) ○, CH_4; ●, CD_4. Reproduced with permission from Kellner and Bell (*76*).

where the pressures are in atmospheres. Power rate law constants were given for TONs for each of the hydrocarbons to C_{10}. For the paraffins at 480 K, the H_2 exponent was ~1.35 through C_4 and then decreased to ~1.0. The CO power increased from about −1.0 for CH_4 to a constant value of −0.3 at C_6. For the olefins, the H_2 exponent decreased from 0.8 at C_2 to 0.2 at C_{10}, and the CO power increased from −0.7 to 0.0 from C_2 to C_{10}. In the author's opinion, these rate equations are neither very accurate nor particularly useful. The kinetics were explained by a mechanism involving the dissociation of CO and interactions of adsorbed methylene and methyl groups. Infrared spectra similar to those on supported Ru by King (*81*) were obtained. For methanation with Ru on fused alumina plates, the rate was inversely proportional to the partial pressure of CO and the activation energy was 27 kcal/mol (*2*).

Dalla Betta *et al.* (*35*) observed constant initial TONs for the FTS on Ru on alumina with average size of Ru particles varying from less than 10 to 91 Å. At 240°C the H_2 and CO orders were 1.28 and −0.47, respectively, and at 300°C, 1.70 and −0.93. The results suggest possible inadequacies of the power rate law. Activation energies were 17 kcal/mol. For tests in a Berty reactor on Ru on alumina at elevated pressure and 235°C, Pannell *et al.* (*110*) obtained power rate law exponents of 0.22 and −0.32 for H_2 and CO, respectively.

Table 9
Ruthenium Supported on TiO_2 and Al_2O_3[a]

Catalyst	Pressure, atm	Conversion of CO, %	Products, wt %								
			C_1	$C_2^=$	C_2	$C_3^=$	C_3	$C_4^=$	C_4	C_5	C_{6+}
2% Ru–TiO_2	1	1	26	12	5	20	8	14	4	10	1
	9.8	7	14	2	4	14	7	16	4	17	22
5% Ru–Al_2O_3	1	2	47	5	8	21		6	1	6	5
	9.8	10	24	1	5	12	4	6	5	12	32

[a] Temperature, 262–274°C. Adapted from Vannice and Garten (*142*).

Vannice and Garten (*142*) found that TiO_2 (Cab-O-Ti) was a better support for Ru than silica, alumina, and carbon with respect to activity and yields of C_{2+} hydrocarbons and olefins. Table 9 compares catalysts supported on alumina and TiO_2 at 1 and 9.8 atm using 1 H_2 + 1 CO feed. A potassium-promoted 1.3% Ru on alumina was prepared by impregnating the alumina with a solution of $K_2Ru_3(CO)_{12}$; it had moderate activity and produced hydrocarbons (*96*).

Everson *et al.* (*56*) used the Engelhard pelleted 0.5% Ru on γ-alumina in a Carberry reactor at 8–16 atm of 3 H_2 + 1 CO. The tests were of 2-h duration at 200–300°C, and the objective was to explore conditions under which the product did not exceed C_{12}. At a given temperature, the molecular weight decreased with decreasing feed rate or increasing conversion. Possibly this relationship arises from the use of a nonstoichiometric feed ratio, the H_2/CO ratio increasing with conversion. Catalytic activity decreased continually during the 2-h tests, and a carbonaceous deposit formed on the catalyst in amounts up to 5.2 wt %. The surface area of the catalyst decreased from 275 m^2/g for fresh material to 140 m^2/g for the sample with a 5.2 wt % carbonaceous deposit.

In a second paper, Everson *et al.* (*55*) studied the effect of thermal sintering on the activity and selectivity of a pelleted 0.5% Ru on alumina in the FTS in a spinning-basket reactor at 8 and 12 atm of 2.3 H_2 + 1 CO gas. The catalyst (Johnson–Matthey) was sintered for 30 min in oxygen before a standard reduction at 250°C for 1 h (e.g., the original catalyst had a dispersion of Ru of 78% and an average crystallite size of 13 Å). Oxygen treatment at 250°C gave a dispersion of 54% and a crystallite size of 19 Å; at 350°C, 36% and 29 Å; and at 500°C, 16% and 64 Å. The yield in the C_5–C_{10} range and the growth constant α increased with increasing crystallite size (Table 10). Also shown in Table 10 are kinetic constants for the original and the most severely sintered catalyst.

Dalla Betta and Shelef (*36*) made *in situ* measurements of infrared

Table 10
Effect of Dispersion on the Fischer–Tropsch Synthesis on Ruthenium[a]

Selectivity experiments[b]					*Kinetic tests, order of reaction*		
Sintering temperature, °C	*Dispersion of Ru, %*	*Crystallite size, Å*	α^c	*C_5–C_{10} selectivity*[d]	*H_2*	*CO*	*E, kcal/mol*
None	78	13	0.65	33.7	0.84	−0.76	28.2
200	70	15	0.66	33.2	—	—	
300	47	22	0.67	36.5	—	—	
500	16	64	0.73	44.7	1.15	−0.71	24.1

[a] Reproduced with permission from Everson *et al.* (*55*).
[b] Conditions: 250°C, 12 atm of 2.3 H_2 + 1 CO, and conversion of 22%.
[c] Slope of a plot of the logarithm of moles of product in a carbon-number fraction as a function of carbon number.
[d] Weight percentage of total hydrocarbons.

spectra during methanation on 5% Ru on alumina. At 250°C, the main feature was a C—O stretching band at ~2000 cm^{-1} observed in pure CO and H_2 + CO mixtures. Bands for hydrocarbons, formate, and water adsorbed on the support were also found.

Vannice *et al.* (*143*) studied Ru–Fe alloys supported on silica (Cab-O-Sil) at metal concentrations of 6 wt % or less and used Mössbauer spectroscopy to characterize the iron. Turnover numbers were at a maximum in the range 100–95% Ru. In the FTS at ~250°C with 3 H_2 + 1 CO, maximum yields of olefins in the C_2–C_4 range were obtained from 65 to 35 atom % Ru in Ru–Fe.

Ott *et al.* (*107*) examined Ru–Fe alloy powders by X-ray photoelectron and secondary-ion mass spectrometry. The first layer of the surface was greatly enriched in Fe. In the FTS at 1 atm of 3.3 H_2 + 1 CO gas and 300°C, pure Ru was the most active, followed by a minimum and a broad maximum between 75 and 33 atom % Ru. Addition of 3 atom % Fe increased olefin production sharply, and the olefin yields remained large until the Fe content was increased beyond 70 atom %. The same authors (*108*) examined the Ru powder and the 97 Ru : 3 Fe alloy after exposure to 3 H_2 + 1 CO at 1 atm and 300°C by X-ray photoelectron and secondary-ion mass spectroscopy. The metal surfaces were covered by less than one-fourth monolayer of carbon. Alloys containing a larger fraction of Fe accumulated carbon more rapidly; the deposition of carbon occurred on the Fe.

Miura and Gonzalez (*97*) characterized a series of Pt–Ru bimetallic clusters on Cab-O-Sil silica, 0.3 mmol metal per gram, and tested them in the FTS with 3 H_2 + 1 CO at 1 atm. Higher hydrocarbons were produced only by preparations containing less than 50% Pt. For smaller amounts of Pt, the yields of CH_4 increased with the amount of Pt, but the composition of the C_2–C_5 hydrocarbons was more or less independent of the Pt content. *In situ* infrared studies suggest that CO chemisorbed on Pt sites was transferred to Ru sites before reacting.

Tamaru (*138*) studied reactions of H_2 + CO on Ru on silica at 150°C. Infrared studies showed that the surface was heavily covered with CO and only sparsely by growing hydrocarbon chains. From the size of bands for CH_3 and CH_2, the chain length was estimated. When CO was removed from the gas phase, the adsorbed hydrocarbon chains were rapidly converted mostly to CH_4. Evacuation of an operating catalyst gave CO and H_2 but no hydrocarbons, which apparently were converted to surface carbon. If synthesis was performed with H_2 + ^{13}CO and this gas changed to H_2 + ^{12}CO, ^{13}C was found in the hydrocarbons for some time, suggesting the presence of ^{13}C on the surface and in growing chains. Treatment with ^{13}CO yielded surface ^{13}C, which was found in the hydrocarbons when ^{12}CO + H_2 was passed over the catalyst.

Ekerdt and Bell (*52*) made simultaneous measurements of kinetics and infrared spectra on catalyst of Ru on Cab-O-Sil in a differential reactor. At ~275°C, the hydrogen and CO orders were 1.5 and −0.6, and the activation energy was 24 kcal/mol. The primary feature of the spectra was a band at 2030 cm^{-1} associated with chemisorbed CO. The position and intensity of this band were not affected by the partial pressure of CO or the H_2/CO ratio. The intensity decreased with increasing temperature and when the catalyst was treated with H_2. This structure was not considered to be an intermediate in the reaction. The operating catalyst contained carbon in amounts corresponding to several monolayers. This carbon was also removed when the catalyst was treated with pure H_2, but it was also not considered to be an intermediate in the reaction.

Kellner and Bell (*77*) made *in situ* infrared measurements during the FTS on Ru on alumina. The surface was heavily covered with CO as a linear absorbed species, a diadsorbed $Ru(CO)_2$ form, and a bridged structure [Ru(CO)Ru]. The bridged form was present in only minor amounts, and the $Ru(CO)_2$ species in moderate amounts. The diadsorbed structure did not hydrogenate rapidly. The coverage of the linear form followed the Langmuir equation

$$\theta = KP/(1 + KP), \tag{3}$$

where

$$K = 1.1 \times 10^{-9} \exp(25{,}000/RT) \quad \text{atm}^{-1}. \tag{4}$$

The surface was nearly completely covered under most conditions; the value of K is large, 392 at 200°C.

Table 11
Effect of Copper on a 1% Ruthenium on Silica Gel[a]

		Order		
% Cu	*E, kcal/mol*	*H_2*	*CO*	*Turnover number at 600 K, molecules per Ru site per second*
0	20.6	1.16	−0.43	5.75×10^{-2}
0.016	21.6	0.66	−0.19	1.49×10^{-2}
0.063	21.1	0.89	−0.36	3.34×10^{-2}
0.32	23.1	0.64	−0.29	3.46×10^{-3}
0.63	22.6	0.75	0.14	1.11×10^{-3}

[a] Adapted from Bond and Turnham (*19*).

Bond and Turnham (*19*) studied the effect of Cu in 1% Ru on silica gel on the rate of the H_2 + CO reaction at ~0.1 atm. Aqueous solutions of $RuCl_3$ and $Cu(NO_3)_2$ were impregnated simultaneously. Data in Table 11 show that the rate decreases substantially with increasing Cu, the H_2 exponent decreases, the CO order increases (becomes more positive), but the activation energy remains constant. Interpretations of these data in terms of the Vannice–Ollis reaction scheme are considered in Chapter 5.

Three papers considered carbon-number distributions from Ru catalysts (*38, 106, 110*); these are discussed in Chapter 5. Also included in the next chapter is a paper by Biloen *et al.* (*18*) on the incorporation of surface carbide and a negative H_2/D_2 isotope effect on supported Ru reported by Kellner and Bell (*74*).

Gupta *et al.* (*61*) observed an increased activity in the hydrogenation of CO_2 when 1.8% Ru on alumina and 1.8% Ru on 13X zeolite were exposed to γ rays from ^{60}Co of intensity of 0.63 Mrad/h. The activity increased during radiation for 2% CO_2 in H_2, and when the radiation was removed the activity decreased following a first-order decay. The activation energy decreased during radiation, from 13.8 to 7.7 kcal/mol for the catalyst on alumina and from 7.3 to 4.2 kcal/mol for the preparation on the zeolite. Pichler and Firnhaber (*114*) found that very active catalysts for the polymethylene synthesis were produced by irradiating RuO_2 with γ radiation from ^{60}Co. RuO_2, either in air or evacuated, was exposed to 4×10^5 rads/h for 65 h. The irradiated oxide reacted with nonane at room temperature to produce finely divided Ru metal plus CO_2 and H_2O.

4.2.2 Rhodium, Palladium, Platinum, Iridium, and Osmium

The hydrogenation of CO and CO_2 on a polycrystalline Rh foil was studied by Sexton and Somorjai (*128*) in a versatile ultrahigh-vacuum apparatus permitting cleaning of the sample and performing Auger elec-

Table 12
Variation of Product Distribution with H_2/CO Ratio and Temperature on Rhodium Foils[a]

Temperature, °C	*Product*[b]	*H_2/CO Ratio* 1:2	3:1	9:1
250	C_1	65	84	93
	$C_2^=$	16	9	4
	C_2	9.8	3	2
	C_{3+}	9.2	4	1
300	C_1	77	89	95
	$C_2^=$	13	7	2
	C_2	4	2	2
	C_{3+}	6	3	1
350	C_1	83	94	98
	$C_2^=$	12	3	0
	C_2	1	2	2
	C_{3+}	4	1	0.2

[a] Reproduced with permission from Sexton and Somorjai (*128*).
[b] C_1, Methane; $C_2^=$, ethylene; C_2, ethane; C_3, propane.

tron and other spectroscopies in the ultrahigh-vacuum mode and catalytic studies at 1 atm or higher pressures. Reaction products, activation energies, and TONs for $H_2 + CO$ on the foil at ~1 atm were nearly the same as on the supported catalysts of Vannice (*140*), as shown in Table 3. Some higher hydrocarbons were produced (Table 12). For the $H_2 + CO_2$ reaction, the product was essentially all CH_4, and this reaction had a larger TON than the hydrogenation of CO (Table 13). In thermal desorption, peaks at 250 and 700°C were found for CO introduced as CO, $H_2 + CO$, or $H_2 + CO_2$. In the synthesis reactions, the Rh is covered by carbon, and the bulk Rh contains carbon and oxygen. In similar experiments at 6 atm,

Table 13
Comparison of Activities for Methanation of the CO–H_2 and CO_2–H_2 on the Initially Clean Rhodium Surface at 700 torr[a]

	3 H_2 + 1 CO	*3 H_2 + 1 CO_2*
Specific rate at 250°C, turnover number (CH_4 molecules/site · sec)	0.017	0.12
Specific rate at 300°C	0.13	0.36
Activation energy, kcal	24 ± 2	16 ± 2

[a] Reproduced with permission from Sexton and Somorjai (*128*).

Table 14
Selectivity of Lanthanum Rhodate[a]

	Temperature, °C			
Products, wt %	*225*	*225*	*300*	*300*
CH_4	16	12	20	16
C_{2+}	7	6	8	9
Methanol	33	40	12	17
Ethanol	19	13	9	14
Acetaldehyde	15	23	43	31

[a] Adapted from Watson and Somorjai (*146*). 1 H_2 + 1 CO at 6 atm.

a polycrystalline Rh foil and a 111 Rh single crystal had the same TON (*25*).

Watson and Somorjai (*145*) found that RhO_2 and Rh_2O_3 were reduced rapidly to the metal in synthesis gas at 6 atm and 250–350°C, but $Rh_2O_3 \cdot 5\, H_2O$ was fairly stable under these conditions. At 300°C, the metal produced mostly CH_4 and only small amounts of higher hydrocarbons and no oxygenates. The hydrated oxide yielded larger amounts of C_{2+} hydrocarbons and sizable quantities of acetaldehydes and propionaldehydes. Large yields of propionaldehyde were obtained when ethylene was added to the feed to the hydrated oxide; on the metal, ethylene addition only increased the higher hydrocarbons slightly. Although the hydrated oxide remained selective for producing oxygenates for several hours, it too was reduced and lost its selectivity. Lanthanum rhodate ($LaRhO_3$), prepared by heating a mixture of La_2O_3 and $Rh_2O_3 \cdot 5\, H_2O$ at 1100°C in air, was highly selective, as shown in Table 14 (*146*). X-Ray photoelectron and Auger spectroscopy showed that the catalyst had a reactive carbonaceous layer, and most of the Rh was in the +1 oxidation state plus some metal. For the formation of methanol, associative adsorption of CO was postulated, but dissociative adsorption of CO for formation of other molecules.

Hungarian workers showed that supported Rh catalysts interacted with $H_2 + CO_2$ to produce infrared bands for formate and CO (*132*). The hydrogenation of CO_2 on Rh on alumina was measurable above 170°C, produced only methane, and had a activation energy of 16 kcal/mole. The $H_2 + CO_2$ reaction was eight times as fast as the H_2 + CO reaction. Titania was the most effective support, silica the least. In a paper on the FTS, steady-state TONs at 275°C with 1 atm of 3 H_2 + 1 CO in terms of moles CH_4 produced per site per second were 62, 11, 3, and 4 for preparations supported on TiO_2, Al_2O_3, MgO, and SiO_2, respectively (*133*). The activation energy and the power rate law exponents for H_2 and CO were 18.3 kcal/mol, 0.75, and −0.88; 24.0, 0.9, and −0.42; 23.7, 0.8, and 0.50; and 22.6, 0.57, and −0.20 for catalysts containing TiO_2, Al_2O_3, MgO, and SiO_2, respectively. Moderate yields of higher hydrocarbons were obtained from the catalysts supported on TiO_2 and Al_2O_3. When ethylene was added to the feed, the yield of propylene was increased; similarly, added propylene gave increased butenes.

Surface carbon was measured on the Rh catalysts by stopping the H_2 + CO, purging the system with He, and then introducing H_2. At the temperature of the previous synthesis, CH_4 was produced for the first few minutes at a rate greater than the rate of the synthesis. Part of this surface carbon hydrogenated at room temperature, and most of the carbon was removed by 400°C. At the start of the process, the amount of surface carbon increased in the same way as activity, but the increase of surface carbon continued after the activity had leveled out and had begun to decrease. *In situ* infrared spectroscopy showed the presence of linearly bonded CO, but at a lower frequency than the Rh–CO species. The existence of bands of adsorbed formate ion was proved by using deuterium instead of hydrogen in the tests. The formate was not believed to be an intermediate in the synthesis. One possibility is that the formate is adsorbed on the support.

In a study of supported catalysts containing 2.3% Rh, the activities of H_2 + CO and H_2 + CO_2 reactions at 1 atm varied with the support in the following way: $ZrO_2 > Al_2O_3 > SiO_2 > MgO$ (*65*). Methane was the principal product from H_2 + CO and the exclusive product from H_2 + CO_2, except at higher temperatures where small amounts of CO were also produced.

Bhasin *et al.* (*16*) hydrogenated CO over Rh and Rh–Fe supported on silica gel (3 to 6 mesh) in a silver-plated Berty reactor at pressures above 25 atm. Important products were methanol, ethanol, acetaldehyde, and acetic acid. Typical catalysts contained 2.5% Rh and different amounts of Fe; operating conditions were 70 atm and 300°C, and Fig. 9 shows the effect of Fe content; here, CH_3COX denotes acetic acid plus acetaldehyde. CH_4 yields were substantial and decreased to a minimum at an atom fraction of Fe of ~0.3. Larger amounts of Fe increased the production of C_{2+} hydrocarbons. The ratio of the formation of ethanol to the formation of acetaldehyde is a linear function of the rate of formation of methanol (Fig. 9). Soufi (*134*), using Rh black, obtained results that are similar in some respects to the present data.

Poutsma *et al.* (*120*) found that supported Pd produced sizable yields of methanol under conditions for which methanol is thermodynamically possible, as well as modest amounts of methyl formate and traces of ethylene

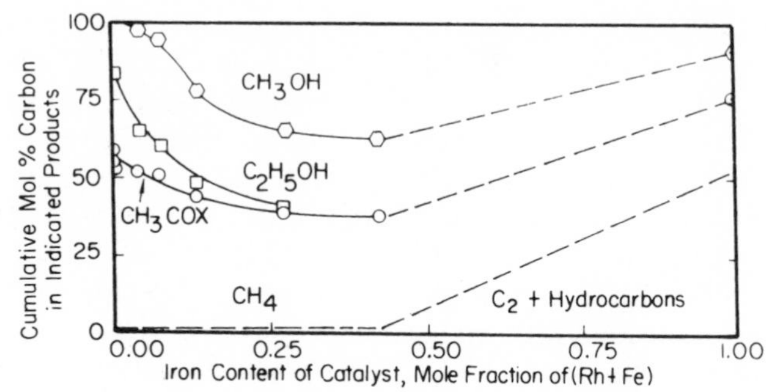

Fig. 9 Selectivities of Rh–Fe on silica-gel catalysts. Reproduced with permission from Bhasin *et al.* (*16*).

glycol. Supported Pt and Ir also produced these molecules, but at ~5% of the rate on Pd. In tests with 4.6% Pd on silica in a Berty reactor at 290°C and 12 atm, the ratio of production of CH_4/CH_3OH was 0.06 and the methanol formation was 0.2 mol/liter · h. At 290°C and 28 atm, the methanol rate increased to about 0.35. In some fixed-bed tests, the ratio CH_4/CH_3OH was less than 0.001.

An alkali-promoted 1.4% Ir on alumina was prepared by impregnating the alumina with a solution of $KIr(CO)_4$; its activity at 1 atm was less than 1% that of a similar Ru preparation, and the products were 72 wt % CH_4, 15 ethylene, and 13 ethane (*96*). Deeba *et al.* (*40*) prepared a magnesia-supported Os catalyst by reaction of MgO with $Os_3(CO)_{12}$. The infrared spectra of the catalyst before and after use in the FTS suggested the presence of a trinuclear osmium carbonyl structure. After oxidation in air, the spectra was that of a mononuclear osmium carbonyl. Both the original and oxidized catalysts were active in the FTS at 38 atm of 4 H_2 + 1 CO at 300°C, and produced C_2 and C_3 hydrocarbons in addition to CH_4.

4.3 Cobalt and Nickel Catalysts

Cobalt and nickel as metals are active catalysts in the FTS, resembling Ru in many ways. One difference is that Co and Ni form carbonyls more readily than Ru; volatile carbonyl formation often limits the operating pressure, particularly for Ni. Cobalt and Ni can also be converted to the carbides Co_2C and Ni_3C, which have low activity compared with the metals. The carbides are not important, because they are apparently not produced under conditions normally used for the FTS.

The development of Co and Ni catalysts to about 1950 was reviewed by

Anderson *et al.* (*5, 135*). The supported low-density Co and Ni catalysts developed by Fischer and co-workers led to the 100 Co : 18 ThO_2 : 100 kieselguhr and finally to the 100 Co : 5 ThO_2 : 8 MgO : 200 kieselguhr used in the commercial plants in Germany from 1936–1945 (*59*). These catalysts operated effectively in the 1940 fixed-bed reactors at 1 or 12 atm. Apparently, no chemical changes occur in the catalyst during synthesis. At 1 atm, the pores of the catalyst slowly fill with wax, which is liquid at synthesis temperatures. Hydrogen or solvent treatments at 7- to 14-day intervals removes the wax and reactivates the catalyst. At 7 to 12 atm, the pores of the Co catalyst fill with liquid hydrocarbons after a short time and remain filled throughout the life of the catalyst. In the pressure synthesis, regeneration steps are usually not needed and are not effective.

Since 1950, there has not been very much research on Co catalysts. On Ni most of the investigations have been made under methanation conditions. Pertinent papers on methanation are included in the discussion here.

4.3.1 Cobalt

A chapter by Eidus and Bulanova (*51*) on the effect of promoters on Co : kieselguhr (100 : 100) catalysts provides a bridge between old and new work on this type of catalyst. Table 15 presents a summary of the 1930 work of Fischer and Koch (*57*) and the Russian extension of the catalyst development studies directed toward replacing ThO_2 with other dioxides. The German work led to Co : ThO_2 : kieselguhr = 100 : 18 : 100. The subsequent development of the Co : ThO_2 : MgO : kieselguhr (100 : 5 : 8 : 200) used in the German commercial plants was not included in reference *57*. In Table 15, all preparations contain 100 Co : 100 kieselguhr + promoters, except for preparations marked with an asterisk (which contained 200 kieselguhr : 100 Co), and these tests were made at atmospheric pressure with 2 H_2 + 1 CO gas. Usually, Na_2CO_3 was a better precipitating agent than K_2CO_3. Catalysts containing ZrO_2–MgO and TiO_2–MgO gave yields of liquids at least as large as those of ThO_2–MgO.

Pichler *et al.* (*116, 117*) analyzed by capillary gas chromatography products from the atmospheric-pressure synthesis on a Co : ThO_2 : kieselguhr (100 : 18 : 100) catalyst at 190°C. Some of the data of this paper are pertinent to the mechanism of chain growth and are reported in Chapter 5. The carbon-number distribution for total hydrocarbons in Fig. 10 has the familiar pattern of a minimum at C_2 and a maximum at C_4 in the mole percentage plot. Detailed analyses for hydrocarbons in the C_6–C_7 range are given in Table 16. Most of the isomers were separated and all of the sample accounted for. The product contained largely straight-chain and monomethyl paraffins and olefins, and only traces of dimethyl- and

Table 15
Effects of Promoters in Cobalt–Kieselguhr Catalysts[a]

Precipitant	Promoters, parts by weight per 100 Co	Synthesis: Temp., °C	Synthesis: Gasoline + oil, g/m^3	Synthesis: Gasoline/oil ratio
Work of Fischer and Koch (*57*)				
K_2CO_3	None	210	34	17
K_2CO_3	20 Al	215	8	22
K_2CO_3	20 Cr	215	11	15
K_2CO_3	20 Zn	215	30	20
K_2CO_3	20 Mg	215	35	10
K_2CO_3	10 Mn	215	47	7
K_2CO_3	15 Mg	205	88	1.4
Na_2CO_3	15 Mn	190	105	1.0
K_2CO_3	12 ThO_2	190	94	0.6
K_2CO_3	18 ThO_2	190	100	0.7
K_2CO_3	24 ThO_2	190	20	0.6
K_2CO_3	48 ThO_2	190	2	

Precipitant	Promoters, parts by weight per 100 Co	Synthesis at 190°C: Gasoline + oil, g/m^3	Synthesis at 190°C: Gasoline/oil ratio
Work of Eidus and Bulanova (*51*)			
K_2CO_3	None	21	7
K_2CO_3	14 ZrO_2	54	0.5
K_2CO_3	18 ZrO_2	62	0.5
K_2CO_3	12 ZrO_2	60	1.8
K_2CO_3	10 ZrO_2 + 6 MgO	46	0.4
K_2CO_3	6 ZrO_2 + 10 MgO	93	1.1
K_2CO_3	6 ZrO_2 + 10 MgO[b]	113	0.8
Na_2CO_3	6 ZrO_2 + 10 MgO[b]	127	0.6
Na_2CO_3	18 TiO_2	80	0.8
Na_2CO_3	6 TiO_2 + 10 MgO	116	0.7
Na_2CO_3	6 TiO_2 + 10 MgO[b]	115	0.7
Na_2CO_3	10 TiO_2 + 6 MgO[b]	108	0.9
K_2CO_3	6 TiO_2 + 10 MgO[b]	86	0.6
K_2CO_3	6 ThO_2 + 10 MgO[b]	104	0.6
K_2CO_3	18 CeO_2[b]	16	0.9
Na_2CO_3	10 CeO_2 + 6 MgO[b]	71	0.6
Na_2CO_3	10 HfO_2 + 6 MgO[b]	76	0.8

[a] All catalysts contain 100 Co : 100 kieselguhr except where noted. All tests were at atmospheric pressure.
[b] In these preparations, Co : kieselguhr was 100 : 200.

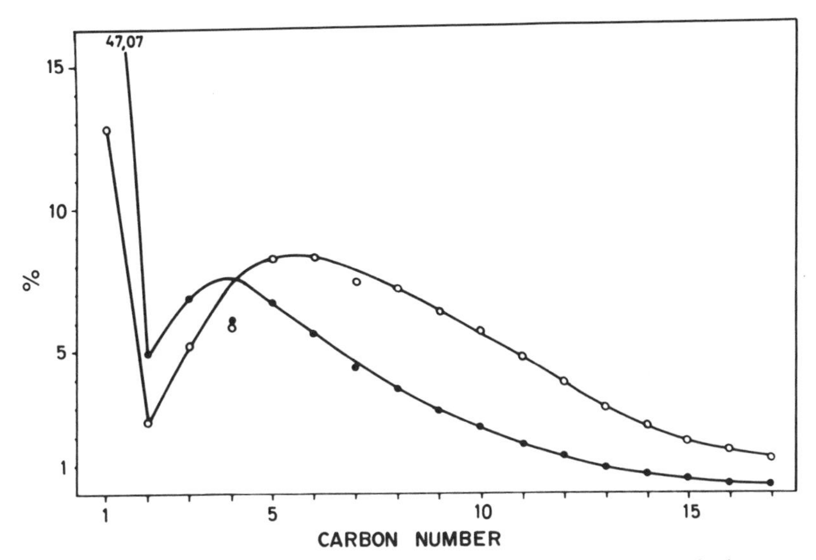

Fig. 10 Carbon-number distribution for hydrocarbons from the atmospheric-pressure synthesis on a standard Co–ThO_2–kieselguhr catalyst at 190°C. ○, Weight basis; ●, molar basis. Reproduced with permission from Pichler *et al.* (*117*).

Table 16
Detailed Analyses of C_6 and C_7 Hydrocarbons from Cobalt[a]

		Olefins, double-bond position				
Structure[b]	*Paraffins*	*1*	*2*	*3*	*4*	*5*
Normal C_6	56.0 }	4.0	{ 16.7	4.0		
2M5	8.1 }		{ 2.9	0.8	A	
3M5	5.1	A	1.3			
Normal C_7	55.7	1.8	10.6	5.9		
2M6	B	0.5 }	4.0		{ 0.6	C
3M6	C	B }			{ 0.9	1.6
23DM5	0.4				0.1	
24DM5		0.1				

3-Ethyl pentane = 0.2; ΣA = 1.1; ΣB = 6.5; ΣC = 11.1

[a] Expressed as a weight percentage. Adapted from Pichler and Schulz (*116*).
[b] M, Methyl; DM, dimethyl (e.g., 23DM5, 2,3-dimethylpentane).

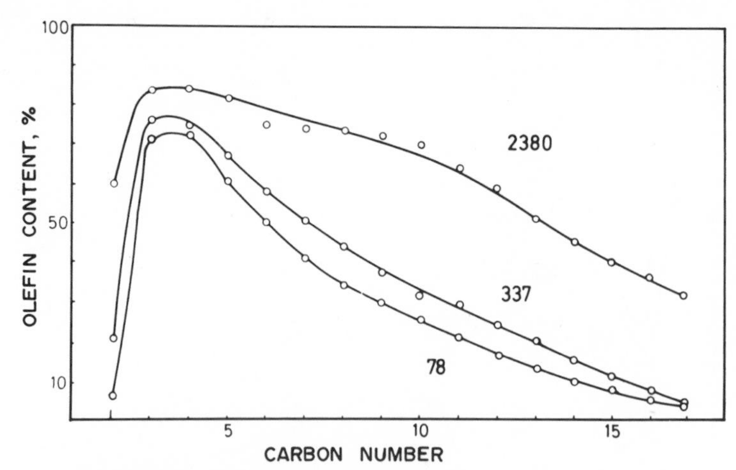

Fig. 11 Olefin content as a function of carbon number for three hourly space velocities. Reproduced with permission from Pichler *et al.* (*117*).

ethyl-substituted species. Naphthenes, aromatics, diolefins, and acetylenes were not found (*118*). Transolefins were produced in larger amounts than cis, usually by a factor of 2.

Olefins, particularly α olefins, seem to be a primary product of the synthesis on Co. The percentage of olefin in carbon-number fractions passes over a maximum at about C_3–C_4 and then diminishes (Fig. 11). The fraction of olefin increases with increasing space velocity, that is, with decreasing conversion; the space-velocity or conversion effects seem to be more pronounced at larger carbon numbers. The percent of α olefin in a given carbon number also increases with increasing space velocity, that is, with decreasing conversion (Fig. 12).

Primary alcohols, C_1 to C_5, were found in the water layer in very small amounts ($<$0.2 wt %). The alcohol yield was independent of space velocity (and conversion). Alcohols were considered to be by-products and not the primary product.

In the 1970s, workers at American universities sought Co catalysts for producing C_2–C_4 olefins for petrochemical feed stocks. These tests were usually made at temperatures from 200 to 300°C, significantly higher than those in the 1940 German work. Dent and Lin (*41*) tested many Co catalysts in a Berty reactor with a feed of 3 H_2 + 1 CO + 6 He at 7.8 atm. After a conditioning period, the catalyst was held for 4-h intervals at 200, 250, 350, 250, 300, and 225°C. The most active catalysts were Co–Cr_2O_3–kieselguhr and Co–ZrO_2–kieselguhr; however, Co–MnO–alumina–K_2O catalysts seemed best for producing olefins. Typical results are shown in Fig. 13. Alkali increased the olefin yield from these catalysts.

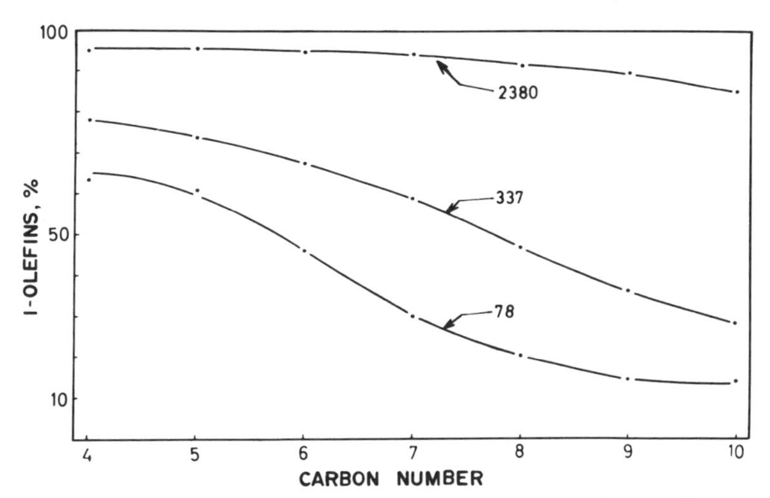

Fig. 12 1-Olefins as fraction of total olefins as a function of carbon number for three hourly space velocities. Reproduced with permission from Pichler *et al.* (*117*).

Borghard and Bennett (*20*) tested a Harshaw Co on silica (100 Co : 94 SiO_2) at 250°C and 20 atm of 2 H_2 + 1 CO gas. About 90% of hydrocarbon was gaseous and 10% liquid. For this catalyst, 2 H_2 + 1 CO was better feed than 1 H_2 + 1 CO or 3 H_2 + 1 CO.

Yang *et al.* (*149*) studied Co–Cu–alumina catalysts; the optimum C_2–C_4 yield was obtained with a Co : Cu ratio of 3 : 7. Alkali, potassium or sodium, was not a promoter for this type of catalyst. The C_2–C_4 fraction was about half of the hydrocarbons. The rate equation $r = kp_{H_2}p_{CO}^{-1/2}$ fits the kinetic data satisfactorily for tests in which the temperature, total pressure, and ratio H_2/CO were changed. For a 14 Co : 1 La_2O_3 : 100 Al_2O_3 catalyst at 5–8 atm and 215°C in a Berty reactor, Pannell *et al.* (*110*) found power rate exponents of 0.55 and −0.33 for H_2 and CO, respectively; the rate of CO_2 production was low. Poisoning tests of an alkalized Co–ThO_2–kieselguhr catalyst (*91*) are described in Chapter 6.

Agrawal *et al.* (*1*) studied kinetics of methanation on Co on alumina in three pseudo–steady-state regions: as clean Co, carbon-deactivated Co, and sulfur-poisoned Co. Turnover numbers, power rate exponents, and activation energies are given in Table 17 for each of the regions. In these experiments, the initial period of relatively constant high activity lasted for 30 h on a feed containing 0.5% CO, and shorter periods on feed with more CO. Then the activity decreased more rapidly over the next 2 or 3 days to a new steady state at least 50-fold lower than the first. The sulfur-poisoned state was obtained with synthesis gas containing 13 ppb H_2S.

Based on other evidence, Agrawal *et al.* (*1*) believed that a mechanism

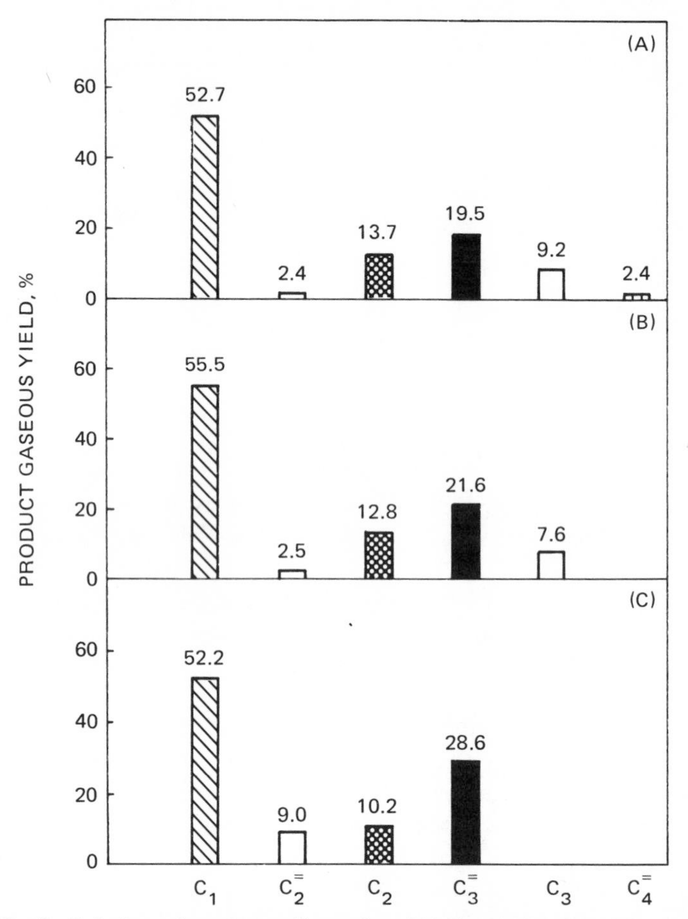

Fig. 13 C_1–C_4 hydrocarbons from a Co catalyst (100 Co : 10 Mn : 400 Al_2O_3 : 3.6 K_2O) at 7.8 atm of 3 H_2 + 1 CO. (A) 231°C. (B) 218°C. (C) 205°C. Reproduced with permission from Dent and Lin (*41*). Copyright 1979 American Chemical Society.

involving a surface-carbon intermediate with the addition of the first hydrogen to the carbon as the rate-determining step was most appropriate:

$$r = k_1 p_{CO} p_{H_2}^{1/2}/(1 + k_2 p_{CO})^2. \qquad (5)$$

The value of the constant k_2, the adsorption equilibrium constant for CO, decreased substantially from the fresh, to the carbon-deactivated, to the sulfur-poisoned states, apparently reflecting a decrease in the heat of adsorption of CO. Equation (5) does not represent data for the fresh catalyst very well, and it was suggested that two or more sets of active sites may be present.

Table 17
Kinetics of Methanation on Cobalt on Alumina in Different Activity Regions[a]

Measure	*Fresh catalyst*	*Carbon-deactivated catalyst*	*Sulfur-poisoned catalyst*
Turnover number at 400°C, s^{-1}	10.0	0.10	0.001
Activation energy, kcal/mol	28 ± 2	16 ± 2	16 ± 2
CO Exponent	−0.24	0.3–1.0	0.3–1.0
H_2 Exponent	0.5	0.5	

[a] Adapted from Agrawal *et al.* (*1*).

Palmer and Vroom (*109*) found that the activity of clean Co and Ni foils increased by factors of 100 and 10, respectively, if the final reduction in H_2 in an alternate oxidation–reduction surface-cleaning procedure was at 525–600 K rather than the usual 800–900 K. Residual oxygen remaining after the lower-temperature reduction was postulated to increase the activity. Ignatiev and Matsuyama (*64*) performed similar experiments using a single crystal of Co and Auger electron spectrometry. The oxygen on the specimen was quickly removed in H_2 at 800–900 K. Reduction was very much slower at 525–600 K; however, the authors concluded that all of this oxygen would be removed after a reasonable exposure to H_2 + CO. The enhanced activity was attributed to the production of very active "small metal clusters" from the reduction of the surface oxide at 525–600 K.

Nakamura *et al.* (*103*) studied alloys of composition Fe_3Co, FeCo, and $FeCo_3$ prepared by reduction of oxides precipitated from solutions of nitrates. A bcc structure was observed for all of the alloys. Auger electron spectroscopy showed that Fe accumulated at the surface. The Fe–Co alloy had the largest activity in H_2 + CO reactions. All of the alloys produced large yields of C_2 and C_3 hydrocarbons (Fig. 14). Pure Fe produced the most higher hydrocarbons and pure Co the least.

4.3.2 Nickel

In the early 1930s, when commercial development of the FTS was first considered seriously, Fischer contended that the use of Co was impractical. An effort was made to prepare useful Ni catalysts, and a number of compositions were tried in the following order: 100 Ni : 18 ThO_2 : 100 kieselguhr; 100 Ni : 20 MnO : 100 kieselguhr; and 100 Ni : 25 MnO : 10

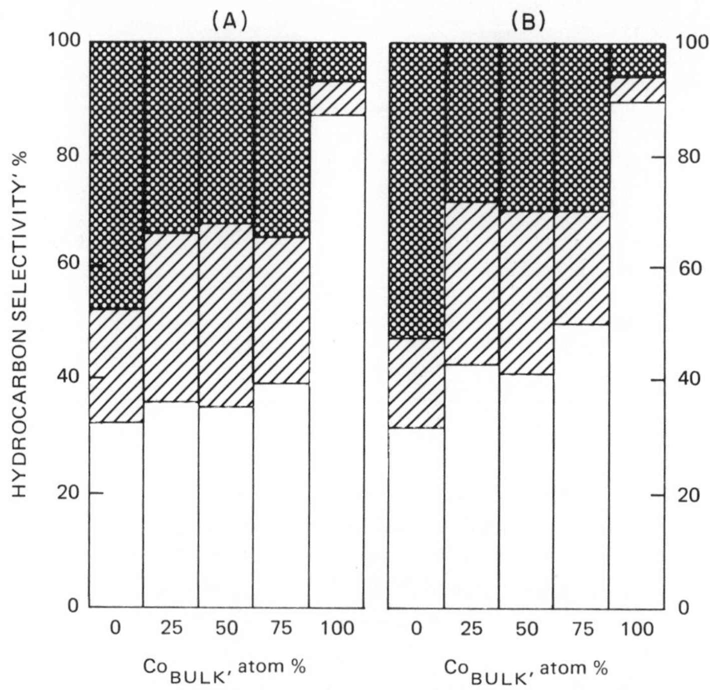

Fig. 14 Selectivity of Fe–Co alloys at two temperatures (A, 523 K; B, 548 K). □, CH_4; ▨, $C_2 + C_3$; ▩, C_{4+}. Reproduced with permission from Nakamura *et al.* (*103*).

Al_2O_3 : 100 kieselguhr. The general experience was that each of these preparations performed moderately well in small reactors, but in pilot plants and larger units the catalyst life was short, product quality poor, and Ni recovery low. At least some of these difficulties resulted from the inadequacy of the early reactors in removing the heat of reaction. Cobalt catalysts containing 100 Co : 5 ThO_2 : 8 MgO : 200 kieselguhr were used in the commercial plants. Unsuccessful attempts were made later to replace part of the Co in the standard catalysts with Ni. This early work is reviewed in reference *135*. Since 1950 a wide variety of catalyst types has been studied; however, the objective has often been the production of CH_4. The literature of methanation is voluminous, and these papers are frequently not pertinent to the FTS. Therefore, only a few papers on methanation are considered here.

Shalvoy *et al.* (*129*) examined unreduced coprecipitated Ni–SiO_2 containing ~50% Ni by X-ray photoemission spectroscopy as well as NiO, $Ni(OH)_2$, and $NiSiO_3$ standards. The spectra of the supported samples suggested that they were composed primarily but not entirely of amorphous $NiSiO_3$; however, X-ray diffraction indicated NiO as the principal

Table 18
Activity and Selectivity of Supported Nickel Catalysts[a]

Catalyst	*Activity at 205°C*		*Temp., °C*	*CO Conversion, %*	*Products, mol %*				
	CO converted, mol/sec · g Ni	*CO converted, mol/sec · g catalyst*			C_1	C_2	C_3	C_4	C_5
1.5% Ni on TiO_2	8.35	0.11	251	13	58	14	12	8	7
10% Ni on TiO_2	22.8	2.28	243	24	50	9	25	8	9
8.8% Ni on η-Al_2O_3	1.6	0.14	230	3	81	14	3	2	—
16.7% Ni on SiO_2	2.36	0.39	220	3	92	5	3	1	—
20% Ni on graphite	0.06	0.01	234	25	87	7	4	1	—

[a] 3 H_2 + 1 CO at 1 atm. Adapted from Vannice and Garten (*142*).

phase. Depending on the support and the way the catalyst is made, particularly the thermal treatment, the interaction of nickel oxide with the support may vary widely. In extreme cases, reduction to Ni metal may be difficult or impossible.

Bartholomew *et al.* (*14, 15*) prepared Ni catalysts supported on silica, alumina, and titania and characterized them by chemisorption of H_2 and CO. Tests were made at 1.4 atm with hot-gas-recycle-type feed, 1 CO + 4 H_2 + 95 N_2, at high flows and 500, 525, and 550 K. The ratio of the production of C_{2+} to that of CH_4 increased with dispersion of Ni for alumina- and silica-supported preparations. Turnover numbers as a function of dispersion followed no clear-cut trend. Activation energies of 17–28 kcal/mol were found for catalysts supported on alumina, 19–27 on silica-supported samples, and ~20 on titania-supported Ni.

Vannice and Garten (*142*) tested the series of supported Ni catalysts shown in Table 18. Those supported on TiO_2 were significantly more active and produced more higher hydrocarbons than catalysts supported on η-Al_2O_3, SiO_2, and graphite. The preparations on titania did not chemisorb H_2 and CO as expected, possibly because of strong interactions between the TiO_2 and Ni. For this reason, TONs are not reported. The titania catalysts were more stable thermally and produced carbonyl less rapidly than other catalysts. Their products did not follow the usual linear logarithm of moles versus carbon number (Fig. 15); products from other catalysts followed this linear relationship.

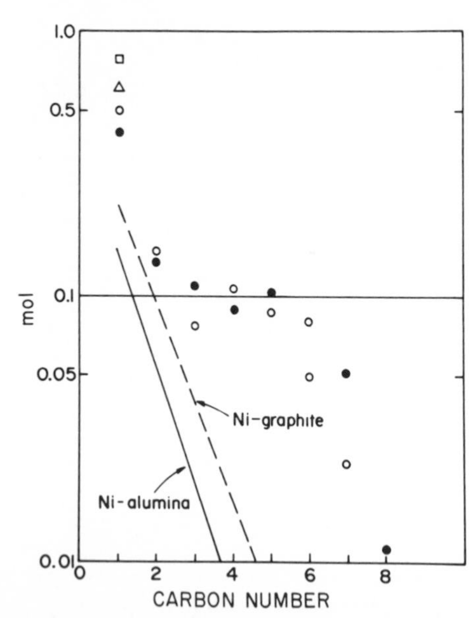

Fig. 15 Products from two 10% Ni-on-TiO_2 catalysts (○ and ●) do not follow the usual logarithmic relationship. Products from 42% Ni on alumina (□) and 20% Ni on graphite (△) have the normal distribution pattern, as indicated by the straight curves; only points of C_1 are shown for the last two catalysts. Adapted from Vannice and Garten (*142*).

Kao *et al.* (*70*) studied in an ultrahigh-vacuum apparatus the strong support interactions of a single crystal of TiO_2 on which Ni was evaporated. Photoemission studies indicated that the Ni particles (on TiO_2) were negatively charged. The activity of the evaporated Ni increased with average thickness of the deposit to a maximum of 5 Å; the maximum activity of Ni on TiO_2 was more than three times that of a Ni(111) single crystal. In the FTS at 80 torr, Ni on TiO_2 produced more higher hydrocarbons than Ni(111). Activation energies of 25.2 and 26.7 kcal/mol were obtained on the supported and single-crystal Ni.

Bhatia *et al.* (*17*) studied supported Ni catalysts for methanation. For Ni exchanged or impregnated on Y zeolite, the optimum activity was obtained for Ni crystallites of 140–160 Å. Nickel on η-alumina was more active in methanation than Ni on zeolite Y.

Nickel-on-silica catalysts were shown to have constant activity per unit weight during a test, but the activity per unit Ni area increased, and the crystallite size, as indicated by magnetic measurements, also increased (*139*). Tests were made at 208°C at 0.8 atm of 10 H_2 + 1 CO feed. Although the formation of $Ni(CO)_4$ was thermodynamically impossible from bulk Ni, it was postulated that a carbonyl process led to the transport of

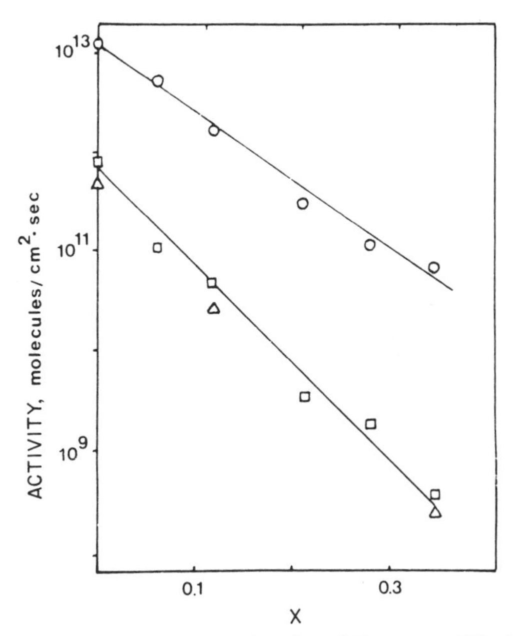

Fig. 16 Activities of Ni–Cu on silica as a function of Cu content X for 4 H_2 + 1 CO gas at 250°C. ○, CH_4; □, C_2H_6; △, C_3H_8. Reproduced with permission from Dalmon and Martin (*37*).

Ni from small to large crystallites. No Ni was lost. The independence of rate on surface area and crystallite size suggests the reaction takes place on large crystallites.

Dalmon and Martin (*37*) studied H_2 + CO reactions on Ni on silica and Ni–Cu on silica; the latter catalysts formed homogeneous alloys with surface and bulk compositions being about equal. Activities were given for production of CH_4, C_2H_6, and C_3H_8 as a function of copper content (Fig. 16). The activities were proportional to $(1 - X)^N$, where X is the atom fraction Cu/(Ni + Cu); N was 12 for CH_4 and 20 for C_2H_6 and C_3H_8. Previous work had suggested that the intermediate in the reaction involved four Ni atoms, Ni_3C + NiO, but apparently larger assemblies are required for high activity, particularly for producing higher hydrocarbons. The crystallite size of Ni on silica was varied 25 to 250 Å. The production of both CH_4 and C_2H_6 had maxima at ~100 Å with a flat maximum for CH_4. Ethane production was low for 25 Å and increased more than 100-fold to the maximum at 100–110 Å.

In 1934, Fischer and Meyer (*58*) prepared skeletal catalysts from alloys of 50 wt % Co or Ni and 50% Al or Si; the Al or Si of the alloys was "extracted" with aqueous NaOH. Similar Si alloys containing half Ni and half Co gave catalysts producing larger yields of liquid hydrocarbons at 190–200°C and atmospheric pressure than either the Co or Ni catalysts. The large density of metal in these preparations led to overheating problems; they definitely were not suitable for the early commercial reactors.

The catalysts prepared from aluminum alloys are known as Raney catalysts (*121*). Raney Ni is a widely used catalyst for liquid-phase hydrogenation and is available commercially as fine powder in its activated state. Raney Co, Fe, and Cu are also available. Catalysts based on Si alloys are not used widely.

Commercially activated Raney Ni as a fine powder was effective in methanation of 4 H_2 + 1 CO_2 or 3.3 H_2 + 1 CO gas at 350–450°C and 21.4 atm, yielding nearly equilibrium conversions to CH_4 at feed rates of ~5 liters (S.T.P) per gram per hour (*9*). Carbon deposition in 11 days of methanation was small; the used catalyst contained less than 0.6 wt % carbon. Lee (*87*) obtained TONs at 211°C in terms of CH_4 produced of 7×10^{-3} and 1.2×10^{-3} s^{-1} for the hydrogenations of CO_2 and CO, respectively, on Raney Ni. Overall activation energies for the methanations of CO_2 and CO were 25 and 35 kcal/mol; the value for hydrogenation of CO is ~10 kcal/mol larger than on other Ni catalysts.

The Raney Ni alloys usually contain 50–60 wt % Al. These alloys are brittle solids and can be crushed and sieved to appropriate particle sizes. The alloy can also be cast into platelets, cylinders, and other shapes appropriate to the requirements of particular reactors. However, the activation in aqueous alkali causes a marked decrease in mechanical strength. Processes have been developed at the U.S. Bureau of Mines, now the Department of Energy, for flame-spraying Raney alloy on the inside or outside of metal tubes or on platelets for use in tube-wall or hot-gas-recycle reactors. The surface of the metal was first roughened slightly by sandblasting. The Raney alloy powder was fed through a H_2–O_2 metallizing gun to produce an alloy layer ~0.6 mm in thickness. The alloy was then activated by treatment with aqueous NaOH at temperatures not exceeding 50°C (*111*).

Wallace and co-workers (*30, 34, 53, 66*) studied H_2 + CO reactions over a variety of intermetallic compounds containing Fe-group metals, such as RNi_5 (R = Er, La, Th, U, or Zr), $CeCo_5$, $CeFe_2$, $ThFe_5$, $ErFe_2$, and $ErFe_3$. The crushed alloys were not porous, but as temperature increased synthesis reactions began. The water that was produced oxidized the alloying element, and the surface area, chemisorption of CO, and activity increased. This activation step can be accomplished in a more controlled way if the oxidation is performed with O_2 at 350°C. For example, in the pretreatment with O_2, $ThNi_5$ was converted to Ni supported on ThO_2, and the surface area and CO chemisorption increased from 0.1 to 12–18 m^2/g and 0.15 to 20–100 μmol/g, respectively. These preparations were active; TONs of 6–11 × 10^{-3} s^{-1} at 205°C. The Ni–ThO_2 produced 20% C_2–C_4 and 80% CH_4 at 200°C.

Melts of mischmetal, an alloy containing Ce, La, Pr, Nd, and Ni, were also very active in methanation (*13*). The metal melts containing 10–50% mischmetal and the remainder Ni were activated by heating in 3 H_2 + 1 CO at 10 atm for several hours. The temperature was then decreased to 200–250°C. The activated samples had surface areas as large as 37 m^2/g and CO chemisorption of 300 μmol/g. X-Ray diffraction showed the presence of rare-earth oxides and Ni_3C. Turnover numbers at 275°C were as large as 150 s^{-1}. Luengo *et al.* (*88*) showed that $CeCo_2$ and $CeNi_2$ were active in pulsed methanation tests of CO and CO_2, and $CeAl_2$ had a moderate activity. On the practical side, these alloy catalysts should be very expensive. Available testing data suggest few if any advantages over conventional catalysts.

Detailed studies of the hydrogenation of CO and CO_2 on Ni have been made under methanation conditions. Sughrue and Bartholomew (*136*) used hot-gas-recycle conditions with Ni on a Corning Corderite monolith, 36 square channels per square centimeter, as well as Ni on alumina, in a Berty differential reactor. Parameters of the power rate law varied widely and irregularly with temperature from 200 to 350°C, the CO exponent from −0.87 to −0.04, and the H_2 power from 0.66 to 1.04. No single fundamental rate equation was useful over the entire temperature range, and three equations based on two-site mechanisms, each valid over part of the temperature spread, were given. The authors interpreted this situation as follows: from 200 to 250°C, the rate-determining step was the adsorption of H_2 with strongly sorbed CO as an inhibitor; from 250 to 300°C, CO dissociation and hydrogenation of surface carbon are the rate-determining steps; and above 300°C, the rate-determining step is the hydrogenation of surface carbon. CH_4 did not inhibit the rate, but H_2O did decrease the rate, irreversibly below 300°C. The effect of H_2O could be included in the equations by adding kp_{H_2O}/p_{H_2} to the terms in the denominator, or by multiplying the whole equation by $1/(1 + kp^n_{H_2O})$, where k and n are constants. A rate equation covering the entire temperature range was obtained by combining two of the expressions of limited validity.

Weatherbee and Bartholomew (*147*) measured rates of hydrogenation of CO_2 on Ni on silica in a single-pass tubular reactor, usually at low conversions of CO_2. From 227 to 327°C, power law exponents for H_2 increased from 0.35 to 0.56, and for CO_2 they decreased from 0.24 to 0.07. Their best fundamental rate equation was based on the dissociation of CO as the rate-determining step with oxygen and CO as the most abundant adsorbed species:

$$r = k_1 p_{CO_2}^{1/2} p_{H_2}^{1/2}/[1 + K_1(p_{CO_2}/p_{H_2})^{1/2} + K_2(p_{CO_2}p_{H_2})^{1/2} + K_3 p_{CO}]^2. \quad (6)$$

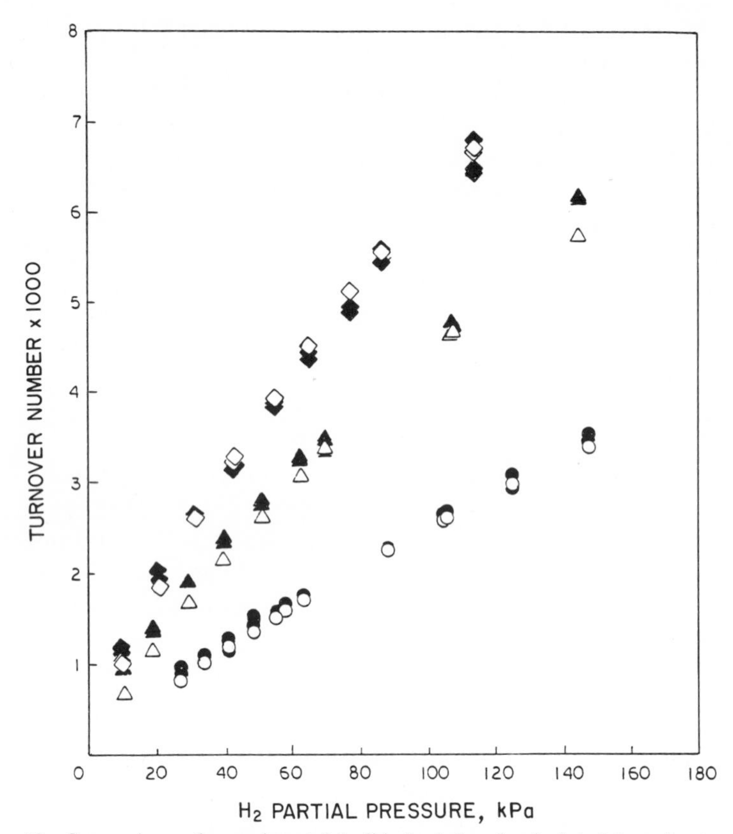

Fig. 17 Comparison of experimental (solid glyphs) and calculated (open) rates of CO methanation on Raney Ni. Temperature, 220.5°C. CO pressures (kPa): ◇, 6.9; △, 13.1; ○, 32.1. From Lee (*87*).

The dissociation of CO_2 to adsorbed CO and oxygen was postulated to be rapid and in "quasi-equilibrium" with CO_2 and CO in the gas phase. At each temperature, an "equilibrium" concentration of CO was obtained that was independent of feed composition, CO_2 conversion, and space velocity. The concentrations of CO observed in tests at ~1.5 atm were 0.003, 0.028, and 0.120% CO at 227, 277, and 327°C, respectively. CO was a strong inhibitor for methanation, if its concentration exceeded the equilibrium value mentioned before.

Lee (*87*) determined kinetics of methanation of CO and CO_2 on Raney Ni in a fixed-bed, single-pass reactor maintained "differential" by keeping the conversion less than 5%. The use of flow controllers for the reactants permitted experiments in which the partial pressure of one feed component was held constant while the partial pressure of the other com-

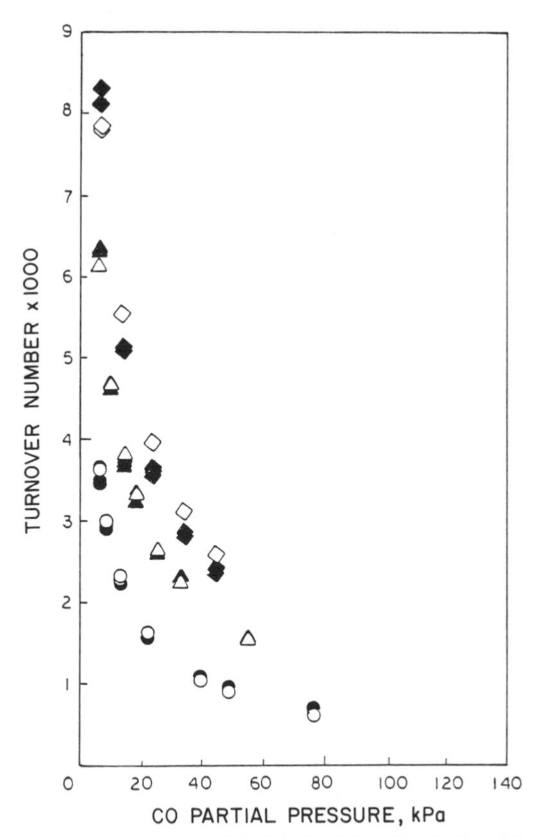

Fig. 18 Comparison of experimental (solid glyphs) and calculated (open) rates of CO methanation on Raney Ni. Temperature, 220.5°C. H_2 pressures (kPa): ○, 42.7; △, 86.9; ◇, 136.5. From Lee (*87*).

ponent was varied. For the hydrogenation of CO_2, CH_4 was the only hydrocarbon produced; for CO, traces of C_2 and C_3 hydrocarbons were also obtained. Typical results are shown in Figs. 17 and 18 for H_2 + CO and Figs. 19 and 20 for H_2 + CO_2. In the H_2 + CO reactions, the rate increased linearly with H_2 at constant CO, and the rate decreased sharply with increasing CO at constant H_2. For H_2 + CO_2, the rate increased initially with increasing H_2 at constant CO_2, but seemed to approach a constant rate at high H_2. At constant H_2, the rate increased initially with increased CO_2, but the rate passed over a maximum and finally decreased. In these reactions, CO was a strong inhibitor, H_2O a moderate inhibitor, and CH_4 a diluent. With low H_2/CO_2 ratios, moderate concentrations of CO were found, but the CO "equilibrium" relationship reported by Weatherbee and Bartholomew (*147*) did not seem to hold. The

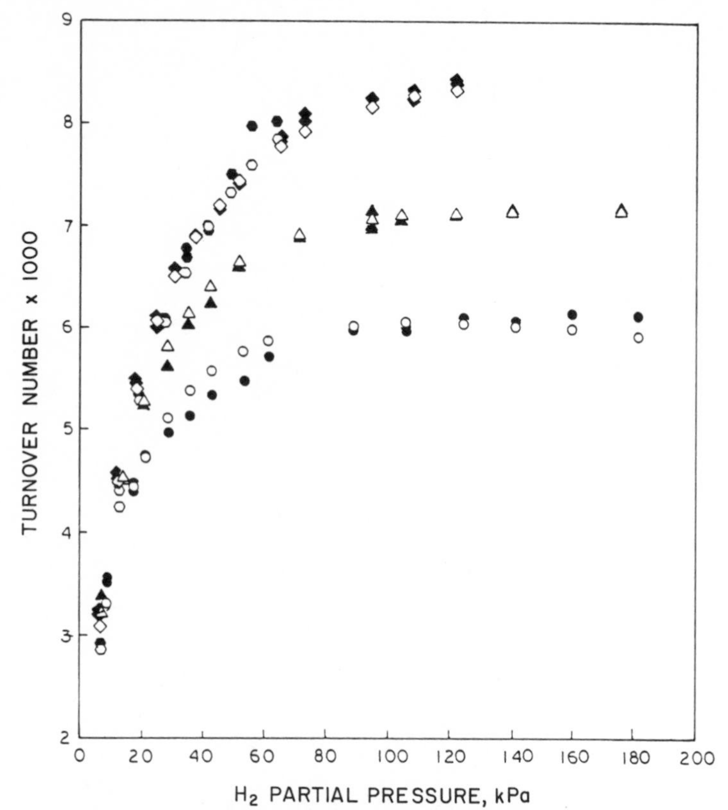

Fig. 19 Comparison of experimental (closed glyphs) and calculated (open) rates of CO_2 methanation on Raney Ni. Temperature, 210.9°C. CO_2 pressures (kPa): ○, 13.8; △, 27.6; ◇, 68.9; ⬡, 137.9. From Lee (*87*).

concentration of CO was a function of the feed ratio H_2/CO_2, as well as temperature in Lee's experiment (*87*).

Rate equations were derived and interpreted by the Langmuir–Hinshelwood methods. The best equation for the hydrogenation of CO was

$$r = k_1 p_{H_2} p_{CO}^{1/2} / [1 + (K_2 p_{H_2})^{1/2} + (K_3 p_{CO})^{1/2}]^3, \qquad (7)$$

where k_1 is a rate constant, and K_2 and K_3 are adsorption equilibrium constants. The slow step was assumed to be $\cdot H + :CH \rightarrow \cdot CH_2 + 2\cdot$, a three-site reaction. CO and H_2 are reversibly and dissociatively adsorbed, the CO to adsorbed carbon and oxygen. The oxygen is rapidly removed by hydrogen. Adsorbed hydrogen and carbon are the abundant surface species and are represented in Eq. (7) by the terms in the denominator.

For CO_2, the same slow step was used, and CO_2 was postulated to

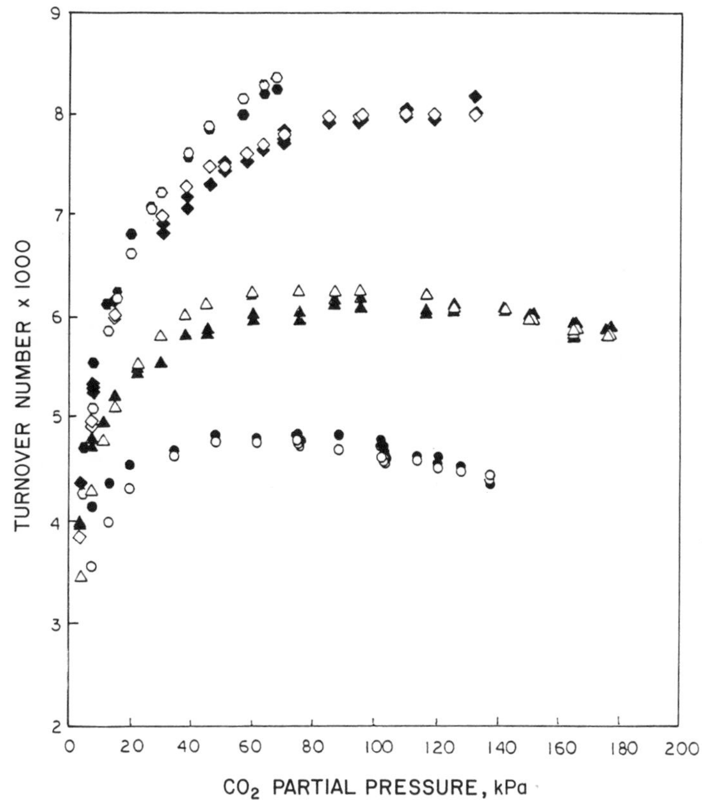

Fig. 20 Comparison of experimental (closed glyphs) and calculated (open) rates of CO_2 methanation on Raney Ni. Temperature, 210.9°C. H_2 pressure (kPa): ○, 13.8; △, 27.6; ◇, 68.9; ⬡, 137.9. From Lee (*87*).

adsorb reversibly and dissociatively to two adsorbed oxygens and a surface carbon, leading to

$$r = k_1 p_{H_2} p_{CO_2}^{1/3} / [1 + (K_2 p_{H_2})^{1/2} + (K_3' p_{CO_2})^{2/3}]^3. \tag{8}$$

The third term in the denominator for adsorbed carbon should have the exponent $\frac{1}{3}$, but $\frac{2}{3}$ provides a better fit of the data. For both feeds, H_2O inhibited the reaction reversibly, but CH_4 was a diluent. If the water is also reversibly and dissociatively adsorbed, a term in the denominator for oxygen atoms may be included as $K_4 p_{H_2O}^{1/3}$.

The ability of Eqs. (7) and (8) to describe methanation is shown by a comparison of the calculated curves with the experimental points in Figs. 17, 18, 19, and 20. Equilibrium constant K_2 has the same significance in Eqs. (7) and (8), and about the same values were obtained from both hydrogenations. The Raney Ni sintered during the long kinetic tests, and the values of the rate constant k_1 often decreased with increasing temperature.

Lee (*87*) also determined reducible carbonaceous material on his catalysts by the following steps, all at the previous synthesis temperature: (a) stopping the flow of synthesis gas, (b) purging the catalyst with He for ~1 h, (c) passing in pure H_2, and (d) measuring the CH_4 produced. On changing to pure H_2, a slight increase in catalyst temperature and a rapid evolution of CH_4 were observed; this initial production rate of CH_4 was larger than in the previous period of methanation. After 1 h of methanation of CO, the carbon content of the catalyst corresponded to 10–20 monolayers, but for H_2 + CO_2 only ~10% of a monolayer. The carbon content increased with increasing partial pressures of CO and CO_2 and decreased with increasing partial pressures of H_2 and H_2O, other factors being constant. Carbon also increased with increasing synthesis temperature. These data suggest, but do not prove, that surface carbon may be an intermediate in methanation.

Pulsed tests on the formation and reaction of adsorbed carbon and carbides, such as those of Wentrcek *et al.* (*148*), on Ni are described in Chapter 5.

4.4 Iron Catalysts

In the FTS, carbides and oxides of Fe can be formed, and carbides, nitrides, and carbonitrides of Fe can be used as catalysts, as described in Chapter 3. The synthesis on Fe is usually done in two ranges of temperature: (a) below 280°C for fixed-bed reactors and for the slurry and oil-circulation processes, in which the catalyst is submerged in oil, and (b) above 320°C for fluidized and entrained units, the high temperatures being required to minimize the production of wax, which causes agglomeration of the fine particles.

Water is the principal primary oxygenated product, but a subsequent water–gas shift proceeds at about the same rate as the primary process. For this reason it is usually desirable to express rate and selectivities in terms of total moles of H_2 + CO consumed; this quantity is not affected by the water–gas shift. Thus, the relative usage of H_2 and CO, the moles of H_2 consumed divided by the moles of H_2 + CO converted, varies widely with conversion of H_2 + CO, as shown in Fig. 21 for nitrided iron operated at several pressures; this curve is typical of all Fe catalysts operated at temperatures less than 280°C (*8*).

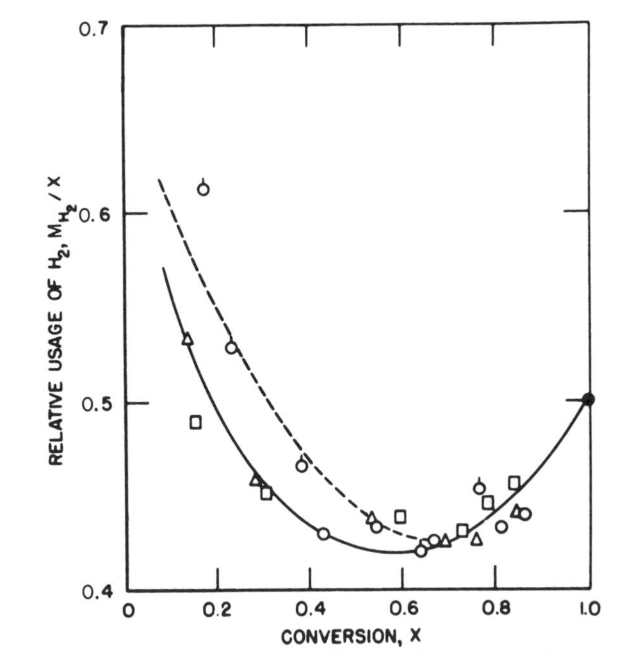

Fig. 21 Relative usage of H_2 as a function of conversion and operating pressure. 1 H_2 + 1 CO gas at 240°C. Pressures (atm): ♁, 7.8; △, 11.2; □, 14.6; ○, 21.4. From Anderson *et al.* (*8*).

The relative usage of H_2 extrapolates to about 0.67 (2 H_2 : 1 CO) at a conversion of zero, and with increasing conversion decreases rapidly to a minimum of 0.43 (0.75 H_2 : 1 CO) at a conversion of ~0.6 and then increases and approaches the solid point that arises from the requirement that if all of the H_2 + CO is consumed, the gas must be consumed in the same ratio as the feed.

According to Dry (*44*), in fluidized or entrained reactors operating at 330°C the water–gas shift is very rapid with respect to synthesis, and the water–gas shift is in thermodynamic equilibrium throughout the bed.

Bureau of Mines studies (*7, 8, 71, 72*) delineated the principal features of kinetics in the lower temperature regime. The differential reaction rate varied as $(1 - X)$ to the 1 or $\frac{1}{2}$ power, where X is the conversion of H_2 + CO, and as operating pressure p to the 0.94 power. The rate at zero conversion was proportional to $p_{CO}^{0.66} p_{H_2}^{0.34}$, and the overall activation energy was 20. kcal/mol. These results led to a useful empirical equation (*10*)

$$-\ln(1 - X) = A[\exp(-E/RT)]p/S, \tag{9}$$

where S is the feed rate of synthesis gas.

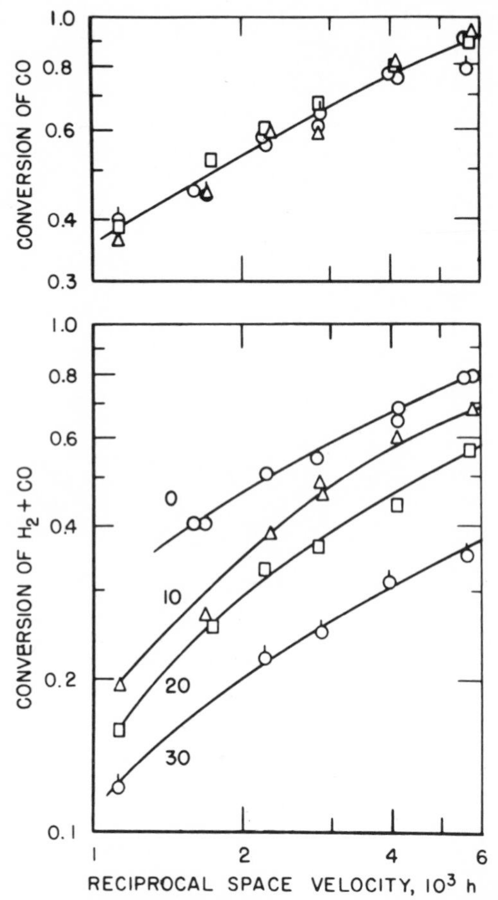

Fig. 22 Logarithmic plots of conversion of CO and H_2 + CO as a function of reciprocal space velocity in H_2O-addition tests with 1 H_2 + 1 CO at 21.4 atm and 240°C. Water added (mol %): ○, 0; △, 10; □, 20; ó, 30. From Anderson *et al.* (*8*).

Karn *et al.* (*8, 72*) studied the effect of water vapor on the FTS on nitrided Fe. Water was an inhibitor, but the rate promptly regained its previous value when the addition of water was stopped. The conversion of H_2 + CO decreased monotonically with increasing added water (Fig. 22), but the rate of the water–gas shift increased with added water so that the conversion of CO remained constant. For 30 mol % added water, there was a net production of H_2. Thus, the rate of consumption of CO did not change with water addition; however, the fraction of CO reacting with H_2O increased. In diluent tests, Ar and CH_4 had about the same effect; CO_2 was a weak inhibitor and H_2O a strong inhibitor.

All of the kinetic effects just described were incorporated into a semi-empirical rate equation (*7*) that was based on successive additions of a C_1 species to a growing chain. The C_1 units may either add to the growing chain or be destroyed by reaction with H_2O. The final equation is

$$r = ap_{H_2}^{0.6}p_{CO}^{0.4} - fr^{0.5}p_{H_2O}^{0.5}, \qquad (10)$$

where a and f are constants. Equation (10) represented moderately well kinetic data for nitrided Fe at 21.4 atm and 225–240°C with feeds having H_2/CO ratios of 2, 1, 0.7, and 0.25 and data from water-addition tests. The constants a and f had activation energies of 20.2 and 18.5 kcal/mol.

Dry (*43, 44, 49*) found that with reduced Fe at 225–265°C and very low conversions, the rate was proportional to the partial pressure of H_2, and at higher conversions, an earlier U.S. Bureau of Mines equation (*4*)

$$r = kp_{H_2}p_{CO}/(p_{CO} + ap_{H_2O}) \qquad (11)$$

represented the kinetics from both fixed- and fluidized-bed reactors satisfactorily. In the low-temperature range, the activation energy of the rate constant k was 15.0–16.8 kcal/mol, and for the high temperature range, 6 kcal/mol. The latter value suggests that high-temperature processes are severely limited by diffusion despite the small particles used.

Iron has perhaps the greatest tendency to produce elemental carbon of the usual metals for the FTS. In addition to clogging fixed beds, carbon deposition causes disintegration of catalyst particles to fines. Apparently, carbon nuclei form inside the Fe crystallites and grow to an extent that the particles are broken by the expanding carbon deposit. In a properly designed fixed-bed reactor that has no "hot spots," carbon deposition is usually not a serious problem below 275°C. In the fluidized and entrained reactors at 330°C, carbon deposition always proceeds at a moderate or larger rate.

Some workers have mapped regions in an equilibrium C–H–O diagram where carbon deposition can occur. These studies have not been very useful, presumably because the carbon was deposited early in the process and not at equilibrium. Manning and Reid (*93*) reported that carbon deposition did not occur on Fe_3O_4 and Fe_3C.

Dry (*44*) reviewed SASOL data on carbon deposition. At 325°C, water or hydrogen-containing molecules increased the rate of the Boudouard reaction, $2\ CO = C + CO_2$, N_2 had no effect, and NH_3 decreased the rate. Paraffinic and olefinic hydrocarbons did not affect the rate. For reduced, fused catalysts containing different structural promoters or none at all, the rate of carbon deposition was proportional to the surface area. Other promoters also did not change the rate of carbon deposition. This process increases sharply with increasing alkali concentration, and at a given alkali content adding acidic oxides such as silica decreased the rate.

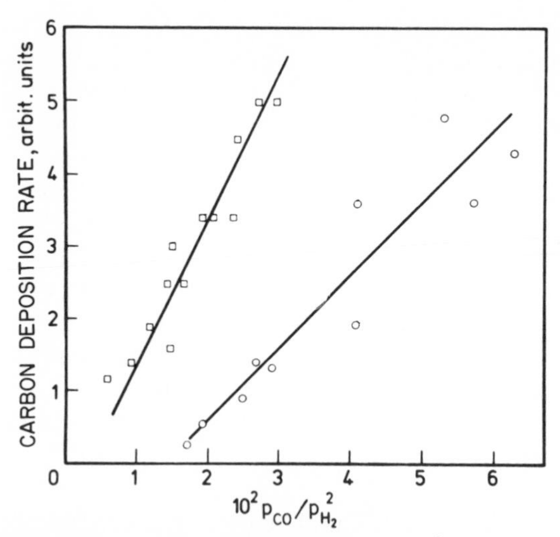

Fig. 23 Rate of carbon deposition as a function of $p_{CO}/p_{H_2}^2$ for two different catalysts. Reproduced with permission from Dry (*44*).

In the synthesis, carbon deposition increases linearly with $p_{CO}/p_{H_2}^2$, as shown in Fig. 23 (*44*). This relationship holds over a variety of situations. Dry (*44*) noted that the factor $p_{CO}/p_{H_2}^2$ properly correlated carbon deposition, but that the ratio H_2/CO did not. For example, the carbon deposition decreased as the operating pressure was increased and as the recycle ratio was decreased from 4.0 to 1.2, whereas in both cases the ratio H_2/CO was nearly constant. Carbon deposition seemed to be independent of the partial pressure of CO_2.

Dry (*44*) derived the factor $p_{CO}/p_{H_2}^2$ by a reaction scheme in which chemisorbed CO either forms a CHOH intermediate or dissociates to carbon and oxygen. The carbon is either hydrogenated to CH_4 or aggregates to produce elemental carbon. The hydroxymethylene intermediate can either grow to long-chain molecules or produce CH_4, and the oxygen atom reacts with H_2 or CO.

The preparation of Fe catalysts is described in the next several pages, starting with the Ruhrchemie precipitated-Fe catalyst that was originally used in the fixed-bed reactors at SASOL. The procedure is typical of several arising from German work in the 1940s, which has been reviewed (*5, 135*). Impregnated-Fe catalysts and precipitated preparations containing large amounts of supports or carriers have generally not been good Fischer–Tropsch catalysts, possibly because of the interaction of the carrier or support with the alkali promoter. For our discussion of impregnated-Fe catalysts, the presentation here follows in part a review by

Guczi (*60*). Dry (*44*) described more recent work on fused–iron oxide catalysts, and the reader is referred to the earlier reviews (*5, 135*).

Alkali, usually K_2CO_3, is the most important promoter for Fe in the author's opinion; however, in the 1970s, only Dry (*44*), Arakawa and Bell (*11*), and McVicker and Vannice (*96*) have made FTS studies on alkali promoters. Earlier work in this area has been reviewed (*5, 135*). Arakawa and Bell (*11*) prepared Fe on alumina with different concentrations of alkali by mixing an aqueous solution of ferric nitrate and potassium nitrate with alumina (Alon C, Cabot Corp.), drying, calcining in air at 350°C, and finally reducing in H_2 at 400°C for 2 days. All of the catalysts contained 20 wt % Fe and K : Fe weight ratios from 0 to 0.4. The larger alkali concentrations were more than 10 times those used in previous tests, and the reasonable performance of the catalyst in these fixed-bed tests at 1 atm was probably the result of the large amount of acidic alumina in the catalyst.

Hydrogen at 110°C and CO at 23°C chemisorbed in about equal amounts, which decreased from ~45 to 20 mmol/g as the K/Fe ratio increased from 0 to 0.2. At 260°C, the rate of consumption of CO, and the production of CO_2 and individual hydrocarbons up to C_5 expressed as TONs, all passed over a maximum at a K/Fe ratio of 0.022 or 0.065. The paper by Arakawa and Bell (*11*) demonstrated that *TON can be misleading*. The measures of productivity listed before, if calculated per unit weight of catalyst, *all decreased monotonically* with increasing K/Fe. The activity of conventional fused or precipitated catalysts per unit weight passed over a maximum at a K/Fe of 0.005–0.01. Arakawa and Bell (*11*) did not explore this concentration range.

With increasing potassium, the yield of CH_4 and higher paraffins decreased and olefins increased. Water was the primary oxygenated product, and CO_2 was produced by a subsequent water–gas shift. The ratio CO_2/H_2O in products increased with increasing K/Fe; the water–gas shift was increased by alkali. Carburization of Fe in synthesis gas proceeded more rapidly on alkalized catalysts.

The Ruhrchemie catalyst was precipitated from a nearly boiling solution of nitrates containing 40 g Fe and 2 g Cu per liter by adding a hot solution of Na_2CO_3 with vigorous stirring (*44*). Addition of the Na_2CO_3 was completed in several minutes when the pH reached 7–8. The precipitate was washed with hot distilled water in a filter press to remove Na^+ to a low level. The precipitate was slurried in water and impregnated with a solution of potassium water glass to provide 25 g SiO_2 per 100 g Fe. Enough aqueous HNO_3 was added to remove the excess potassium to give 0.5 g K_2O per 100 g Fe after filtering. The filter cake was predried, extruded, and finally dried to ~10 wt % H_2O.

Table 19
Alkali in Precipitated Iron: Activity and Wax Yields[a]

Catalyst type	*K_2O Level*[b]	*Hard-wax selectivity*[c]	*Activity*
Unsupported ferric oxide	0	5	26
	1.0	34	47
	1.6	41	50
	2.0	53	53
	3.0	63	40
SiO_2-promoted Fe_2O_3	12	18	112
	16	20	19
	21	30	85
	24	38	83
	32	44	75

[a] Adapted from Dry (*44*).
[b] Relative quantities.
[c] Weight percentage of hydrocarbons boiling above 500°C.

The structure of the catalyst is determined by the original precipitating conditions, and the SiO_2 from the water glass interacts with the hydrous ferric oxide. The silica stabilizes the iron oxide gel, and 25 SiO_2 : 100 Fe gave a large area in the original state, 390 m^2/g, after 58% reduction in H_2, 270 m^2/g (*44*). The silica makes the reduction of the ferric oxide more difficult.

In the synthesis at 200–250°C, the metallic Fe in the partly reduced catalyst is rapidly converted to Hägg carbide. Further reduction and carbiding continue slowly for several days and are accompanied by an increase in activity. Gas composition and temperature should be carefully adjusted so that the catalyst attains its optimum activity promptly without permanent damage (*44*). Chapter 3 presents numerous examples of composition changes of Fe catalysts in this lower-temperature operating range from Bureau of Mines work.

Dry (*44*) studied the effect of promoters and supports on the performance of precipitated Fe of the type just described; unfortunately, only relative concentrations were provided for some components. In all cases, K_2O was an important promoter, and often the optimum alkali content is changed by the addition of a support. Silica was the best support in terms of both activity and wax production. For activity, $SiO_2 > ZnO > Al_2O_3 >$ none $> Cr_2O_3 > MgO$, and for wax production, SiO_2, $Al_2O_3 > Cr_2O_3 >$ none $> MgO > ZnO$. Table 19 shows that as the K_2O level is increased, a maximum activity is achieved at relatively low concentrations, whereas the wax yields still increase at higher K_2O levels. Acidic oxides such as silica (including kieselguhr) and alumina sequestered the alkali, and larger concentrations were required. These SASOL data are in general agreement with older work.

Iron catalysts containing a substantial fraction of support or carrier have generally been ineffective FTS catalysts, and highly dispersed supported Fe is difficult to prepare. Guczi (*60*) has summarized work on supported Fe and supported Fe alloys of Pt and Ru. Mössbauer spectroscopy is the prime research tool for studying these catalysts. Usually, if the iron oxide is highly dispersed on the support, it cannot be reduced to Fe^0. On the other hand, the poorly dispersed oxide is in the form of large crystallites; these oxides do reduce, but to large Fe crystallites. In acidic solution (pH = 1), Fe^{3+} adsorbs weakly on silica gel or alumina. In less acid situations (pH = 3), Fe^{3+} interacts strongly, but part of it hydrolyzes to a basic ion. Drying and calcination conditions are also important in determining the nature of the ferric oxide and its reducibility.

In catalysts supported on silica and alumina, the iron oxide migrates before reduction and the metal crystallites grow after formation. On magnesia, this migration does not occur and both large and small metal particles are obtained. With coprecipitated Fe_2O_3–MgO, it is also possible to obtain small metal crystallites.

An alternate preparation involves the impregnation of compounds containing Fe^0 as in carbonyl clusters such as $Fe_3(CO)_{12}$; however, this procedure is complicated by the oxidation of Fe^0 from the cluster decomposition by surface hydroxyls and traces of H_2O. Brenner and Hucul (*21*) obtained high dispersions of Fe by impregnating γ-alumina with solutions of $Fe(CO)_5$, $Fe_2(CO)_9$, or $Fe_3(CO)_{12}$. The temperature-programmed decompositions of these preparations suggested that the different carbonyls might lead to different catalytic properties. Activations above 300°C led to oxidation of the Fe.

Hugues *et al.* (*63*) impregnated $Fe(CO)_5$, $Fe_3(CO)_{12}$, and $[HFe_3(CO)_{11}]^-$ dissolved in pentane on alumina and magnesia. On dehydroxylated supports, the carbonyls decomposed at temperatures less than 200°C to superparamagnetic Fe particles with diameters less than 15 Å. Hydroxylated supports gave mixtures of metal and oxides. The catalysts often produced large yields of ethylene and propylene initially, but the selectivity changed to a more normal pattern after a few hours. McVicker and Vannice (*96*) prepared potassium-promoted Fe catalysts by impregnating alumina with solutions of $K_2Fe(CO)_4$ or $KFe(CO)_2 \cdot C_2H_5$. In the atmospheric-pressure synthesis, some of these preparations were 40–50 times as active as conventional catalysts. The ratios of alkali to Fe were enormous; two typical preparations contained 5.5% K, 3.9% Fe, and 2.2% K, 3.2% Fe.

Vannice *et al.* (*69, 144*) studied Fe catalysts supported on a variety of activated carbons at atmospheric pressure. The Fe content varied from 2 to 7%. Turnover numbers for some of the carbon-supported catalysts were as much as eight times larger than those for Fe on alumina. The C_2 + C_3 fraction usually was about half olefinic. The products were mostly gaseous hydrocarbons, and the amount of CH_4 was ~50%. Carbon was a good support for Fe; iron oxides on these carbons were readily reducible. For catalysts containing ~5% Fe, one on a graphitic carbon black with a surface area of 56 m^2/g, had relatively large Fe crystallites (600 Å) and normal ferromagnetic properties. The other on Carbolac-1 color black with an area of 950 m^2/g had small Fe crystallites (~20 Å) and superparamagnetic properties (*68*). In this study, the catalysts were examined by Mössbauer spectroscopy, conventional magnetic methods, and chemisorption.

Shultz *et al.* (*131*) showed that the "carbon" deposited by the action of CO or synthesis gas on various forms of Fe is an active, long-lived catalyst. The source of the activity is crystallites of Fe imbedded in the carbon. These experiments were made in a carbon steel reactor of 1.5 cm internal diameter that was loosely packed with glass wool, or in a similar stainless steel reactor with ~0.5 g of Fe catalyst or electrolytic Fe dispersed on glass wool to occupy 50 cm^3 of reactor volume. These catalysts were operated at 300–325°C at 21.4 atm of 1 H_2 + 1 CO at a space velocity of 300 h^{-1} based on the reactor volume. In one test, the carbon containing only 0.07 g Fe operated at 316°C with constant activity for 89 days, and the products were typical of the fluidized or entrained reactors.

The Guczi review of impregnated Fe catalysts also described work on supported Pt–Fe and Ru–Fe (*60*). The Fe became more easily reducible when Pt or Ru was added, and alloys were formed to some extent. Clarke and Creamer (*33*) have also reviewed alloy catalysts. Dilute, supported Fe, whether or not containing alloying elements, apparently has not been useful for the FTS. In addition, in the presence of the large amount of support, alkali promoters are usually ineffective.

Fused magnetite catalysts of the type used in the ammonia synthesis are effective in the FTS. These catalysts are prepared by melting iron oxides plus promoters in an electric furnace or by burning Fe in oxygen. The latter method produces catalysts of low silica content. Structural promoters that yield samples of high surface areas after reduction include Al_2O_3, MgO, CaO, TiO_2, and ZrO_2. Alkali, almost always K_2O, increases activity and is known as a chemical promoter. Part of the structural promoter is incorporated in the magnetite as spinels, but usually a substantial fraction of these oxides plus impurities in the original magnetite, particularly silica, plus most of the K_2O are often found in slaglike inclusions or envelopes that surround the magnetite crystallites (*28, 29, 43, 44, 104*). Preparations containing large amounts of silica require larger amounts of

Table 20
Effect of Alkali in the High-Temperature (330°C) Fischer–Tropsch Synthesis on Fused Catalysts[a]

Alkali	*Amount, on relative molar basis*	*Activity, conversion of CO + CO_2 (%)*	*Selectivity[b] to*		*C_3H_6/C_3H_8 Ratio*
			CH_4	*Alcohols + ketones*	
K_2O	0	63	41	0.3	0.2
	1.75	90	20	1.7	3.0
	2.1	90	15	2.6	5.7
	2.65	92	13	3.4	7.6
	3.4	94	12	3.8	11
	3.95	93	10	4.0	12
Na_2O	0	63	41	0.3	0.2
	2.3	78	15	2.7	1.8
	4.5	85	13	2.0	2.3
	6.8	84	14	2.5	2.0
	9.1	87	16	2.7	1.7

[a] Adapted from Dry (*44*).
[b] On a percentage carbon atom basis.

the alkali promoters. In the oxide form, the catalyst has no porosity and a very low surface area, and it must be reduced in a high flow of H_2 at 450–500°C before use. The reader is referred to a review by Dry (*44*) and the original papers of SASOL research on the effect of promoters on lattice parameters and surface areas and chemisorption of CO and CO_2 on reduced catalysts (*31, 32, 42, 45–48*). These studies generally agree with older work and probably represent the largest number of promoters and compositions examined by a single research group.

For the fused catalysts as well as other Fe catalysts, K_2O is the important promoter, and the presence of "acidic" oxides (Al_2O_3, TiO_2, and SiO_2) decreases the effectiveness of the alkali so that larger concentrations are required for the same effect. Fused catalysts used at low temperatures (240–270°C) required about the same amount of alkali promoter as other Fe catalysts (*5*).

Fused catalysts used in entrained or fluid-bed reactors at 330°C have different promoter requirements, because the rate is great and large surface areas are not needed. The reaction is usually diffusion controlled and only a modest surface area is necessary. K_2O seems to be the necessary promoter in the high-temperature syntheses (Table 20); however, Na_2O is surprisingly effective under these conditions.

Satterfield and Huff (*125*) determined carbon-number distributions of products from a stirred slurry reactor at 234–269°C and 7.9 atm. A reduced, fused-Fe catalyst (170- to 270-mesh) was slurried in a C_{28} paraffin. A typical largely olefinic product with some oxygenates was obtained. Semilogarithmic plots of moles of hydrocarbon and oxygenates as a function of carbon number were linear, including the low-carbon-number end (Fig. 24). The plots were nearly identical over the entire temperature range and for several feed compositions and conversions, from 59 to 96%. For different operating conditions, the relative amounts of hydrocarbons and oxygenates changed, but the total molar amount was unchanged. The product contained only a very small amount of branched molecules.

Deckwer *et al.* (*39*) studied a 1 Fe : 6 Mn catalyst, prepared in a unique continuous-precipitating unit, in a slurry reactor at 12 atm and 288–303°C, with the production of C_2–C_4 olefins as one of the goals of the work. For a CO-rich feed ($CO/H_2 = 1.65$) and space velocities above 600 h^{-1}, yields of C_2–C_4 olefins were ~58 g/m_n^3 converted. The selectivity was relatively independent of temperature and of feed ratio as long as CO/H_2 was 1.35 or more. Plots of the carbon-number distribution according to the semilogarithmic equation gave $\alpha = 0.675$, about the same as found by Satterfield and Huff (*125*) at lower temperatures. Earlier, Kölbel and Ralek (*82*) obtained similar results for Fe–Mn catalysts.

Both groups (*39, 125*) presented analyses of kinetics in the slurry reactor, taking the rate as $r = kp_{H_2}$, and accounting for mass-transfer resistances by the liquid. The ratios of mass-transfer resistance to reaction resistance were usually ~0.7, but decreased to 0.4 at the highest temperature (*39*). For one set of data, the activation energy of the rate was 26 kcal/mol, and that of the overall process, 19 kcal/mol. Activation energies for the reaction of 26, 17, and 22 kcal/mol were obtained for slurry tests of Fe–Mn catalyst from continuous precipitation, Fe–Mn batch precipitation, and Fe–Cu continuous precipitation, respectively.[1]

Pichler and Schulz (*116*) presented detailed analyses by capillary gas chromatography of products from the SASOL fixed-bed and entrained reactors. Table 21 presents the analysis for C_6 and C_7 fractions of hydrocarbons. For both products, the major unsaturated molecules were terminal olefins. The fraction of branched molecules was small for the fixed-bed product and moderate for the fluid-bed product. Isomer distributions are considered in detail in Chapter 5. As shown in Table 22, naphthenes and aromatics were found in small amounts in the products from the fixed-bed, low-temperature process of SASOL and in substantially larger quantities in the products of the entrained, high-temperature unit (*118*). The carbon-number distributions for Fe are similar to those for Co (Fig. 10), but the yield of CH_4 is usually smaller and the C_2 yield often larger. For

[1] Further work on high Mn/Fe catalysts is described in the Addendum, p. 293.

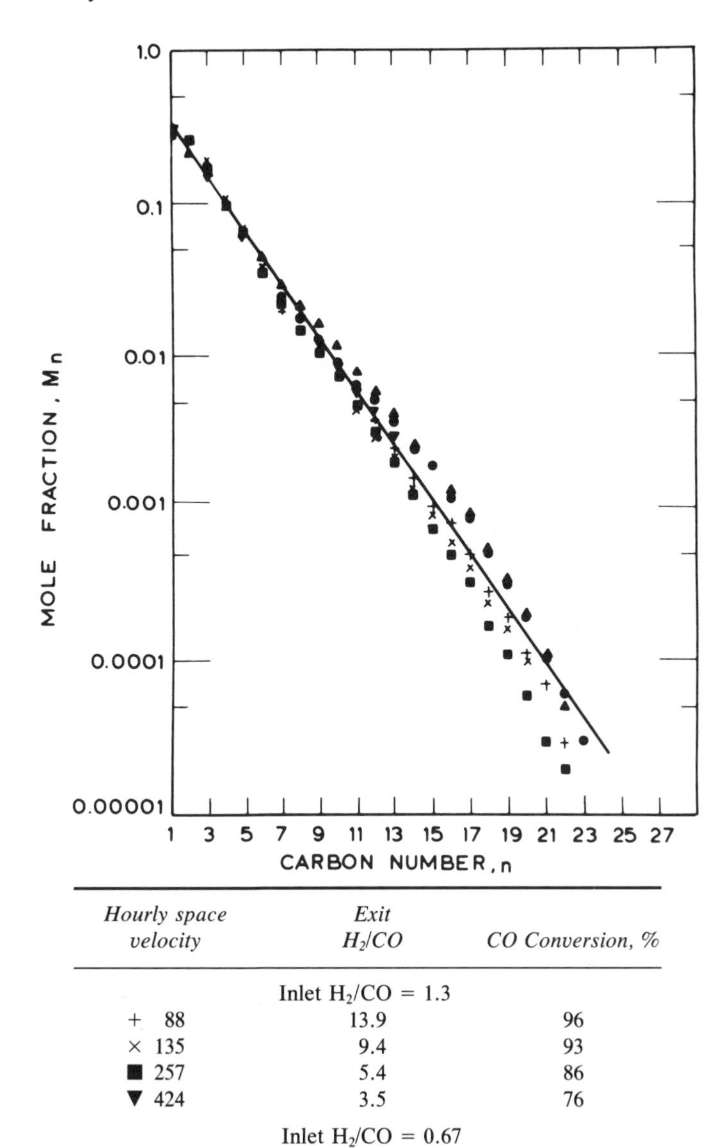

Hourly space velocity	*Exit H_2/CO*	*CO Conversion, %*
	Inlet H_2/CO = 1.3	
+ 88	13.9	96
× 135	9.4	93
■ 257	5.4	86
▼ 424	3.5	76
	Inlet H_2/CO = 0.67	
● 107	1.2	79
▲ 209	0.85	59

Fig. 24 Application of Flory equation to slurry products at 269°C. $\alpha = 0.67 \pm 0.02$. Reproduced with permission from Satterfield and Huff (*125*).

Table 21
Detailed Analyses of C_6 and C_7 Hydrocarbons from Iron[a]

Structure[b]	Paraffins	Olefins, position of double bond: 1	2	3	4	5
	Weight percentage					
Fixed bed, 220–240°C						
Normal C_6	47.3 }	28.0	{ 15.8	1.5		
2M5	1.9 }		{ —	0.2	A	
3M5	2.5	A	0.2			
23DM4	0.2					
Methylcyclo-pentane	0.2			$\Sigma A = 2.2$		
Normal C_7	46.3	28.3	15.4	1.2		
2M6	B	0.9 }	0.1		{ 0.3	C
2M6	C	B }			{ 0.2	0.8
23DM5	0.3	0.1				
24DM5		0.05				
3-Ethylpentane = 0.2; napthenes = 0.4; toluene = 0.1; $\Sigma B = 2.1$; $\Sigma C = 2.6$						

Structure[b]	Paraffins	Olefins, position of double bond: 1	2	3	4	5
	Weight percentage					
Entrained bed, 320°C						
Normal C_6	17.9 }	41.8	{ 16.8	2.1		
2M5	2.8 }		{ 0.2	0.9	A	
3M5	3.4	A	1.3			
23DM4	0.1					
Methylcyclopentane = 0.2; benzene = 0.2; $\Sigma A = 11.6$						
Normal C_7	13.3	36.9	12.0	1.5		
2M6	B	3.8 }	0.8		{ 2.7	C
3M6	C	B }			{ 0.6	5.0
23DM5	0.3	0.4				
24DM5		0.4				
34DM5		0.8				
3-Ethylcyclopentane = 0.7; naphthenes = 0.8; toluene = 3.5; $\Sigma B = 6.7$; $\Sigma C = 7.7$						

[a] Adapted from Pichler and Schulz (*116*).
[b] M, Methyl; DM, dimethyl (e.g., 23DM5 = 2,3 dimethylpentane).

Table 22
Naphthenes and Aromatics in SASOL Products[a]

Compound[b]	*Fixed bed*	*Entrained*	*Compound*	*Fixed bed*	*Entrained*
Napthenes			Aromatics		
CP	0.14	0.66			
MCP	0.57	3.55	Benzene	0.003	0.24
ECP	0.25	2.56			
12 DMCP[c]	0.34	1.95			
13 DMCP[c]	0.30	2.10			
MCH	0.08	0.36	Toluene	0.14	3.45
*n*PrCP	0.21	2.37			
1M2ECP	0.05	0.52			
1M3ECP[c]	0.37	3.00			
123TMCP[c]	0.10	0.55			
124TMCP[c]	0.15	0.88			
ECH	0.04	0.35	Ethylbenzene	0.04	2.03
13 + 14 DMCH	0.02	0.13	*m*-Xylene	0.03	0.78
12 DMCH		0.04	*p*-Xylene	0.01	0.44
			o-Xylene	0.22	1.91
			o-Ethyltoluene	0.12	1.24
			m- and *p*-Ethyltoluene	0.03	1.59
			n-Propylbenzene	0.03	1.66

[a] Weight percentage in carbon-number fraction. Adapted from Pichler *et al.* (*118*).
[b] C, Cyclo; P, pentane; M, methyl; E, ethyl; D, di; H, hexane; *n*Pr, *n*-propyl; T, tri.
[c] These fractions were separated into cis and trans isomers.

products from a slurry reactor (Fig. 24), both C_1 and C_2 fall on the linear logarithmic plot, when total moles of hydrocarbons plus oxygenates are used in the graph.

In the FTS under certain circumstances, the ratios H_2O/H_2 and CO_2/CO become large enough to oxidize metallic Fe, iron carbides, and iron nitrides to Fe_3O_4. In addition to operation at high conversions, oxidation can also occur at the center of porous catalyst particles, if diffusion-in-pore limitations increase the concentration of products above those in the bulk gas stream and decrease the concentration of reactants. Early work in the <280°C regime suggested that only the outer 0.1 mm of the particle is active in the FTS because of diffusional resistance (*10*). This outer zone is reduced and carbided, and the core of the particle is oxidized to magnetite. Chapter 3 describes Bureau of Mines studies of the change of composition of Fe catalysts during synthesis. Under some conditions, the Fe catalysts converted to carbides, nitrides, and carbonitrides oxidized at a slower rate than did reduced Fe.

More recent investigations generally agree with the older studies, but they benefit from the availability of Mössbauer spectroscopy for analysis of Fe phases. Unlike magnetic analysis and X-ray diffraction, Mössbauer spectroscopy is not seriously affected by the presence of small crystallites.

Raupp and Delgass (*122*) studied impregnated 10 Fe : 90 SiO_2 (Cab-O-Sil), 10 Fe : 90 MgO, and 5 Fe : 5 Ni : 90 SiO_2 by Mössbauer spectroscopy. Vacuum drying of the moist catalyst at mild conditions led to significantly smaller crystallites of ferric oxide, but after drying the crystallites are relatively stable. In the FTS, the reduced sample was used with 3.3 H_2 + 1 CO at 250°C and, presumably, 1 atm. In the initial part of the synthesis on Fe–SiO_2, small crystallites were converted to ϵ- and ϵ'-carbides, and larger crystallites on SiO_2 and small particles on MgO contained χ-$Fe_{2.5}C$. The reduced, supported Ni–Fe alloy did not form carbides. *In situ* spectroscopic and kinetic measurements showed that the rate in the initial part of the FTS is directly proportional to the amount of bulk carbide present. More higher hydrocarbons were produced as the iron carbides increased. Hydrogenation of bulk carbides was slower than either carbiding or the FTS, and hydrogenation of surface and bulk carbides yielded only CH_4.

Niemantsverdriet *et al.* (*105*) followed the formation of carbides in the FTS at 1 atm in a precipitated ferric oxide and iron catalysts supported on a TiO_2–CaO mixture using Mössbauer spectroscopy and X-ray diffraction. After synthesis for 1 day at 160°C, an unknown carbide (Fe_xC) and a hexagonal carbide [ϵ'-$Fe_{2.2}C$, according to Amelse *et al.* (*3*)] were formed; in 1 day at 190°C, Fe_xC, ϵ'-$Fe_{2.2}C$, and Hägg carbide (χ-$Fe_{2.5}C$) were found; and in 1 day at 240°C, 35% of the Fe was present as ϵ'-$Fe_{2.2}C$, and 65% as χ-$Fe_{2.5}C$. In 3 h at 350°C, 12% of the Fe was found as ϵ'-Fe_2C, 47% as χ-$Fe_{2.5}C$, and 41% as cementite. In 1.5 h at 450°C, all of the Fe was converted to cementite. In these tests, the conversions were held below 5% and magnetite was not found in the catalyst. The activity in the FTS at 240°C increased with carbide content up to ~8.5% carbon, which occurred at ~5 h. The activity then decreased. The decrease was attributed to elemental carbon formation, but it may well have been the result of accumulation of waxes.

Le Cäer *et al.* (*85*) made a critical review of the characterization of Fe carbides by Mössbauer spectroscopy. *In situ* Mössbauer measurements were made during the treatment of a reduced 10% Fe on alumina with 1 H_2 + 1 CO at 250°C. Patterns for magnetic and superparamagnetic ϵ-carbides were obtained with the amount of superparamagnetic component, attributable to fine particles, increasing with time. The carburization was postulated to involve migration of carbon atoms along defects in the initial particles, forming smaller crystallites of carbide. Adsorption measure-

Table 23
Changes of Catalyst during Synthesis[a]

Time of synthesis, h	*Activity, CH_4/min · g Fe*	*Surface composition from X-ray photoemission spectroscopy, arbitrary units (C/Fe)*	*Phases present*[b]	*Total C/Fe, H_2 treatment*	
				To 250°C	*To 500°C*
Unreduced sample					
0	0	0.26	Fe_2O_3	0	0
1.5	285	0.95	M,χ	0.21	0.24
5.5	335	1.25	M,χ	0.72	0.92
15	250	2.45	χ,M	1.13	1.53
Reduced sample					
0	—	—	α	—	—
0.25	310	4.4	α, $Fe_{20}C_9$	—	—
1.0	205	—	—	1.0	1.08
10	70	3.5	$Fe_{20}C_9$	—	—
24	40	2.8	$Fe_{20}C_9$	1.09	1.17

[a] Adapted from Reymond *et al.* (*123*).
[b] Phases from X-ray diffraction in decreasing order of intensity of patterns: Fe_2O_3, α-Fe_2O_3; M, magnetite; χ, χ-Fe_2C (χ-Fe_5C_2?, Hägg carbide); α, metallic Fe; $Fe_{20}C_9$, ϵ'-$Fe_{2.2}C$(?).

ments indicated an increase in the area of the Fe part of the catalyst. As mentioned in Chapter 3, the areas of reduced, fused catalysts did *not* increase substantially in conversion to χ-carbides or ϵ-nitrides.

Raymond *et al.* (*123*) followed composition changes in the atmospheric-pressure synthesis on α-Fe_2O_3 prepared from the decomposition of $Fe(NO_3)_3 \cdot 9\ H_2O$ at 160–200°C. One portion of α-Fe_2O_3 was heated briefly in He at 250°C and another was reduced in H_2 at 250°C for 16 h; both samples were tested at 250°C in 9 H_2 + 1 CO in a differential reactor. The raw catalyst became active during the synthesis; the rate attained a maximum at 5 h and then decreased. The reduced sample was active initially, and the rate decreased with time. Table 23 summarizes activity and composition from X-ray photoelectron spectroscopy, phases from X-ray diffraction, and total CH_4 generated by H_2 treatment of the used catalyst at 250°C and, subsequently, at 500°C. The raw catalyst was reduced rapidly to magnetite and, at a slower rate, was converted to χ-carbide. The carbide became the major phase by 15 h. The reduced sample was converted to "$Fe_{20}C_9$" [ϵ'-$Fe_{2.2}C$? (*3*)] at a moderate rate to become the only phase observed after 10 h. The selectivities of the two

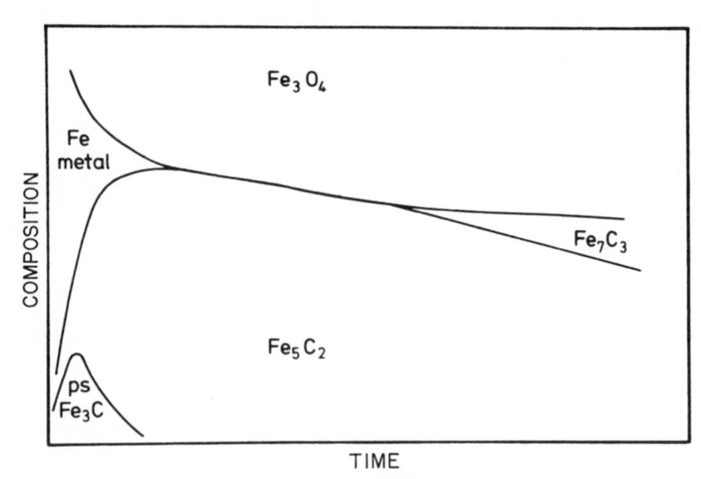

Fig. 25 Schematic composition changes in the high-temperature synthesis starting with catalysts of 100% metallic Fe. The time scale is not given; it depends on alkali content. Reproduced with permission from Dry (*44*).

catalysts were about the same, which differed markedly from the selectivity of precipitated-Fe catalysts used directly in the synthesis and those pretreated in H_2 (see Fig. 16 in Chapter 3).

In the fluidized and entrained processes at ~325°C and 20–30 atm, a similar set of composition changes occurs in the reduced-Fe catalysts, as shown in Fig. 25 (*44, 92, 116*). The metallic Fe in the reduced catalyst was rapidly converted to magnetite and Hägg carbide plus the transient presence of an unstable carbide, called "pseudocementite," with an X-ray diffraction pattern similar to Fe_3C. After several days, the Eckstrom and Adcock carbide (*92*) appeared; this carbide was originally designated as FeC but has subsequently been assigned the composition Fe_7C_3 (*62*). Eckstrom and Adcock (*50*) reported that 90% of the Fe was changed to this carbide in synthesis at 27 atm and 360°C, but Dry (*44*) found that the concentrations in the SASOL commercial reactors never reached this level. At SASOL, in 5-cm internal diameter fixed, fluidized-bed reactors, the Eckstrom–Adcock carbide was not found unless the tests were operated at pressures exceeding 60 atm.

Krebs *et al.* (*83*) studied the FTS on a 6 × 6 × 1 mm platelet of magnetite and on Fe foil in an ultrahigh-vacuum apparatus in which the sample could be moved from the vacuum apparatus to a small atmospheric-pressure reactor and back again without contamination. The catalyst were examined by Auger electron and by X-ray photoelectron spectroscopies at various stages of reduction and surface cleaning; however, few data were reported on the composition of the surface after FTS. Data from the synthesis tests were typical of alkali-free Fe at 1 atm. The rate

Table 24
Composition of C_5 Fraction from Potassium-Promoted Fused-Iron Catalyst Systems at 325°C and 12 bars[a]

Component	*Hydrocol*	*Fe/zeolite*
Isopentane	3.5	32
n-Pentane	7.9	11
1-Pentene	67.2	3
2-Pentene	5.8	18
2-Methyl-1-butene	3.5	19
3-Methyl-1-butene	11.1	2
2-Methyl-2-butene	0.7	58
Unsaturates, wt % of total	88.5	57
Research octane number, C_5 − 200°C	68.5	92

[a] Reproduced with permission from Caesar *et al.* (*24*).

increased with time of reduction of the Fe_3O_4 in H_2 at 500°C to a maximum at 2 h and then decreased. At the start of the tests the product was largely paraffinic, but olefins increased and paraffins decreased with time so that after 10 h the product was methane and alkenes. Unreduced Fe_3O_4 and a clean Fe foil had low activity in the FTS at 300°C. Reducing the Fe_3O_4 and oxidizing and reducing the foil increased activity substantially; the larger rates were attributed to larger surface areas of the treated specimen.

Kikuchi *et al.* (*79*) studied reactions of graphite lamellar catalysts prepared from $FeCl_3$. Reduction in H_2 at 200°C for a week reduced $FeCl_3$ to $FeCl_2$ and to metallic Fe. In reaction of 3 H_2 + 1 CO at 20 atm and 400°C, the selectivity varied with amount of $FeCl_2$ remaining and interlayer spacing. Catalysts were not very active; ~1 cm^3 (S.T.P.) of CO was consumed per gram catalyst per minute.

Workers at Mobil (*26*) reported a process utilizing the shape-selective zeolite ZSM5 for converting methanol to an aromatic gasoline in high yields. This zeolite was combined with a potassium-promoted ammonia-synthesis catalyst by mixing intimately 12- to 35-mesh powders in the ratio of one Fe catalyst to four zeolite (*24*). The results were remarkable in that the mechanical mixtures of Fe and zeolite were more active than the Fe catalyst alone. Their products were limited to the gasoline range and contained large yields of branched hydrocarbons, as shown in Table 24, where data from the literature are also given. In tests at 330°C and 12 atm of 2 H_2 + 1 CO gas at a space velocity of 3300 h^{-1}, the mixture gave conversions of CO and H_2 of about 95 and 50% compared with 25–75 and 16–47% for Fe alone. The authors postulated that the zeolite somehow intercepts the intermediates in the chain growth and converts them to highly branched molecules that do not grow further. An alternate explana-

Table 25
Comparison of Products from Iron on Alumina and Iron on ZSM5 at 35 atm of 1 H_2 + 1 CO Feed at 371°C[a]

Product	Distribution, wt %	
	3% Iron on alumina	*3% Iron on ZSM5*
C_1	30.7	33.4
C_2	17.5	10.8
C_3	17.1	17.3
C_4	6.7	19.3
C_{5+}	14.3	19.2
Oxygenates	13.7	0
Olefins in product, %	35.6	5.1
Aromatics in C_{5+}, %	0	25.2

[a] Reproduced with permission from Chang *et al.* (*27*).

tion is that the ZSM5 refines the iron Fischer–Tropsch products, converting it to a branched aromatic gasoline. A second Mobil paper (*27*) compared products from 3% Fe impregnated on alumina with 3% Fe on ZSM5 in the FTS at 371°C, 35 atm of 1 H_2 + 1 CO, and low conversions. As shown in Table 25, the products were similar, except that the oxygenates and olefins from the Fe on alumina were replaced by aromatics from Fe on ZSM5. Nearly all of the aromatics were in the range C_6–C_{10}.

German workers (*98*) made somewhat similar tests of ZSM5 and the less acidic silicalite with the Mn-rich Fe catalyst described previously. The zeolites were initially exchanged with aqueous HCl. Single-stage tests were prepared by milling and mixing fine powders of the catalyst and pelleting this mixture. For two-stage tests the zeolite was placed at the outlet of the reactor, preceded by a wad of glass wool and an equal volume of Fe catalyst. The data in Table 26 indicate a large increase in branched isomers for the silicalite test, with both modes of operation giving similar results. The two-stage ZSM5 test was about the same as the silicalite tests, but in the single-stage test the aromatics increased largely at the expense of branched isomers. A mechanism was postulated in which for the single-stage processes, C_3 and C_4 olefins in "equilibrium" with the Fe catalyst were rapidly converted by the adjacent zeolite to branched molecules and aromatics. The rapid depletion of olefins in the gas phase increased the production of olefins. In the two-stage operation, olefin formation was not accelerated. On passing through the zeolite bed, the hydrocarbons (particularly the olefins) were isomerized to branched olefins, and the branched species were converted to aromatics.

Table 26
Tests of Manganese–Iron plus Zeolites[a]

Catalyst arrangement	*Mn/Fe*	*Mn/Fe–Silicalite*		*Mn/Fe–ZSM5*	
		Single stage	*Two stage*	*Single stage*	*Two stage*
Space velocity, h^{-1}	450	340	340	350	360
Conversion of $H_2 + CO$	0.80	0.87	0.88	0.78	0.86
Yields, g/m^3 fed					
CH_4	33.0	37.2	26.1	42.2	31.5
C_2–C_4 Olefins	39.1	18.5	21.0	0.3	8.7
C_2–C_4 Paraffins	31.4	39.5	33.1	42.4	35.0
C_{5+}	61.0	89.0	106.2	76.4	105.7
Composition of C_{5+} hydrocarbons, wt %					
Branched isomers	33.9	78.7	78.0	42.4	71.0
Straight-chain isomers	66.1	15.7	22.0	11.6	19.0
Aromatics	0	5.6	0	45.9	10.0
Olefins	62.0	59.9	60.0	0	43.0
Paraffins	38.0	34.5	40.0	54.1	47.0

[a] About 1.5 CO + 1 H_2 feed at 310°C and 12 atm. Adapted from Moller *et al.* (*98*).

4.5 Molybdenum and Tungsten Catalysts

A modest amount of research has been done on the hydrogenation of oxides of carbon on Mo, but there are only a few studies on W. As shown in Table 2, W and Re catalysts have low activity. The moderate activity of Mo catalysts and their possible resistance to sulfur poisoning have led to occasional detailed studies.

As shown in Table 1, Mo and W have many characteristics similar to Fe-group elements, except that these metals oxidize more readily than Fe and form very stable carbides and nitrides (see Tables 5, 9, and 10 in Chapter 2). MoO_3 reduces rapidly to MoO_2 at 350°C in H_2 or synthesis gas. The equilibrium constants (H_2O/H_2 or CO_2/CO) for MoO_2 and Fe_3O_4 both increase with increasing temperature (see Table 5 in Chapter 2); however, MoO_2 requires temperatures ~100°C higher than Fe_3O_4 for these ratios to have the same values. An oxycarbide of Mo has been described (*86*).

Saito and Anderson (*124*) prepared and tested a number of Mo compounds with moderate surface areas in the hydrogenations of CO and CO_2 at atmospheric pressure and 350°C (Table 27). The authenticity of the

Table 27
Comparison of Results for Hydrogenation of CO_2 and CO[a]

	Initial				*After 10 h*		
	Rate, μmol/g · min	*Specific rate,[b] μmol/min · m²*	$(C_2 + C_3)/C_1$[c]	*WGS*[d]	*Rate, μmol/g · min*	$(C_2 + C_3)/C_1$[c]	*WGS*[d]
3.7 H_2 + 1 CO_2							
MoO_2	8.5	0.11	0.01	28.2	7.0	0.00	44
MoS_2	1.1	0.05	0.00	10,800.0	1.3	0.00	8500
Mo	3.6	0.50	0.01	96.5	6.2	0.02	39
Mo–carbide	43.8	6.26	0.14	24.3	15.0	0.10	27
Mo–nitride	1.1	0.16	0.00	57.2	1.6	0.00	65
Ni	285.0	238.0	0.00	180.0	265.0	0.00	168
Fe	>143.0	>17.9	>0.32	<51.4	20.0	0.03	600
3.1 H_2 + 1 CO							
MoO_2	45.9	0.65	0.07	20.0	45.9	0.07	20
MoS_2	7.9	0.34	0.04	0.3	6.2	0.06	0.3
Mo	61.5	8.20	0.03	21.0	37.0	0.18	18
Mo–carbide	181.0	24.8	0.18	18.5	15.0	0.21	13
Mo–nitride	37.4	5.12	0.23	21.3	44.0	0.14	18
Ni	504.0	520.0	0.01	0.1	325.0	0.01	0.1
Fe	>93.0	>11.0	>0.30	>3.4	24.0	0.00	0.0

[a] Reproduced with permission from Saito and Anderson (*124*).
[b] The initial rate of hydrocarbon formation per unit surface area of a catalyst at 350°C and 1 atm.
[c] Distribution of hydrocarbons produced.
[d] Water–gas shift WGS = ($[CO_2][H_2]$)/($[CO][H_2O]$). At equilibrium at 350°C, WGS = 21.0.

Table 28
Tests of Impregnated Molybdenum Catalysts at 21 atm[a]

	Support					
Conditions and products	*Silica–alumina cracking catalyst*		*Low-area silica–alumina*		*Charcoal*	
Mo, %	11.2	11.2	7.8	8.2	36.7	36.7
Feed gas, H_2/CO	1	3	1	1	1	3
Temperature, °C	355	401	339	400	404	420
Conversion of H_2 + CO, %	54	41	67	58	87	77
Hydrocarbon distribution, wt %						
C_1	68	80	24	63	76	97.5
C_2	22	16	14	22	19	1.2
$C_3 + C_4$	10	4	24	9	5	1.3
C_{5+}	0	0	38	6	0	0
Usage ratio, H_2/CO	1.0	1.5	0.8	1.0	1.0	1.9

[a] Adapted from data of Shultz *et al.* (*130*).

compound was checked by powder X-ray diffraction; the diffraction patterns of the particular compound always predominated, with other diffraction lines from traces of the starting material.

On the Mo compounds, CO was hydrogenated about five times as fast as CO_2. For both CO and CO_2 hydrogenations, metallic Ni was several times more active than the Mo samples, and usually the metallic Fe was more active than the Mo also. Metallic Mo, Mo_2C, Mo_2N, and metallic Fe had the largest tendency to produce higher hydrocarbons. In 10 h of synthesis, the carbided sample lost more than half of its activity, but the activity of the other preparations increased. The active Mo_2C was prepared by carburization with 40 H_2 + 1 C_3H_8; reduced Mo carburized with CO had low activity. MoS_2 had the lowest activity of the materials compared.

The tendency of the water–gas shift to proceed as a preliminary reaction in the $H_2 + CO_2$ reactions and as a subsequent reaction for the H_2 + CO synthesis is shown by the equilibrium constant designated in Table 27 as WGS, where WGS = 21 for equilibrium at 350°C. All of the Mo compounds except MoS_2 were active water–gas shift catalysts, and the Mo catalysts were more active in 1 H_2 + 1 CO gas than in 3 H_2 + 1 CO.

In the earlier Bureau of Mines work (*130*), the tendency of catalysts containing Mo to perform well in CO-rich feed gases was again shown (Tables 28–30). Impregnated catalysts prepared with ammonium molyb-

Table 29
Tests of Coprecipitated Molybdenum–Alumina at 21 atm[a]

Conditions and products	*Promoter*			
	Alumina		*Alumina*[b]	
Mo, %	28.3	28.3	26.8	26.8
Feed gas, H_2/CO	1	3	1	3
Temperature, °C	417	407	394	420
Conversion of H_2 + CO, %	66	66	57	41
Hydrocarbon distribution, wt %				
C_1	71	83	72	85
C_2	20	14	20	12
$C_3 + C_4$	9	3	8	3
Usage ratio, H_2/CO	1.1	1.8	1.1	1.5

[a] Adapted from Shultz *et al.* (*130*).
[b] Mo converted to a carbonitride of composition $MoC_{0.57}N_{0.33}$.

date (Table 28) were not very active, and the product usually was entirely gaseous hydrocarbons. Similar results were obtained with coprecipitated Mo–alumina catalyst (Table 29). A carbonitride having the composition $MoC_{0.57}N_{0.33}$ was prepared by reduction in H_2 followed by carbidizing in CO and subsequent nitriding in NH_3, all steps at 600°C. The carbonitride was no more active than the reduced catalyst. Mixed, precipitated catalysts also containing Ni or Co were very much more active than Mo alone, but the presence of Mo made it possible for these preparations to operate properly with CO-rich feed.

Solutions of carbonyls of Mo or W might be a useful impregnant for porous solids, and if the carbonyl were decomposed in the proper way, the metal might be obtained. Brenner *et al.* (*22*) reported that supported $Mo(CO)_6$ and $W(CO)_6$ may be decomposed below 200°C in a temperature-programmed reaction system and that the metal was partly oxidized. If the impregnated solids were heated to 600°C, the oxidation numbers of the metals were large (4.2–4.9). McLane and Anderson (*95*) tested catalysts prepared in this way. $Mo(CO)_6$ was dissolved in sodium-dried benzene, and the steps in the impregnation were performed in an inert atmosphere. The carbonyl in the catalyst was decomposed in He or H_2 and the rate of heating varied. The most active catalysts were obtained at a heating rate of 200°C/h up to 500°C. Not all of the carbonyl was decomposed on the support; a Mo mirror developed on the glass at the outlet of the catalyst tube. Catalysts of low activity were obtained at lower heating rates, 50°C/h, and those heated at 20°C/h were inactive.

Table 30
Tests of Nickel–Molybdenum and Cobalt–Molybdenum Catalysts at 21 atm[a]

Conditions and products	Compositions					
	Ni/Mo = 0.27 by weight				*Co/Mo = 2.68 by weight*	
Feed gas	1 H_2 + 1 CO	2 H_2 + 1 CO	3 H_2 + 1 CO	4 H_2 + 1 CO_2	2 H_2 + 1 CO	3 H_2 + 1 CO
Temperature, °C	257	275	293	350	300	299
Conversion of H_2 + CO, %	76	75	77	96[b]	75	68
Hydrocarbon distribution, wt %						
C_1	64	83	84	99.7	75	77
C_2	21	14	12	0	12	10
$C_3 + C_4$	15	3	4	0.3	10	10
C_{5+}	0	0	0	0	3	3
Usage ratio, H_2/CO	0.9	1.2	2.0	3.9[b]	1.2	1.8

[a] Adapted from Shultz *et al.* (*130*).
[b] Conversion of $H_2 + CO_2$ and H_2/CO_2 usage.

Table 31
Initial Activities of Supported Molybdenum Catalysts[a]

Phase in fresh catalyst	*Support*	*Temperature, °C*	*Initial rate, μmol/g · min*	
			CH_4	CO_2
$Mo(CO)_6$[b]	SiO_2	400	7.5	13
$Mo(CO)_6$[b]	13X	450	0	15
MoO_3	None	450	0.8	4.6
MoO_3[c]	MgO	400	0	17.0
	Al_2O_3	450	15	25.5
	SiO_2	450	61	60.0
	Charcoal	450	32	27

[a] $2\ H_2 + 1\ CO$ feed at 1 atm. Reproduced with permission from McLane and Anderson (*95*).

[b] The samples were prepared to contain 15% $Mo(CO)_6$ by weight. A substantial part of the carbonyl was lost in the decomposition step.

[c] These preparations contained 20% by weight of MoO_3.

The best activity at 400°C of catalysts prepared from carbonyls was 7.5 μmol/g · min (Table 31). A 13X zeolite similarly impregnated with $Mo(CO)_6$ was inactive. The low activity of the carbonyl catalysts may be attributed to large losses of carbonyl during heating and a substantial degree of oxidation of the metal by hydroxyls from the surface and by traces of water evolved in heating.

Table 31 includes for comparison preparations impregnated with ammonium paramolybdate. These samples were reduced in H_2 in a temperature-programmed reactor with water vapor monitored by a thermal-conductivity cell (*95*). Reduction was obtained by heating the catalyst at a rate of 200°C/h to 500°C and holding for 80 min at 500°C. Silica was the best support, followed by charcoal and alumina.

Copper was added to the Mo supported on SiO_2, based on the idea that factors increasing the ease of reduction of MoO_2 might improve the catalyst. Initial activities (Table 32) indicated that the rate of production of CH_4 at 450°C increased to a maximum when 1 part of CuO per 20 MoO_3 by weight was added. For all of these catalysts, the water–gas shift was more rapid than the synthesis. Addition of Fe-group oxides to Mo catalysts also increased the rate (Tables 33 and 34). Some higher hydrocarbons were produced, but only traces above C_2. The addition of NiO and CuO nearly doubled the rate. The last column in Table 33 gives the calculated rate of CO_2 production if all of the oxygen were converted to CO_2. The observed rates for CO_2 were only slightly smaller than the calculated ones, suggest-

Table 32
Initial Activities of Molybdenum–Copper–Silica[a]

	Initial rates, μmol/g · min			
	2 H_2 + 1 CO		*1 H_2 + 1 CO*	
Weight ratio, CuO/MoO_3	*CH_4*	*CO_2*	*CH_4*	*CO_2*
0	45	48	43	49
	36	42	38.5	54
	61	60	—	—
0.025	51	61	—	—
0.05	90	82	89	108.5
	89	68	97	125
0.087	48.5	58	53.5[b]	82[b]
0.5	51	49.5	55	75
1.0	43	49	61	93
Cu/SiO_2[c]	0	0		

[a] Catalysts contain 20 MoO_3 + 80 silica. Tests at 450°C and 1 atm. From McLane and Anderson (95).
[b] At 480°C.
[c] Blank test.

ing that the water–gas shift must be at or near equilibrium. At 9 atm in tests of similar catalysts supported on charcoal (Table 34), NiO again was best Fe-group cocatalyst. The tests of the catalysts containing Fe indicated that silica is a better support than charcoal.

In an attempt to increase the yields of C_2 and C_3, a variety of preparations of Mo on charcoal were prepared; the preparations contained Fe in moderate amounts, and Cu and K_2O or Na_2O in small concentrations. In

Table 33
Metal Oxides Increase Activity of MoO_3 on Silica[a]

		Initial rates at 450°C, μmol/g · min			
Composition of catalyst	*Conversion of H_2 + CO, %*	*CH_4*	*C_2H_6*	*CO_2*	*CO_2 (calc.)*
MoO_3–NiO	38	102	14	127	130
MoO_3–CuO	34	89	14	108.5	117
MoO_3–Fe_2O_3	23	68	9	80	86.5
MoO_3–CoO	14	46	9	59	63
MoO_3	14	43	7	49	57

[a] 1 H_2 + 1 CO at 1 atm and 450°C. Catalysts of the type 20 MoO_3 + 80 silica gel + 1 metal oxide. From McLane and Anderson (95).

Table 34
Tests of Molybdenum on Charcoal Promoted with Iron-Group Oxides[a]

Promoter	Conversion of $H_2 + CO$, %	Rate, μmol/g · min		
		CH_4	C_2H_6	CO_2
NiO	83	670	60	780
CoO	41	256	31	285
Fe_2O_3	21	110.5	17	145
Fe_2O_3[b]	31	151	23	150

[a] Catalyst of the type 20 MoO_3 + 80 charcoal + 1 metal oxide, 1 H_2 + 1 CO at 9 atm and 450°C. Reproduced with permission from McLane and Anderson (*95*).
[b] In this catalyst, silica was used as a support.

accord with Dry's result (*44*) for Fe catalysts, Na_2O and K_2O were equally effective in operation at 350 or 450°C. Table 35 presents data for a 5-day test at 450°C; no difficulties were encountered in this test in a fixed-bed reactor. The activity of the catalyst increased slowly with time, but the amount of olefins decreased. The products contained ~50 wt % C_2 and C_3 hydrocarbons.

Bridgewater *et al.* (*23*) studied 10% Mo on acid-washed coconut charcoal at 1.6–16 atm. In some instances, charcoal impregnated with ~12% K_2CO_3 was used; these preparations had low activity. Catalysts were pretreated in two ways: (a) heated in N_2 to 500°C for 16 h and then in H_2 at 450°C for 16 h to give samples containing MoO_2; (b) heated in H_2 for 16 h at 800°C to Mo_2C, the carbon being furnished by the support. The cata-

Table 35
Test of a Promoted Molybdenum–Iron–Charcoal Catalyst[a]

Days on stream	Conversion of $H_2 + CO$, %	Rates of production, μmol/g · min					$C_2 + C_3$, wt %
			C_2		C_3		
		CH_4	Paraffin	Olefin	Paraffin	Olefin	
1	58	148	55	16	15	17.5	62
2	60	171	60	14	14	14	61
3	63	187	62	9	15	13	55
4	65	205	67	5	16	10	52
5	69	239	70	1	21	5.5	48

[a] 20 MoO_3 + 10 Fe_2O_3 + 1 CuO + 1 Na_2O + 68 charcoal, 1 H_2 + 1 CO gas at 9 atm and 450°C. From McLane and Anderson (*95*).

Table 36
Tests of Commercial Cobalt Molybdate[a]

Catalyst	*Conversion, %*	*Selectivity, % CO to*				
		CO_2	*CH_4*	*C_2*	*$C_3 + C_4$*	*C_{5+}*
Without K_2O	43	50.3	26.2	17.5	10.4	0
With K_2O	45.5	41.2	14.9	15.2	9.5	19.2

[a] 15 atm of 2.1 H_2 + 1 CO feed; hourly space velocity, 200; and 302°C. Adapted from Madon *et al.* (*90*).

lysts pretreated by methods (a) or (b) had about the same activity and selectivity but were very much less active than Mo on silica.

To obtain improved thermal efficiency, methanation should be done at as high a temperature as the catalyst can tolerate. For this purpose, Araki *et al.* (*12, 137*) developed a 20 NiO : 25 MoO_3 : 55 magnesium aluminate catalyst that survived 12 days of methanation in a 45 H_2 + 15 CO + 40 CH_4 feed at a space velocity of 15,000 h^{-1}, 60 atm, and 650°C. The temperature was decreased periodically to 300, 400, and 500°C to determine if the activity had decreased; the CH_4 in the dry gas from the synthesis at 400°C decreased from 85 to 75% in 12 days. The catalyst was prepared by hot, wet mixing and grinding together of the support and ammonium paramolybdate plus nickel nitrate for several hours. The wet material was dried at 500°C and reduced in H_2 at 700°C for 15 h before use. During the 12 days of methanation, the catalyst composition did not change; phases present were a Ni–Mo alloy, Mo_2C, and $MgAl_2O_4$.

Workers at Dow (*99–102*) and Exxon (*90*) studied Mo catalysts, particularly their resistance to poisoning by sulfur compounds. Because the poisoning was a major aspect of these studies, it seems appropriate to describe these researches in detail in Chapter 6, and to summarize briefly the studies on pure feed here.

Madon *et al.* (*90*) made synthesis tests of a commercial Co–Mo catalyst on alumina, Co : Mo : alumina = 3 : 11 : 86 with and without the addition of a substantial amount of K_2CO_3 (3.45 wt % K_2O in the total catalyst). Typical results for the reduced catalysts are given in Table 36. The alkali promoter increased C_{5+} and olefins, and decreased CH_4. In other studies, a commercial MoS_2 powder (surface area, 2.2 m^2/g) impregnated with 3% KOH produced large yields of C_{5+} at 20 atm of 1.97 H_2 + 1 CO at 350°C, but the activity was only moderate; conversion of H_2 + CO was 52% at an hourly space velocity of 179. Selectivities (percentage CO converted to) were CH_4, 2.7; C_2, 1.1; $C_3 + C_4$, 0.7; C_{5+}, 90.4; CO_2, 5.1.

Table 37
Selectivity in Catalyst Tests[a]

	Test				
Conditions and products	*1*	*2*	*3*	*4*	*5*
Temperature, °C	333	336	351	353	389
Pressure, atm	20.5	33.5	33.4	34.7	34.1
Feed ratio, H_2/CO	0.76	0.77	0.77	0.98	0.98
Hours on stream	45	69	139	163	107
Conversion of H_2 + CO, %	56	57	61	64	75
Weight percentage					
CH_4	25	26	30	32	35
C_2–C_5	62	54	61	57	64
Liquids	13	19	9	11	2

[a] Adapted from Murchison and Murdick (*101*).

Tests were also made of 10% WO_3 on alumina and 3% NiO : 10% WO_3 on alumina, both with and without 3.3% K_2O. Activity was low; conversions of H_2 + CO of 10.9, 7.8, and 16.1% were obtained for the alkalized WO_3 on alumina, and NiO–WO_3 on alumina without and with alkali, respectively, at 450°C, an hourly space velocity of 1 H_2 + 1 CO of 115, and 20 atm. At these conditions for W on alumina, C_{5+} was 69% of converted CO, but for the catalyst containing Ni the selectivities to C_{5+} were less than 1.5%.

Murchison and Murdick (*99–102*) described studies of Mo catalysts designed to increase the yield of LPG. For Mo on carbon, increasing alkali to 4% K_2O increased yields of C_2–C_4 largely at the expense of CH_4. For alkali carbonates in equimolar amounts, Na_2CO_3, K_2CO_3, Rb_2CO_3, and Cs_2CO_3 had about the same effect in increasing C_2–C_4 yields. An alkalyzed Mo on carbon had relatively constant activity and selectivity for more than 50 days. Conversion of CO was ~68%, and 60 wt % or more of hydrocarbons was in the range C_2–C_4. Table 37 shows changes in selectivity with operating pressure (tests 1 and 2), temperature (tests 2 and 3, and 4 and 5), and feed ratio (tests 3 and 4). Increasing pressure increased the yield of liquid hydrocarbons at the expense of C_2–C_5. Increasing temperature increased CH_4 slightly and C_2–C_5 moderately. Changing the feed ratio made no sizable changes in selectivity.

Long tests were made in a large pilot plant, a "single commercial tube reactor," on Mo–alkali–carbon catalysts (*99–102*). At 366°C and 35 atm, conversions of both H_2 and CO were more than 56% and the hydrocarbon product contained C_1, 32; C_2–C_4, 62; and C_{5+}, 6 wt %. In another test with 0.8 H_2 + 1 CO containing 10 ppm H_2S, activity was relatively constant and the C_{2+} was greater than 60% of the hydrocarbons for more than 1500 h.

Murchison and Murdick (*99–102*) made kinetic studies in a Berty recycle reactor. The differential reaction rate was given as

$$r = Ap_{H_2}^{0.5}p_{CO} \exp(-22{,}805/RT)/(1 + 1.02p_{CO_2}), \quad (12)$$

where the activation energy is given in calories per mole and the pressure in atmospheres. The term in the denominator was very important in limiting conversions to relatively small values, particularly because with most feeds, CO_2 is the principal oxygenated product.

On the basis of the research cited, Mo catalysts may be useful for producing LPG and other low molecular weight fractions. Despite the large tendency of reduced Mo to oxidize, modest activities and long lives were possible. The catalysts were usually excellent for the water–gas shift, and in the FTS, equilibrium in the water–gas shift was approached in many instances. As a consequence, these catalysts operated effectively on CO-rich feed, and apparently carbon deposition was not a serious problem in tests at 400–500°C. Alkali, K_2CO_3 or Na_2CO_3, was an effective promoter for increasing the average molecular weight of the product, but the activity was usually not increased. Mixed catalyst containing Mo plus Fe-group metals had activities very much larger than preparations of Mo alone, and the selectivities for hydrocarbon production seemed to be determined by the Fe-group metal. The Mo part provided high water–gas shift activity plus the ability to operate on CO-rich feed. Finally, Mo catalysts have at least a modest resistance to poisoning by sulfur compounds, as discussed in Chapter 6.

Molybdenum catalysts generally have relatively low activity compared with Fe-group metals and Ru, and a 100-fold increase in activity would make Mo a serious competitor for commercial operations, especially for synthesis of particular fractions such as C_2–C_4.

References

1. Agrawal, P. K., Katzer, J. R., and Manogue, W. H., *Ind. Eng. Chem. Fundam.* **21,** 385 (1982).
2. Agrawal, P. K., Katzer, J. R., and Manogue, W. H., *J. Catal.* **74,** 332 (1982).
3. Amelse, J. A., Butt, J. B., and Schwartz, L. H., *J. Phys. Chem.* **82,** 558 (1978).
4. Anderson, R. B., *in* "Catalysis" (P. H. Emmett, ed.), Vol. 4, p. 279. Van Nostrand-Reinhold, Princeton, New Jersey, 1956.
5. Anderson, R. B., *in* "Catalysis" (P. H. Emmett, ed.), Vol. 4, Chapter 2. Van Nostrand-Reinhold, Princeton, New Jersey, 1956.

6. Anderson, R. B., *in* "Catalysis" (P. H. Emmett, ed.), Vol. 4, p. 237. Van Nostrand-Reinhold, Princeton, New Jersey, 1956.
7. Anderson, R. B., and Karn, F. S., *J. Phys. Chem.* **64,** 805 (1960).
8. Anderson, R. B., Karn, F. S., and Schultz, J. F., *Bull.—U.S. Bur. Mines* **614,** 46 (1964).
9. Anderson, R. B., Lee, C.-B. and Machiels, J. C., *Can. J. Chem. Eng.* **54,** 590 (1976).
10. Anderson, R. B., Seligman, B., Shultz, J. F., Kelly, R., and Elliott, M. A., *Ind. Eng. Chem.* **44,** 391 (1952).
11. Arakawa, H., and Bell, A. T., *Ind. Eng. Chem. Process Des. Dev.* **22,** 97 (1983).
12. Araki, M., Takaya, H., Ogawa, K., Suzuki, K., Hosoya, T., and Todo, N., *J. Jpn. Pet. Inst.* **25,** 85 (1982).
13. Atkinson, G. B., and Nicks, L. J., *J. Catal.* **46,** 417 (1977).
14. Bartholomew, C. H., Pannell, R. B., and Butler, J. L., *J. Catal.* **65,** 335 (1980).
15. Bartholomew, C. H., Pannell, R. B., Butler, J. L., and Mustard, D. G., *Ind. Eng. Chem. Prod. Res. Dev.* **20,** 296 (1981).
16. Bhasin, M. M., Bartley, W. J., Ellgen, P. C., and Wilson, T. P., *J. Catal.* **54,** 120 (1978).
17. Bhatia, S., Bakhshi, N. N., and Mathews, J. F., *Can. J. Chem. Eng.* **59,** 492 (1981).
18. Biloen, P., Helle, J. N., and Sachtler, W. M. H., *J. Catal.* **58,** 95 (1979).
19. Bond, G. C., and Turnham, B. D., *J. Catal.* **45,** 128 (1976).
20. Borghard, W. G., and Bennett, C. O., *Ind. Eng. Chem. Prod. Res. Dev.* **18,** 18 (1979).
21. Brenner, A., and Hucul, D. A., *Inorg. Chem.* **18,** 2836 (1979).
22. Brenner, A., Hucul, D. A., and Hardwick, S. J., *Inorg. Chem.* **18,** 2836 (1979).
23. Bridgewater, A. J., Burch, R., and Mitchell, P. C. H., *J. Catal.* **78,** 116 (1982).
24. Caesar, P. D., Brennan, J. A., Garwood, W. E., and Ciric, J., *J. Catal.* **56,** 274 (1979).
25. Castner, D. G., Blackadar, R. L., and Somorjai, G. A., *J. Catal.* **66,** 257 (1980).
26. Chang, C. D., Kuo, J. C. W., Lang, W. H., Jacob, S. M., Wise, J. J., and Silvestri, A. J., *Ind. Eng. Chem. Prod. Res. Dev.* **17,** 255 (1978).
27. Chang, C. D., Lang, W. H., and Silvestri, A. J., *J. Catal.* **56,** 268 (1979).
28. Chen, H. C., and Anderson, R. B., *J. Colloid Interface Sci.* **38,** 535 (1972).
29. Chen, H. C., and Anderson, R. B., *J. Catal.* **38,** 161 (1973).
30. Chin, R. L., Elattar, A., Wallace, W. E., and Hercules, D. M., *J. Phys. Chem.* **84,** 2895 (1980).
31. Clarke, J. J., Dry, M. E., Montano, J. J., and van Zyl, W. J., *Trans. Faraday Soc.* **57,** 2239 (1961).
32. Clarke, J. J., Dry, M. E., and van Zyl, W. J., *J. S. Afr. Chem. Inst.* **15,** 11 (1962).
33. Clarke, J. K. A., and Creaner, A. C. M., *Ind. Eng. Chem. Prod. Res. Dev.* **20,** 575 (1981).
34. Coon, V. T., Takeshita, T., Wallace, W. E., and Craig, R. S., *J. Phys. Chem.* **80,** 1878 (1976).
35. Dalla Betta, R. A., Piken, A. G., and Shelef, M., *J. Catal.* **35,** 54 (1974).
36. Dalla Betta, R. A., and Shelef, M., *J. Catal.* **48,** 111 (1977).
37. Dalmon, J. A., and Martin, G., *Proc. Int. Congr. Catal., 7th, 1980* (T. Seiyama and K. Tanabe, eds.), Part A, p. 402. Elsevier, Amsterdam, 1981.
38. Dautzenberg, F. N., Helle, J. N., van Santen, R. A., and Verbeek, H., *J. Catal.* **50,** 8 (1977).
39. Deckwer, W.-D., Serpemen, Y., Ralek, M., and Schmidt, B., *Ind. Eng. Chem. Process Des. Dev.* **21,** 222 (1982).
40. Deeba, M., Scott, J. P., Barth, R., and Gates, B. C., *J. Catal.* **71,** 373 (1981).
41. Dent, A. L., and Lin, M., *Adv. Chem. Ser.* **178,** 47 (1979).

42. Dry, M. E., *Brennst.-Chem.* **50,** 193 (1969).
43. Dry, M. E., *Ind. Eng. Chem. Prod. Res. Dev.* **15,** 282 (1976).
44. Dry, M. E., *in* "Catalysis: Science and Technology" (J. R. Anderson and M. Boudart, eds.), Vol. 1, Chapter 4. Springer-Verlag, Berlin and New York, 1981.
45. Dry, M. E., du Plessis, J. A. K., and Leuteritz, G. M., *J. Catal.* **6,** 194 (1966).
46. Dry, M. E., and Ferriera, L. C., *J. Catal.* **7,** 352 (1967).
47. Dry, M. E., Leuteritz, G. M., and van Zyl, W. J., *J. S. Afr. Chem. Inst.* **16,** 15 (1963).
48. Dry, M. E., and Oosthuizen, G. J., *J. Catal.* **11,** 18 (1968).
49. Dry, M. E., Shingles, T., and Boshoff, L. J., *J. Catal.* **25,** 99 (1972).
50. Eckstrom, H. C., and Adcock, W. A., *J. Am. Chem. Soc.* **72,** 1042 (1950).
51. Eidus, Ya. T., and Bulanova, T. F., *in* "Scientific Selection of Catalysts" (A. A. Balandin, B. A. Kazanskii, V. E. Vasserberg, G. V. Isagulyants, and G. I. Levi, eds.) Izd. Nauka, Moscow, 1966 (transl. by A. Aledjem, pp. 206–213. Daniel Davey & Co., Jerusalem, 1968.
52. Ekerdt, J. G., and Bell, A. T., *J. Catal.* **58,** 170 (1979).
53. Ellatar, A., Wallace, W. E., and Craig, R. S., *Adv. Chem. Ser.* **178,** 7 (1979).
54. Elliott, D. J., and Lunsford, J. H., *J. Catal.* **57,** 11 (1979).
55. Everson, R. C., Smith, K. J., and Woodburn, E. T., *Proc. Int. Mtg. S. Afr. Inst. Chem. Eng., 3rd, Stellenbosch, 1–3 Apr. 1980,* P2D/1.
56. Everson, R. C., Woodburn, E. T., and Kirk, A. R. M., *J. Catal.* **53,** 186 (1978).
57. Fischer, F., and Koch, H., *Brennst.-Chem.* **13,** 61 (1932).
58. Fischer, F., and Meyer, K., *Brennst.-Chem.* **15,** 84, 107 (1934).
59. Frohning, C. D., Rottig, W., and Schnur, F., *in* "Chemierohstoffe aus Kohle" (J. Falbe, ed.), pp. 234–257. Thieme, Stuttgart, 1977.
60. Guczi, L., *Catal. Rev.—Sci. Eng.* **23,** 329 (1981).
61. Gupta, N. M., Kamble, V. S., and Iyer, R. M., *J. Catal.* **66,** 101 (1980).
62. Herbstein, F. H., and Snyman, J. A., *Inorg. Chem.* **3,**(6), 894 (1964).
63. Hugues, F., Bussière, P., Bassett, J. M., Commerenc, D., Chauvin, Y., Bonneviot, L., and Oliver, D., *Proc. Int. Congr. Catal., 7th, 1980* (T. Seiyama and K. Tanabe, eds.), Part A, p. 418. Elsevier, Amsterdam, 1981.
64. Ignatiev, A., and Matsuyama, T., *J. Catal.* **58,** 325 (1979).
65. Iizuka, T., Tanaka, Y., and Tanabe, K., *J. Catal.* **76,** 1 (1982).
66. Imamura, H., and Wallace, W. E., *J. Catal.* **65,** 127 (1980).
67. Jacobs, P. A., Verdonck, J., Nijs, R., and Uytterhoeven, J. B., *Adv. Chem. Ser.* **178,** 15 (1979).
68. Jung, H.-J., Vannice, M. A., Mulay, L. N., Stanfield, R. M., and Delgass, W. N., *J. Catal.* **76,** 208 (1982).
69. Jung, H.-J., Walker, P. L., Jr., and Vannice, M. A., *J. Catal.* **75,** 416 (1982).
70. Kao, C.-C., Tsai, S.-C., and Chung, Y.-W., *J. Catal.* **73,** 136 (1982).
71. Karn, F. S., Shultz, J. F., and Anderson, R. B., *J. Phys. Chem.* **64,** 446 (1960).
72. Karn, F. S., Shultz, J. F., and Anderson, R. B., *Actes Congr. Int. Catal., 2nd, 1960* Vol. II, p. 2439 (1961).
73. Karn, F. S., Shultz, J. F., and Anderson, R. B., *Ind. Eng. Chem. Prod. Res. Dev.* **4,** 265 (1965).
74. Kellner, C. S., and Bell, A. T., *J. Catal.* **67,** 175 (1981).
75. Kellner, C. S., and Bell, A. T., *J. Catal.* **70,** 418 (1981).
76. Kellner, C. S., and Bell, A. T., *J. Catal.* **71,** 288 (1981).
77. Kellner, C. S., and Bell, A. T., *J. Catal.* **71,** 296 (1981).
78. Kempling, J. C., and Anderson, R. B., *Ind. Eng. Chem. Process Des. Dev.* **9,** 116 (1970).

79. Kikuchi, E., Aranishi, Y., Koizumi, A., and Morita, Y., *Proc. Int. Congr. Catal., 7th, 1980* (T. Seiyama and K. Tanabe, eds.), Part B, p. 1436. Elsevier, Amsterdam.
80. King, D. L., *J. Catal.* **51,** 386 (1978).
81. King, D. L., *J. Catal.* **61,** 77 (1980).
82. Kölbel, H., and Ralek, M., *Chem. Ind.* (*London*) **31,** 700 (1979).
83. Krebs, H. J., Bonzel, H. P., Schwarting, W., and Gafner, G., *J. Catal.* **72,** 199 (1981).
84. Kuznetsov, V. L., Bell, A. T., and Yermakov, Y. I., *J. Catal.* **65,** 374 (1980).
85. Le Caër, G., Dubois, J. M., Pijolat, M., Perrichon, V., and Bussière, P., *J. Phys. Chem.* **86,** 4799 (1982).
86. Leclercq, L., Imura, K., Yoshida, S., Barbee, T., and Boudart, M., *in* "Preparation of Catalysts" (B. Delmon, P. A. Jacobs, and G. Poncelet, eds.), p. 627. Elsevier, Amsterdam, 1979.
87. Lee, C. B., Ph.D. Thesis, Dept. of Chem. Eng., McMaster University, Hamilton, Ontario, Canada (1983).
88. Luengo, C. A., Cabrera, A. L., MacKay, H. B., and Maple, M. B., *J. Catal.* **47,** 1 (1977).
89. Lunde, P. J., and Kester, F. L., *J. Catal.* **30,** 423 (1973).
90. Madon, R. J., Bucker, E. R., and Taylor, W. F., *U.S. Energy Res. Dev. Admin.* (*Dep. Energy*), *Final Rep.* Contract No. E(46-1)-8008 (1977).
91. Madon, R. J., and Taylor, W. F., *Adv. Chem. Ser.* **178,** 93 (1979).
92. Malan, O. G., Louw, J. D., and Ferreira, L. C., *Brennst.-Chem.* **42,** 209 (1961).
93. Manning, M. P., and Reid, R. C., *Ind. Eng. Chem. Process. Des. Dev.* **16,** 358 (1977).
94. McKee, D. W., *J. Catal.* **8,** 240 (1967).
95. McLane, R. C., and Anderson, R. B., unpublished data.
96. McVicker, G. B., and Vannice, M. A., *J. Catal.* **63,** 25 (1980).
97. Miura, H., and Gonzalez, R. D., *Ind. Chem. Prod. Res. Dev.* **21,** 274 (1982).
98. Moller, K., Deckwer, W. D., and Ralek, M., *Workshop Zeolites, 1982.*
99. Murchison, C. B., *Pap. Coal Technol. '80 Conf. 1980.*
100. Murchison, C. B., *Int. Conf. Uses Chem. Molybdenum, 4th, 1982.*
101. Murchison, C. B., and Murdick, D. A., *Prepr. Pap. A.I.Ch.E. Meet. 1980.*
102. Murchison, C. B., and Murdick, D. A., *Hydrocarbon Process.* January 1981, p. 159.
103. Nakamura, M., Wood, B. J., Hon, P. Y., and Wise, H., *Prepr. Int. Congr. Catal., 1980* Paper A29 (1980).
104. Nielsen, A., "An Investigation on Promoted Iron Catalysts for the Synthesis of Ammonia," 3rd ed. Gjellerups Forlag, Copenhagen, Denmark, 1968.
105. Niemantsverdriet, J. W., van der Kraan, A. M., van Dijk, W. L., and van der Baan, H. S., *J. Phys. Chem.* **84,** 3363 (1980).
106. Nijs, H. H., and Jacobs, P. A., *J. Catal.* **66,** 401 (1980).
107. Ott, G. L., Fleisch, T., and Delgass, W. N., *J. Catal.* **60,** 394 (1979).
108. Ott, G. L., Fleisch, T., and Delgass, W. N., *J. Catal.* **65,** 253 (1980).
109. Palmer, R. L., and Vroom, D. A., *J. Catal.* **50,** 244 (1977).
110. Pannell, R. B., Kibby, C. L., and Kobylinski, T. P., *Proc. Int. Congr. Catal., 7th, 1980* (T. Seiyama and K. Tanabe, eds.), Part A, p. 447. Elsevier, Amsterdam, 1981.
111. Pennline, H. W., Schehl, R. R., and Haynes, W. P., *Ind. Eng. Chem Process Des. Dev.* **18,** 156 (1979).
112. Pichler, H., *Adv. Catal.* **4,** 289 (1952).
113. Pichler, H., and Bellestedt, F., *Erdoel Kohle Erdgas, Petrochem. Brennst.-Chem.* **26,** 560 (1973).
114. Pichler, H., and Firnhaber, B., *Angew. Chem.* **75,** 166 (1963); *Brennst.-Chem.* **44,** 33 (1963).

115. Pichler, H., Meier zu Köcker, H., Gabler, W., Gärtner, R., and Kioussis, D., *Brennst.-Chem.* **48,** 266 (1967).
116. Pichler, H., and Schulz, H., *Chem.-Ing.-Tech.* **42,** 1162 (1970).
117. Pichler, H., Schulz, H., and Elstner, M., *Brennst.-Chem.* **48,** 3 (1967).
118. Pichler, H., Schulz, H., and Kühne, D., *Brennst.-Chem.* **49,** 344 (1968).
119. Ponec, V., and van Barneveld, W. A., *Ind. Eng. Chem. Prod. Res. Dev.* **18,** 268 (1979).
120. Poutsma, M. L., Elek, L. F., Ibarbia, P. A., Risch, A. P., and Rabo, J. A., *J. Catal.* **52,** 157 (1978).
121. Raney, M., *J. Am. Chem. Soc.* **54,** 4116 (1932).
122. Raupp, G. B., and Delgass, W. N., *J. Catal.* **58,** 337, 348, 361 (1979).
123. Reymond, J. P., Meriaudeau, P., and Teichner, S. J., *J. Catal.* **75,** 39 (1982).
124. Saito, M., and Anderson, R. B., *J. Catal.* **63,** 438 (1980); **67,** 296 (1981).
125. Satterfield, C. N., and Huff, G. A., Jr., *J. Catal.* **73,** 187 (1982).
126. Schulz, H., *in* "Chemierohstoffe aus Kohle" (J. Falbe, ed.), pp. 334–355, 421. Thieme, Stuttgart, 1977.
127. Schulz, H., and Bellstedt, F., *Ind. Eng. Chem. Prod. Res. Dev.* **12,** 176 (1973).
128. Sexton, B. A., and Somorjai, G. A., *J. Catal.* **46,** 167 (1977).
129. Shalvoy, R. B., Reucroft, P. J., and Davis, B. H., *J. Catal.* **56,** 336 (1979).
130. Shultz, J. F., Karn, F. S., and Anderson, R. B., *Rep. Invest.—U.S. Bur. Mines* **6974,** 20 (1967).
131. Shultz, J. F., Karn, F. S., Anderson, R. B., and Hofer, L. J. E., *Fuel* **30,** 181 (1961).
132. Solymosi, F., and Erdöhelyi, A., *Proc. Int. Congr. Catal., 7th, 1980* (T. Seiyama and K. Tanabe, eds.), Part B, p. 1448. Elsevier, Amsterdam, 1981.
133. Solymosi, F., Tombacz, I., and Kocsis, M., *J. Catal.* **75,** 78 (1982).
134. Soufi, F., Doctoral Dissertation, Engler–Bunte Institute, University of Karlsruhe (1969).
135. Storch, H. H., Golumbic, N., and Anderson, R. B., "The Fischer–Tropsch and Related Syntheses." Wiley, New York, 1951.
136. Sughrue, E. L., and Bartholomew, C. H., *Appl. Catal.* **2,** 239 (1982).
137. Takaya, H., Araki, M., Shin, S., Suzuki, K., Hosoya, T., Ogawa, K., and Todo, N., *Proc. Int. Congr. Catal., 7th, 1980* (T. Seiyama and K. Tanabe, eds.), Part B, p. 1442. Elsevier, Amsterdam, 1981.
138. Tamaru, K., *Proc. Int. Congr. Catal., 7th, 1980* (T. Seiyama and K. Tanabe, eds.), Part B, p. 47. Elsevier, Amsterdam, 1981.
139. van Meerten, R. Z. C., Habets, H. M. J., Beaumont, A. H. G. M., and Coenen, J. W. E., *Proc. Int. Congr. Catal., 7th, 1980* (T. Seiyama and K. Tanabe, eds.), Part B, p. 1440. Elsevier, Amsterdam, 1981.
140. Vannice, M. A., *J. Catal.* **37,** 449, 462 (1975).
141. Vannice, M. A., *J. Catal.* **40,** 129 (1975).
142. Vannice, M. A., and Garten, R. L., *J. Catal.* **56,** 236 (1979); **66,** 242 (1980).
143. Vannice, M. A., Lam, Y. L., and Garten, R. L., *Adv. Chem. Ser.* **178,** 25 (1979).
144. Vannice, M. A., Walker, P. L., Jr., Jung, H.-J., Moreno-Castillo, C., and Mahajan, O. P., *Proc. Int. Congr. Catal., 7th, 1980* (T. Seiyama and K. Tanabe, eds.), Part A, p. 460. Elsevier, Amsterdam, 1981.
145. Watson, P. R., and Somorjai, G. A., *J. Catal.* **72,** 347 (1981).
146. Watson, P. R., and Somorjai, G. A., *J. Catal.* **74,** 282 (1982).
147. Weatherbee, G. D., and Bartholomew, C. H., *J. Catal.* **77,** 460 (1982).
148. Wentrcek, P. R., Wood, B. J., and Wise, H., *J. Catal.* **43,** 363 (1976).
149. Yang, C.-H., Massoth, F. E., and Oblad, A. G., *Adv. Chem. Ser.* **178,** 35 (1979).

CHAPTER 5

Mechanism of the Fischer–Tropsch Synthesis

5.1 Introduction

The mechanism of the growth of the carbon chain is the principal subject of this chapter. The choice of this subject was largely based on the importance of product distribution in the practical use of the Fischer–Tropsch synthesis (FTS) as well as on the interests of the author. The reader is referred to a review of FTS by Rofer-DePoorter (*113*) for a more detailed account of the chemisorptions of H_2 and CO and certain elementary reactions than that given here. Simultaneous adsorption of H_2 and CO often suggests the composition of the intermediate (*27, 61, 83, 116, 135a*). Most of the studies by the surface physicists are not considered here, nor are theoretical studies on chemisorption and catalysis, such as those of Kölbel and Tillmetz (*84*).

The reaction schemes that have been proposed for the FTS may be more easily understood and compared by considering their similar and disparate features, as summarized in Table 1 for the initiation and growth of carbon chains. Table 1 shows only growth steps leading to straight chains, and details of bonding are not given. In mechanisms I–IV, CO is bonded to the metal M at the carbon, but in V and VI, at the oxygen. In mechanisms III–VI, the intermediate contains oxygen, and a mole of H_2O is eliminated in each of the growth steps in B–E. Steps I and II lead to the same intermediate. In all of the growth steps, carbons are added to the growing chain one at a time.

In a number of investigations described in this chapter, small amounts of tagged species were introduced in the feed or deposited on the catalyst, and the products were then examined for the tagged atoms. For example, the tagged molecule may be an olefin or alcohol added to the feed, or reactive carbon or methylene groups may be deposited on the catalyst. *Incorporation* is a general term describing the added molecule reacting with the synthesis gas. The tagged molecule may (a) initiate growing chains, (b) be built into the growing chain by adding repeatedly, or (c) react in other ways (e.g., olefins being hydrogenated to paraffins). This latter reaction would not be considered incorporation. We note in passing that incorporation of a particular tagged molecule does not prove unambiguously that the reaction intermediate is a species of this type.

5.2 Detailed Characterizations of Synthesis Products

Products from the FTS are generally not in thermodynamic equilibrium with respect to the reactants nor with respect to other products; therefore, the selectivity data may contain valuable mechanistic information.

Synthesis products from Co and Fe were well characterized up to C_7 or C_8 in older work (*4*) and more recently by Schulz and Pichler and others using capillary gas chromatography up to C_{17}. Detailed analyses of products from Ni catalysts do not seem to be available. As described in Chapter 4, the "polymethylene" from the high-pressure, low-temperature synthesis on Ru has been characterized as well as the low molecular weight fractions. For the synthesis on Ru at moderate pressure and higher temperatures, the available characterizations are often contradictory.

Pertinent characteristics of products from FTS based mostly on data from Co and Fe are as follows:

1. Water is a principal primary product. Most of the CO_2 is produced in a subsequent water–gas shift. Primary alcohols and/or α-olefins may also be primary synthesis products.
2. The carbon-number distribution for hydrocarbons or total products on a molar basis is large for C_1, passes through a minimum at C_2, a maximum at C_3 or C_4, and then decreases monotonically.
3. Monomethyl-substituted hydrocarbons are present in moderate amounts with methyl groups in about equal amounts at all positions along the carbon chain.
4. Dimethyl species are present in significantly smaller amounts than monomethyl. *None of the dimethyls contains quaternary carbons.* Ethyl-substituted molecules are found in about the same amounts as dimethyls.
5. Naphthenes and aromatics are produced in small amounts in fixed-bed Fe and in moderate amounts from fluidized Fe. Cobalt and Ru usually produce none of these molecules.
6. Olefins from Fe are often found in amounts exceeding 50% in each carbon number, and more than 60% of these are α-olefins. The olefin fractions are more-or-less independent of carbon number. For prod-

Table 1
Pertinent Features of Mechanisms for the Fischer–Tropsch Synthesis

Chain initiation		
Surface carbide	**I.**	M–C≡O (O, C, M stacked) ⟶ C—O over M M ⟶ C over M + O over M $\xrightarrow[-H_2O]{H_2}$ CH_2 over M + M
Enolic-1	**II.**	M–C–O (O, C, M stacked) $\xrightarrow{H_2}$ H C OH over M $\xrightarrow[-H_2O]{H_2}$ CH_2 over M
Enolic-2	**III.**	M–C–O (O, C, M stacked) $\xrightarrow{H_2}$ H C OH over M
CO Insertion–1	**IV.**	H over M $\xrightarrow{CO}$ H over M with CO ⟶ HCO over M ⟶ HC—O over M M $\xrightarrow[-H_2O]{H_2}$ CH_2 over M + M
CO Insertion–2	**V.**	M–O–H $\xrightarrow{CO}$ M–O–CH=O $\xrightarrow{H_2}$ M–O–C(H_2)HO $\xrightarrow{H_2}$ M–O–CH_3
Alkoxy	**VI.**	O—C over M M $\xrightarrow{H_2}$ H C H / O / M + M

ucts from Co, both the fraction of total olefins and the fraction of α-olefins are smaller, and both decrease with increasing carbon number.

7. Carbon chains in alcohols are similar to those found in hydrocarbons. Yields of alcohols on a molar basis are maximal at C_2 and decrease monotonically with increasing carbon number. The low yields of methanol are probably the result of thermodynamic limitations.

Chain growth

A.

$$\underset{M}{\overset{H_2}{C}} + \underset{M}{\overset{H_2}{C}} \longrightarrow \underset{M}{\overset{HCH_2}{CH_2}} + M$$

B.

$$\underset{M}{RCOH} + \underset{M}{HCOH} \xrightarrow[-H_2O]{H_2} + \underset{M}{\overset{R}{H_2CCOH}} + M$$

C.

$$\underset{M}{RCH_2} \xrightarrow{CO} \underset{M}{RCH_2}\ CO \longrightarrow \underset{M}{\overset{RCH_2}{CO}} \longrightarrow \underset{M\quad M}{\overset{RCH_2}{C\!-\!O}} \xrightarrow[-H_2O]{H_2} \underset{M}{\overset{RCH_2}{CH_2}} + M$$

D.

$$\underset{M}{O}\!/CH_3 \xrightarrow{CO} \underset{M}{O}\!/\overset{O}{\overset{\|}{C}}CH_3 \xrightarrow{H_2} \underset{M}{O}\!/\overset{OH}{C}CH_3 \xrightarrow{H_2} \underset{M}{O}\!/CH_2CH_3$$

E.

$$\underset{\underset{M}{O}}{RCH} + \underset{\underset{M}{O}}{HCH} \longrightarrow \underset{\underset{M}{O}\ \ \underset{M}{O}}{R\overset{H}{C}\!-\!CH_2} \longrightarrow \underset{M}{O} + \underset{\underset{M}{O}}{\overset{RCH_2}{CH_2}}$$

Data of Pichler *et al.* (*107*) provide detailed analyses of all hydrocarbon isomers from C_4 to C_{12} and for a hydrogenated product (olefins converted to paraffins) from C_4 to C_{17} for products from the SASOL plant and an atmospheric-pressure test with a standard Co–ThO_2–kieselguhr catalyst. The isomer distributions for hydrogenated hydrocarbons in Table 2 include products from the SASOL fixed-bed reactors for constant-temperature operation at 220 and 241°C. The "practical" product is a composite of a number of these units at different temperatures and in different stages

Table 2
Analyses of Hydrogenated Products from Iron and Cobalt Catalysts[a]

		SASOL Products				*Cobalt, 1 atm, fixed bed,*
		Iron, fixed bed			*Entrained iron,*	
Carbon number	*Product*	*220°C*	*Practical*	*241°C*	*320°C*	*190°C*
4	N4[b]	96.4	95.9	96.0	91.6	95.4
	2M3	3.6	4.1	4.0	8.4	4.6
5	N5	94.9	93.1	91.2	80.5	87.8
	2M4	5.1	6.7	8.8	18.8	12.2
	Cyclo 5		0.14		0.7	
6	N6	94.0	90.5	88.3	70.7	80.6
	2M5	2.9	4.6	5.3	14.6	12.5
	3M5	2.5	4.0	5.0	10.1	6.8
	23DM4	0.3	0.3	0.4	0.8	0.1
	Mcyclo 5[c]	0.3	0.6	1.0	3.6	
7	N7	94.1	90.6	86.3	58.7	73.6
	2M6	2.0	3.3	4.4	11.1	11.3
	3M6	2.6	3.9	5.7	16.7	14.3
	23DM5	0.4	0.6	0.8	1.8	0.3
	24DM5	0.02	0.1	0.1	0.4	0.1
	3E5	0.2	0.4	0.5	0.8	0.4
	Naphthenes	0.6	1.0	1.9	7.0	
	Toluene	0.04	0.14	0.3	3.5	
8	N8	93.8	90.1	84.7	53.6	67.9
	2M7	1.8	2.7	3.9	10.4	10.1
	3M7	2.3	3.3	5.0	12.3	12.3
	4M7	0.9	1.2	1.7	5.2	6.8
	23DM6	0.2	0.3	0.5	1.3	0.3
	24DM6	0.07	0.2	0.2	1.0	0.5
	25DM6	0.05	0.1	0.2	0.5	0.4
	34DM6	0.06	0.2	0.3	0.7	
	3E6	0.3	0.6	0.6	1.5	1.7
	234TM5		0.02		0.2	
	2M3E5	0.04	0.05	0.08	0.2	
	Naphthenes	0.35	0.9	1.8	7.9	
	Aromatics	0.17	0.3	1.0	5.2	
9	N9[b]	94.2	90.9	84.3	50.3	63.3
	2M + 4M8	2.8	3.8	6.4	18.4	21.7
	3M8	1.8	2.4	4.3	10.9	11.3
	23DM7	0.1	0.2	0.5	1.1	0.1
	24DM7	0.02	0.06	0.2	0.7	0.6
	25DM + 35DM7	0.2	0.2	0.5	1.9	1.4
	26DM7	0.04	0.1	0.2	1.0	0.2
	34DM + 4E7	0.2	0.4	0.8	2.1	0.6

Table 2 (*continued*)

		SASOL Products				*Cobalt, 1 atm, fixed bed,*
		Iron, fixed bed			*Entrained iron,*	
Carbon number	*Product*	*220°C*	*Practical*	*241°C*	*320°C*	*190°C*
	3E7	0.2	0.4	0.8	1.2	0.8
	Naphthenes	0.3	1.3	1.4	6.8	
	Aromatics	0.08	0.18	0.36	4.5	
10	N10	93.8	92.9	83.8	50.0	60.4
	2M9	1.5	1.7	3.3	8.1	7.9
	3M9	1.8	1.8	3.8	9.8	10.3
	4M9	1.1	1.2	2.5	8.4	10.4
	5M9	0.5	0.6	1.2	4.0	5.0
	23DM + 34DM + 4E8	0.5	0.5	1.2	3.4	2.1
	24DM8	0.02	0.04	0.1	0.6	0.4
	25DM + 35DM8	0.1	0.1	0.6	1.8	1.5
	26DM8	0.1	0.1	0.4	1.5	0.6
	27DM8	0.05	0.1	0.3	0.7	0.2
	36DM8	0.04	0.1	0.3	1.0	0.2
	3E8	0.2	0.3	0.6	1.1	1.0
	Naphthenes + aromatics	0.04	0.11		3.2	
	Unknown peaks[d]	(5) 0.2	(13) 0.4	(12) 1.9	(15) 6.4	
11	N11	94.5	92.7	84.4	51.1	55.9
	2M10	1.4	1.6	3.0	7.6	8.0
	3M10	1.5	2.0	3.5	10.3	9.9
	4M10	1.0	1.1	2.4	7.1	10.5
	5M10	0.8	1.0	2.2	6.7	9.2
	23DM + 34DM9	0.03	0.3	0.4	3.3	
	24DM9	0.03	0.03	0.06	0.5	0.2
	25DM + 35DM9	0.05	0.1	0.4	1.6	0.6
	26DM9	0.04	0.1	0.3	0.9	0.8
	27DM9	0.07	0.1	0.4	1.3	0.6
	28DM + 36DM9	0.08	0.2	0.5	1.5	1.1
	37DM9		0.03	0.2	0.4	0.3
	3E9	0.2	0.3	0.7	1.0	1.2
	4E9	0.03	0.03	0.3	0.5	1.1
	5E9	0.08	0.1	0.4	0.7	0.6
	Unknown peaks	(2) 0.2	(12) 0.3	(5) 0.8	(17) 5.5	
12	N12[b]	94.7	92.5		40.5	55.2
	2M11	1.2	1.5		6.5	6.7
	3M11	1.3	1.7		7.9	9.6
	4M11	0.8	1.1		6.4	9.1
	5M11	0.8	1.0		6.2	8.5
	6M11	0.4	0.6		3.4	4.2
	24DM10				1.9	0.1

(*continued*)

Table 2 (*continued*)

		SASOL Products				*Cobalt, 1 atm, fixed bed, 190°C*
		Iron, fixed bed			*Entrained iron, 320°C*	
Carbon number	*Product*	*220°C*	*Practical*	*241°C*		
	25 + 26 + 35DM10	0.04	0.06		0.8	0.6
	28DM10	0.05	0.1		1.4	0.7
	29 + 37DM10	0.08	0.15		1.6	0.9
	36DM10	0.06	0.09		0.09	0.5
	38DM10		0.06		0.7	0.2
	3E10	0.2	0.2		0.8	1.0
	4E10	0.2	0.2		2.5	0.9
	5E10	0.1	0.2		1.6	0.5
	Unknown peaks[d]	(2)	0.04		(13) 0.04	(18) 15.0
13	N13	94.5	92.8		42.4	48.5
	2M12	1.2	1.5		7.8	6.4
	3M12	1.3	1.6		8.5	8.2
	4M12	0.8	1.0		7.2	8.6
	5M12	0.6	1.0		6.9	8.1
	6M12	0.6	0.9		6.7	7.6
	3E11	0.05	0.2		0.8	0.7
	Unknown peaks	(9) 0.9	(14) 1.0		(21) 19.7	(14) 11.9
14	N14	95.1	92.7		37.1	46.6
	2M13	1.1	1.4		6.2	5.9
	3M13	1.1	1.5		7.0	7.3
	4M13	0.7	0.9		6.5	8.0
	5M13	0.6	0.9		5.5	7.3
	6M + 7M13	0.9	1.2		11.1	10.4
	3E12	0.2	0.2		2.4	0.5
	Unknown peaks	(5) 0.3	(12) 1.2		(20) 24.2	(80) 14.0
15	N15	94.4	93.8		36.7	46.8
	2M14	1.1	1.2		6.1	5.3
	3M14	1.2	1.3		7.8	7.1
	4M14	0.7	0.8		6.3	7.8
	5M14	0.7	0.7		5.6	7.3
	6M14	0.7	0.7		6.0	6.7
	7M14	0.7	0.6		6.5	7.1
	3E13	0.1	0.2		0.4	0.4
	Unknown peaks	(6) 0.4	(11) 0.7		(25) 24.6	(16) 11.5
16	N16[b]	94.7	92.2		30.6	43.7
	2M15	0.9	1.4		6.7	5.4
	3M15	1.0	1.4		9.0	6.8
	4M15	0.6	0.9		6.0	8.5
	5M15	0.6	0.9		6.0	8.5
	6M15	0.6	0.9		5.7	7.0

Table 2 (*continued*)

Carbon number	Product	SASOL Products: Iron, fixed bed 220°C	Practical	241°C	Entrained iron, 320°C	Cobalt, 1 atm, fixed bed, 190°C
	7M + 8M15	0.9	1.1		5.3	11.6
	Unknown peaks[d]	(5) 0.7	(12) 1.2		(21) 30.7	(11) 8.5
17	N17	94.4	91.6		29.7	44.1
	2M16	1.1	1.4		5.1	5.2
	3M16	1.1	1.5		6.9	6.1
	4M16	0.8	0.9		5.4	8.2
	5M16	0.6	0.8		4.9	6.2
	6M16	0.6	0.8		4.1	8.9
	7M + 8M16	1.0	1.8		10.3	21.3
	Unknown peaks	(4) 0.4	(12) 1.2		(25) 33.6	

[a] Adapted from Pichler *et al.* (*107*).
[b] The following notations were used to identify the molecules: N, normal, straight chain; M, methyl; DM, dimethyl; E, ethyl; TM, trimethyl (e.g., 24DM8 represents 2,4-dimethyloctane).
[c] Plus traces of benzene.
[d] Number of peaks (in parentheses) and total amount.

of catalyst life. A significant difference between the fixed-bed Fe products and the SASOL entrained Fe as well as the Co hydrocarbons is that the fraction of normal species remains essentially constant for the fixed-bed Fe products whereas this fraction decreases with increasing carbon number for the other products. Other isomer distributions are given in Chapter 4.

Next, several carbon-number distributions are presented to supplement those in Chapter 4. Carbon-number distributions on a weight basis often have a maximum that is a distinctive feature of the distribution, as shown for the lower molecular weight products from the polymethylene synthesis (*119*) in Fig. 1. In this figure, oxygenates and hydrocarbons are summed on a weight-of-carbon basis for each carbon number, and the ordinate is weight percentage of total carbon, an analog to weight-percentage plots for hydrocarbons. For this example at least, plotting data on a weight-percentage carbon basis seems appropriate as the 100 and 120°C products contain substantial amounts of oxygenates in the C_2–C_7 range, and those from 140°C, no oxygenates. Carbon-number distributions for Co products are given in Fig. 10 of Chapter 4.

Pannell *et al.* (*103*) presented carbon-number distributions for hydrocarbons from Co, Fe, and Ru catalysts in a Berty reactor, a differential

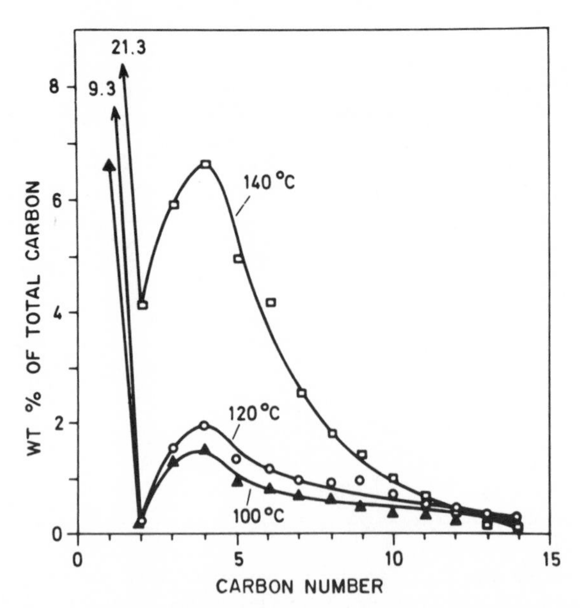

Fig. 1 Carbon number distributions of hydrocarbons + oxygenates on a weight-percentage carbon basis for low molecular weight products from the polymethylene synthesis at 1000 atm in a batch reactor with hexadecane as a suspension medium. Reproduced with permission from Schulz (*119*).

reactor with large internal recycle. This reactor should afford an ideal situation for subsequent reactions of primary products, such as the water–gas shift and hydrogenation and incorporation of olefins. Nevertheless, the water–gas shift proceeded very slowly on Co and Ru. Table 3 gives catalyst compositions, operating conditions, and overall selectivity data, and Figs. 2, 3, and 4 present carbon-number distributions on a weight basis for hydrocarbons. The chain branching in Table 3 for Co, 6–13%, is low compared with data in Table 2, 40% for Co at C_{10}. Maxima in the carbon-number distributions occur at carbon numbers of 6 and 7, 5 and 6, and 3 for Ru, Co, and Fe, respectively. For Ru and Co, the ratio of C_3 olefins to paraffins and the average molecular weight decreased and the branching at C_{10} increased with increasing H_2/CO feed ratio when other conditions were constant. Olefins were the major product from Fe, and some oxygenated molecules were also produced.

Schulz (*118*) provided the only detailed analyses of products from a Ni catalyst. The monomethyl species were largely 2- and 3-methyl, and the proportion of these and other methyl isomers in carbon-number fractions fell off sharply with increasing carbon number (Fig. 5).

Table 3
Product Distributions in a Recycle Reactor[a]

A. 0.5% $Ru–Al_2O_3$ at 235°C					
H_2/CO Ratio at outlet	1.2	1.8	1.8	2.0	
Inlet flow, liters/g·h	8.2	8.2	19.9	8.2	
Pressure, atm	8	8	30	8	
Weight percentage					
C_1	11	16	6	16	
$C_2–C_4$	12	17	5	12	
$C_5–C_{11}$	44	43	59	44	
$C_{12}–C_{18}$	25	19	15	21	
C_3H_6/C_3H_8	1.7	1.0	0.1	0.8	
Percentage branching, C_{10}	3.8	4.4	5.6	4.7	
B. 14% Co–1% $La_2O_3–Al_2O_3$ at 215°C					
H_2/CO Ratio at outlet	0.9	0.9	1.3	2.0	2.0
Inlet flow, liter/g·h	0.48	0.48	0.48	0.48	0.48
Pressure, atm	5.4	8.2	6.8	8.2	6.1
Weight percentage					
C_1	13	11	26	37	35
$C_2–C_4$	10	8	15	18	17
$C_5–C_{11}$	40	36	36	32	35
$C_{12}–C_{18}$	21	28	15	8	8
C_3H_6/C_3H_8	1.0	1.7	0.3	0.1	0.1
Percentage branching, C_{10}	13	6	11	8	13

C. K- and Cr-Promoted iron at 260°C, 7.9 atm
H_2/CO Ratio at outlet, 1.7
Inlet flow, 0.2 liter/g·h
CO Conversion, 55%
Carbon-number distribution, wt %

C_1	11
$C_2–C_4$	29
$C_5–C_{11}$	36
$C_{12}–C_{18}$	15

$C_2H_4/C_2H_6 = 1.8$
$C_3H_6/C_3H_8 = 3.6$

[a] Adapted from Pannell *et al.* (*103*).

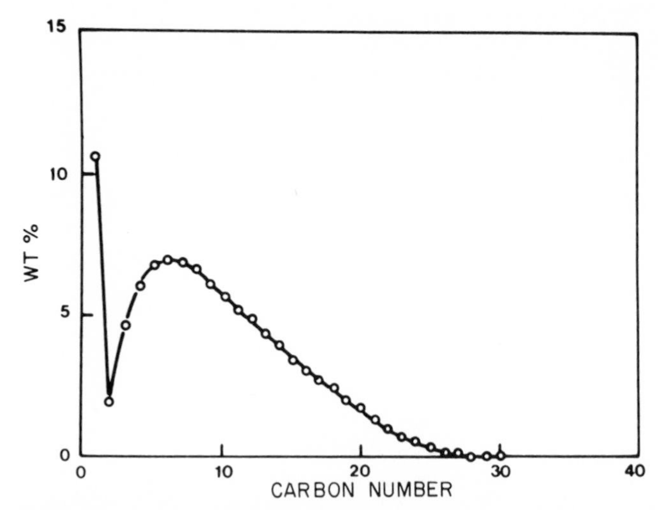

Fig. 2 Hydrocarbons from 0.5% Ru on alumina at 235°C in a recycle reactor with 1.2 H_2 + 1 CO feed at 8 atm and a conversion of CO of 53%. Reproduced with permission from Pannell *et al.* (*103*).

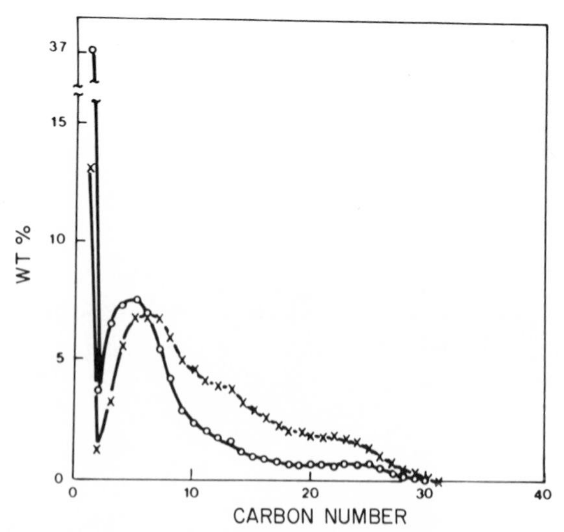

Fig. 3 Hydrocarbons from 14% Co–1% La_2O_3 on alumina at 215°C in a recycle reactor: ○, 8.2 atm of 2 H_2 + 1 CO; X, 5.4 atm of 0.9 H_2 + 1 CO. Reproduced with permission from Pannell *et al.* (*103*).

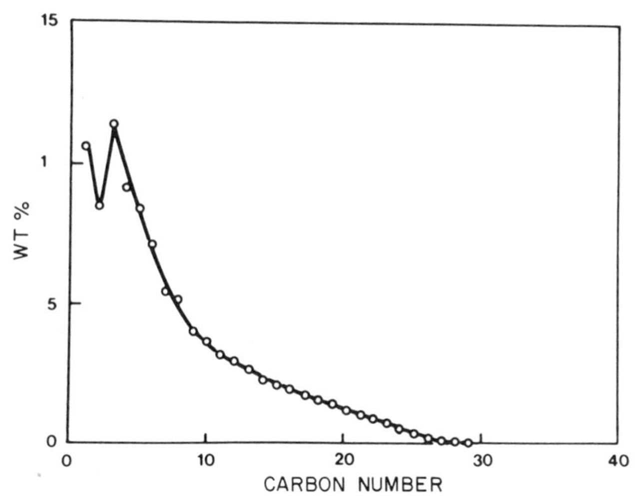

Fig. 4 Hydrocarbons from K- and Cr-promoted Fe at 260°C, 7.9 atm of 1.7 H_2 + 1 CO, in a recycle reactor. Reproduced with permission from Pannell *et al.* (*103*).

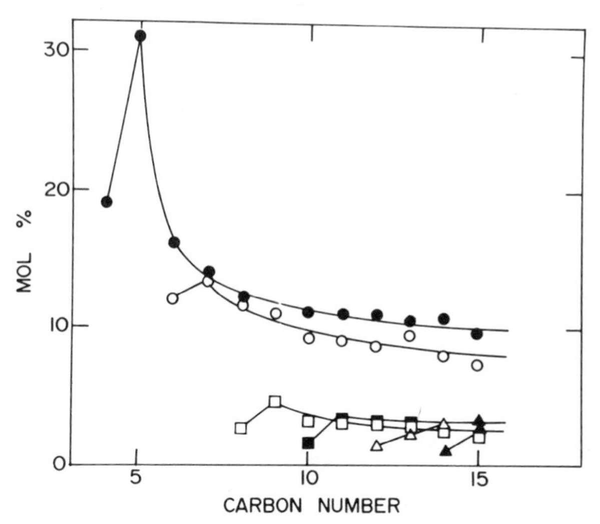

Fig. 5 Methyl (●, 2-; ○, 3-; □, 4-; ■, 5-; △, 6-; ▲, 7-) branched hydrocarbons in carbon-number fractions from the atmospheric-pressure synthesis on Ni. Reproduced with permission from Schulz (*118*).

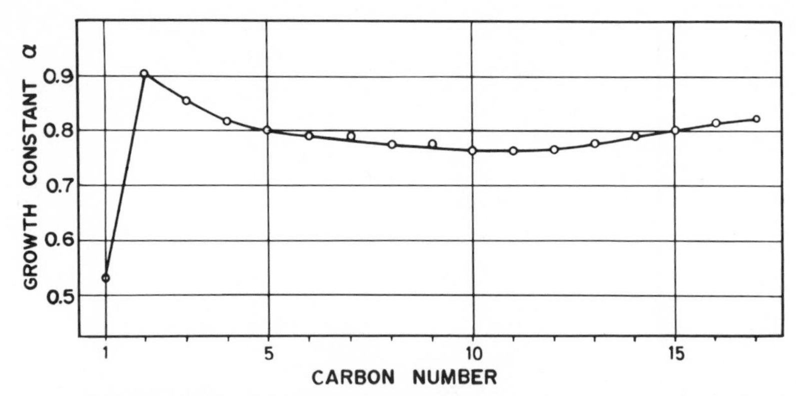

Fig. 6 Chain-growth constant α from Eq. (1) as a function of carbon number for products from the atmospheric pressure synthesis on Co–ThO_2–kieselguhr at 190°C. Reproduced with permission from Pichler *et al.* (*106*).

5.3 Prediction of Carbon-Number Distributions

Most studies of the growth of the carbon chain assume that carbon atoms are added to the growing chain one at a time. In 1946, Herington (*70*) defined a probability β_n that the molecule of carbon number n desorbs from the catalyst rather than grows as

$$\beta_n = \phi_n \Big/ \sum_{n+1}^{\infty} \phi_i = (1 - \alpha_n)/\alpha_n, \tag{1}$$

where ϕ_n is the number moles of carbon number n and α the probability of chain growth. Equation (1) provides an unambiguous description of the chain-growth process, and an analysis of this type was made by Pichler *et al.* (*106*) for the carbon-number distribution from a Co catalyst at 1 atm (Fig. 6).

Friedel and Anderson (*55*) showed that if parameter β or α remains constant over a range of carbon number, in this range $\phi_{n+1}/\phi_n = \alpha$ and

$$\phi_n = \phi_i \alpha^{n-1}. \tag{2}$$

Equation (2) may be derived from a reaction sequence involving one-carbon additions where A_n and G_n are adsorbed and gaseous species

$$\begin{array}{ccccccc} & G_n & & G_{n+1} & & G_{n+2} & \\ & {\scriptstyle k_3}\nearrow & & {\scriptstyle k_3}\nearrow & & {\scriptstyle k_3}\nearrow & \\ \longrightarrow & A_n & \xrightarrow{k_2} & A_{n+1} & \xrightarrow{k_2} & A_{n+2} & \xrightarrow{k_2} \end{array}$$

containing n carbons. For steady-state conditions, a balance at A_{n+1} gives $k_2A_n = (k_2 + k_3)A_{n+1}$ and

$$A_{n+1}/A_n = \phi_{n+1}/\phi_n = k_2/(k_2 + k_3) = \alpha, \tag{3}$$

where k_2 and k_3 are the propagation and termination rate constants. Plots of Eq. 2 gives α, which is related to the ratio k_2/k_3. In passing, we note that if chain propagation stops at, say $n + 1$, the yield of the $n + 1$ species is increased according to

$$\phi_{n+1}/\phi_n = \alpha/(1 - \alpha). \tag{4}$$

Thus, for $\alpha = 0.8$, ϕ_{n+1} is five times larger than it would be if propagation continued to $n + 2$. We also note that the maximum in the carbon-number distribution on a weight basis is given by $n = 1/\ln(1/\alpha)$, where n is the carbon number of the maximum.

Manes (*92*) summed Eq. (2) over groups of carbon numbers and showed how yields of gasoline, diesel oil, and wax changed with α. The diesel-oil fraction was always in the range 20–30 wt % of the total of gasoline + diesel oil + wax, relatively independent of the value of α. If Eq. (2) holds over all carbon numbers, the equation can be rearranged in terms of either the mole or weight fractions of the total product. These forms have no substantial advantage over Eq. (2).

Equation (2) was rediscovered in about 1976 and perhaps impertinently given the name Schulz–Flory equation, honoring the famous polymer chemists (*6, 7, 66, 67, 90, 91*). A more appropriate name might have been the "Bureau of Mines equation." Workers currently call it the Flory equation.

Dautzenberg *et al.* (*35*) examined transients in the chain-growth process at 210°C and 10 atm on Ru on alumina. Synthesis gas, 1 H_2 + 1 CO, was passed over the freshly H_2-treated catalyst for short periods (4–12 min), then H_2 was passed over the catalyst and the temperature increased to 350°C. The temperature was then decreased rapidly to 210°C and the cycle repeated. Products were accumulated from many cycles to obtain sufficient hydrocarbon for analyses.

Equations were developed for the carbon-number distribution in the non–steady-state situation. During the synthesis periods, chains are initiated continuously and grow continuously. The result of the short synthesis time is to limit sharply the amount of high molecular weight molecules (Fig. 7): the carbon chain grows ~1 unit/min. The agreement between experiment and theory is good. In this plot, $x = (k_2 + k_3)t$, where k_2 and k_3 are the rates of propagation and termination and t the time from the start of the synthesis period. For two different catalysts, k_2 was 1.4–1.6 × 10^{-2} s^{-1}, and k_3, 7.3–8.3 × 10^{-4}. These constants in Eq. (3) give $\alpha = 0.95$, the value obtained from carbon-number distributions of steady-state experiments.

Nijs and Jacobs (*98*) studied changes in selectivity in the startup period

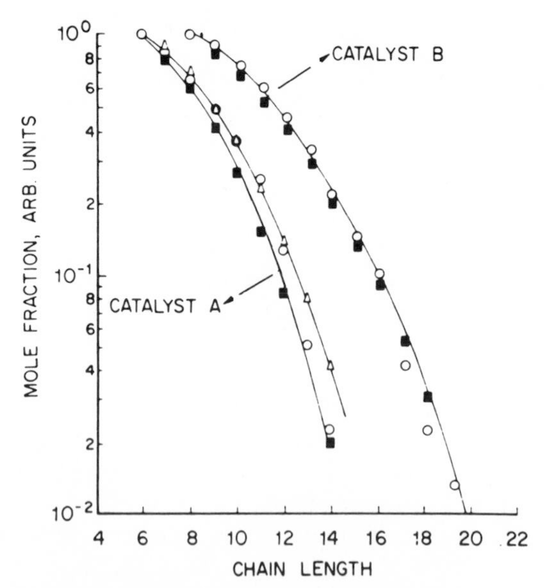

Fig. 7 Carbon-number distributions of hydrocarbons from Ru on alumina for short operating periods. Catalyst A: 8-min pulse; ○, experiment; ■, theory ($x = 7$, $\alpha = 0.95$); △, theory ($x = 8$, $\alpha = 0.95$). Catalyst B: 12-min pulse; ○, experiment; ■, theory ($x = 12$, $\alpha = 0.90$). Reproduced with permission from Dautzenberg *et al.* (*35*).

for supported Ru. The first products were largely CH_4; it was postulated that the catalyst was being "carbided" in this period. More and more larger molecules were produced, and eventually the steady-state selectivity with $\alpha = 0.9$ was obtained.

A number of papers suggest that the carbon-number distribution is related to the pore size of the catalyst or to the size of the metal crystallites, particularly, that the length of the growing chain is limited by the dimensions of the pores or the metal crystallites. Vanhove *et al.* (*130*) prepared 2% Co on aluminas with different pore sizes by impregnating them with solutions of $Co_2(CO)_8$. Products from the sample with a mean pore radius of 300 Å had a maximum in the carbon-number distribution at C_{16}, and another preparation with a mean radius of 65 Å had a maximum at C_3. Catalysts with larger amounts of Co (5%) gave normal carbon-number distributions.

Nijs *et al.* (*100*) performed synthesis tests with 1.5 H_2 + 1 CO at 14 atm and 525 K on 2% Ru on Y-zeolite catalysts. With Ru–NaY, metal particles did not exceed 40 Å, and the distribution of products was cut off sharply at C_{10}. A Ru–LaY catalyst had smaller metal particles; its products did not exceed C_5 or C_6. Nijs *et al.* (*99*) introduced Ru into Y-zeolite

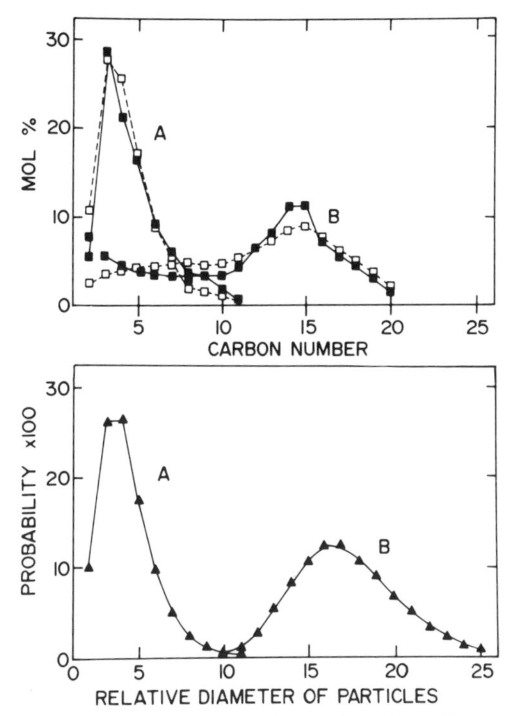

Fig. 8 (Top) Carbon-number distributions for products from Co on alumina with solid points showing experimental and open points calculated data. (Bottom) Metal-particle-size distributions required to fit the data in the top part. Reproduced with permission from Nijs and Jacobs (*97*).

cages, 2.8% Ru in NaY. Growth constants by the Herington method [Eq. (1)] for the products from this catalyst decreased sharply with carbon number. Ruthenium on SiO_2 gave a constant growth parameter.

Nijs and Jacobs (*97*) postulated that the simple chain-growth scheme occurs, but that the chains are terminated at some carbon chain length that is proportional to the size of the metal crystallites in the catalyst. A distribution of metal-particle sizes in the catalyst is postulated, the particle sizes being represented by a log-normal distribution. Typical experimental and calculated carbon-number distributions and the postulated distribution of crystallite diameters are given in Fig. 8. The same growth probability, $\alpha = 0.89$, was used for fitting curves A and B, and the relative mean particle size and its standard deviation were 4 and 1.44 for curve A and 17 and 2.10 for curve B. It was shown earlier that stopping the growing chain at a particular carbon number increases the amount of the last species by a factor of $1/(1 - \alpha)$ compared with the yield expected if chain growth continued. This sudden increase at the end of each of the individual carbon-number distribution curves apparently is smoothed out in calculating the composite carbon-number distribution.

Novak *et al.* (*101*) showed that the incorporation of α-olefins produced in the FTS in a continuously stirred tank-reactor situation led to drastic changes of carbon-number distribution in some cases. Incorporation of olefins is considered in the next section as the source of branched-chain molecules. In addition, values of α that change with bed length in a plug-flow reactor also made significant modifications of carbon-number distribution. Similar problems have been investigated by Collis and Schwarz (*31*).

5.4 Origin and Prediction of Branched Species

The early U.S. Bureau of Mines work on the prediction of isomer distribution postulated reaction steps for the formation of straight-chain and branched species as an integral part of the reaction sequence. The possibility of branched molecules being produced from straight chains by isomerization on acidic components of the catalyst was not considered seriously at that time, because acidic catalysts were not known then. More recent work with Ru on acidic supports suggests that isomerization is possible (cf. reference *77* and Table 8 of Chapter 4).

Another hypothesis has been proposed: the main sequence of chain growth leads only to straight chains, and methyl-substituted species are produced by the incorporation of propylene into the growing chain (*69, 121*). Therefore, this section includes a brief survey of data on the incorporation of olefins. Interactions of olefins plus synthesis gas over Co catalysts illustrate the variety of reactions that can occur depending on reaction conditions and the type of catalyst:

1. The olefins may initiate chains or may become part of a growing chain to produce products typical of the FTS.
2. Ethylene in the presence of H_2 + CO polymerizes to form a "polyethylene." In some instances, the polymerization continues when the CO is removed from the feed.
3. Hydroformylation reactions may occur, producing for example from ethylene, propionic aldehyde as the primary product.

Pichler and Schulz *et al.* (*105, 106, 121*) introduced ^{14}C-tagged olefins and a paraffin in small amounts to the feed to a standard Co–ThO_2–kieselguhr catalyst at 1 atm and 185–190°C, tagged ethylene and propylene to a precipitated-Fe catalyst at 20 atm and 220°C, and tagged ethylene to a fused-Fe catalyst at 20 atm and 320°C (Table 4). On Co, nearly all of the olefins reacted, and ~30% of the ethylene and propylene were incor-

Table 4
Incorporation of Olefins and Paraffins in the Fischer–Tropsch Synthesis[a]

Tracer	Tracer in synthesis gas, vol %	Tracer, % of ^{14}C reacted: Hydrogenated	Incorporated	Hydrogenolysis: Total	Hydrogenolysis: To CH_4	Total reacted
A. Co–ThO_2–kieselguhr, 2 H_2 + 1 CO, SVH = 75,[b] 1 atm, 185–190°C						
[^{14}C]Ethylene	0.30	66.7	29.0	4.3	4.3	>99.95
[1-^{14}C]Propylene	0.29	50.9	31.1	11.7	9.2	93.7
[2-^{14}C]Propylene	0.78	52.0	31.3	9.8	6.4	93.1
n-[1-^{14}C]Hexadecene-1	0.1	79.4	6.3	14.1	3.7	99.8
n-[1-^{14}C]Butane	0.29	—	<0.5	<0.1	<0.1	<1
2-[15-^{14}]Methylpentadecane	0.23	—	0.25	0.45	0.15	0.7
B. Alkalized precipitated iron, 2 H_2 + 1 CO, SVH = 80,[b] 20 atm, 220°C						
[^{14}C]Ethylene	0.21	67.4	9.1	0.1	0.1	76.6
[1-^{14}C]Propylene	0.30	42.2	1.1	0.6	0.06[c]	44.1
C. Alkalized fused iron, fixed bed, SVH = 450,[b] 20 atm, 320°C						
[^{14}C]Ethylene	0.34	65.0	11.5	0.6	0.6	77.1

[a] Adapted from Pichler and Schulz (*105*).
[b] Hourly space velocity.
[c] Probably should be 0.6 (R. B. Anderson).

porated. Products of this reaction were hydrogenated and were then separated by capillary gas chromatography and the radioactivity of the fractions determined.

Figure 9 gives plots of radioactivity per mole as a function of carbon number for straight-chain and branched species from the incorporation of C_2 and C_3 olefins. For C_2 olefins, the radioactivity was larger for even- than for odd-carbon-number molecules for both straight and branched

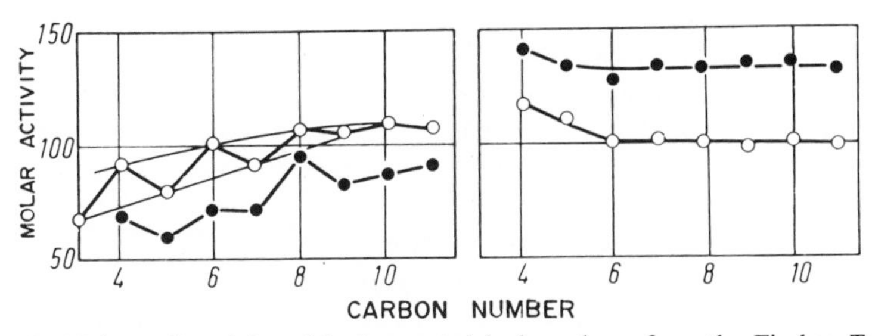

Fig. 9 Molar radioactivity of hydrogenated hydrocarbons from the Fischer–Tropsch synthesis on Co at 1 atm as a function of carbon number. [^{14}C]Ethylene (left) and [1-^{14}C]propylene (right) added to synthesis gas. ○, *n*-Paraffins; ●, monomethyl paraffins. The molar radioactivity of hexane was taken as 100. Reproduced with permission from Pichler and Schulz (*105*).

hydrocarbons. A simple explanation is that part of the even-carbon-number molecules is produced only by the polymerization of ethylene. The odd–even variations are superimposed on a general increase in radioactivity with carbon number. On this basis, the ethylene was postulated to initiate chains and to build into growing chains, as well as to polymerize.

For propylene addition, the radioactivity was independent of carbon number, and the radioactivity for branched molecules was larger than for straight chains. The curves for the product from the incorporation of propylene do not have maxima at intervals of 3. The constancy of the radioactivity as a function of carbon number suggests that the propylene predominately initiated chains and did not appreciably build into growing chains. Only 6% of the hexadecene was incorporated; the hexadecene had a relative radioactivity of 291,000, compared with ~6000 for C_{17}–C_{20}. The C_1–C_{13} hydrocarbons had counts of less than 300; only 14% of the C_{16} was split to smaller molecules.

On the Fe catalysts (Table 4B,C), part of the added olefins did not react, ~10% of the ethylene was incorporated, and in the low-temperature test only 1% of the propylene was built in. Less than 1% of the added olefins was split. On the Co catalyst (Table 4A), less than 1% of the added tagged *n*-butane and 2-methylpentadecane was found to react, and less than 0.5% was incorporated into synthesis products. Earlier work by Kölbel *et al.* (*82*) showed that waxes are not hydrocracked appreciably to CH_4 during the FTS. Higher hydrocarbons containing ^{14}C were produced on the catalyst using H_2 + ^{14}CO feed. The adsorbed hydrocarbons were extracted with toluene, separated, and their radioactivity measured. The hydrocarbons were then readsorbed on the catalyst and the FTS resumed with H_2 + ^{12}CO feed. The $^{14}CH_4$ produced is a measure of the CH_4 produced by cracking the adsorbed higher hydrocarbons. For supported Co and Ni with 2 H_2 + 1 CO gas at 1 atm at temperatures less than 200°C, ~3% of the total CH_4 was produced by cracking larger hydrocarbons. On Co, a larger amount of CH_4 was produced with 3 H_2 + 1 CO than with 2 H_2 + 1 CO feed. For a precipitated Fe containing MgO, CaO, K_2CO_3, and kieselguhr, the CH_4 from cracking was less than 2.7% of the total CH_4 at temperatures up to 285°C. These papers (*82, 105*) indicate that the products are not modified substantially by hydrogenolysis reactions.

In other studies of the standard Co–ThO_2–kieselguhr (*106*) at 1 atm, introducing 1% propylene in the synthesis gas caused no discernible change in the products. With 10% propylene in the feed, about half of it reacted and half of this fraction was hydrogenated to propane. Monomethyl isomers were increased by a factor of 1.3 over a feed without the 10% propylene. Isobutylene, 10% in synthesis gas, had a similar effect in increasing methyl-branched molecules; however, the average molecular weight of the product was also increased.

On the standard Co catalyst, changing the space velocity or conversion alters the selectivity drastically (*106*), including the carbon-number distribution, the amount of total olefins and α-olefins in carbon-number fractions, and the quantity of branched molecules (Fig. 10 and Table 5). In the results shown in Table 5, the proportions of branched hydrocarbons were determined at hourly space velocities of 78, 2380, and 76; increasing space velocity decreased the fraction of branched hydrocarbons by a significant amount. The author has made calculations showing that the mole percentage of propylene at the end of the reactor was ~0.86 and ~0.06 for the low and high space velocities. In the present experiments, the relatively small effects of space velocity on branching compared with the large changes in overall product distribution do not lend credence to the idea that all or even a sizable fraction of methyl branching results from propylene incorporation, and other studies by Pichler (*105*) confirm this notion. Iron catalysts do not incorporate propylene in appreciable amounts, and products from Fe often have a large amount of branching. Compared with Co catalysts, the selectivity of Fe is remarkably insensitive to changes in space velocity, conversion, feed-gas composition, and temperature, as discussed later in this section. We shall return to the origin of branched molecules at the end of this section.

The author's simple chain-growth (SCG) scheme (*124*), developed about 1950, is one of the more useful ways of representing isomer and carbon-number distributions. Here, branching is a part of the primary growth pattern represented by the reaction network shown. Carbon atoms are added one at a time at the end or penultimate carbons at one end

$$\begin{array}{lllllll}
\overset{*}{C}\overset{*}{C} \rightarrow & C\overset{*}{C}\overset{*}{C} \rightarrow & CC\overset{*}{C}\overset{*}{C} \rightarrow & CCC\overset{*}{C}\overset{*}{C} \rightarrow \\
 & & \searrow & \searrow \\
 & & & CC\underset{C}{C}\overset{*}{C} \rightarrow \\
 & \searrow & & \\
 & C\underset{C}{C}\overset{*}{C} \rightarrow & C\underset{C}{C}\overset{*}{C}\overset{*}{C} \rightarrow & \searrow
\end{array} \tag{5}$$

of the molecule, and addition at the penultimate carbon is not permitted if a carbon has already been added to it. For the steady-state situation, the methods described in Section 5.3 give

$$\phi_{n+1}/\phi_n = \alpha/(\alpha + \beta + \gamma) = a \tag{6}$$

for addition to an end carbon, and for addition at the penultimate carbon,

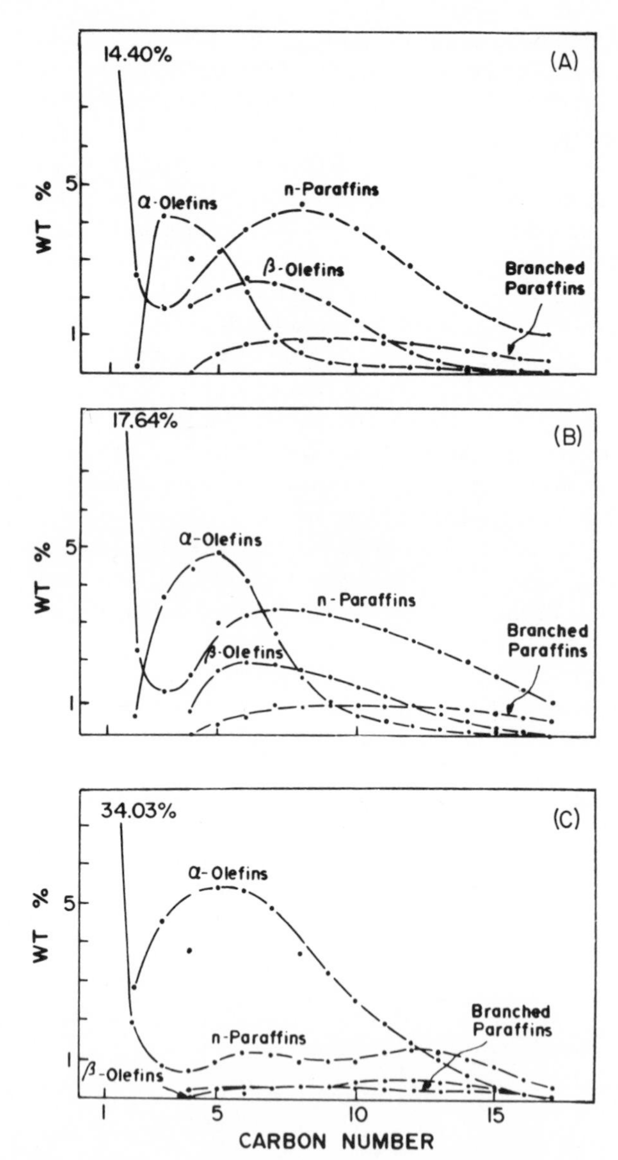

Fig. 10 Effect of hourly space velocity (A, 78; B, 337; C, 2380) on products from the Fischer–Tropsch synthesis on Co–ThO_2–kieselguhr at 1 atm and temperatures from 197 to 210°C. Reproduced with permission from Pichler *et al.* (*106*).

Table 5
Branched Hydrocarbons from a Cobalt Catalyst as a Function of Space Velocity[a]

Hourly space velocity	78	2380	76
Temperature, °C	197	210	207
CO Conversion, %	84.6	10.1	68.7
Carbon chain	*Percentage in carbon-number fraction*		
2-Methylpentane	7.9	6.8	7.8
3-Methylpentane	4.8	4.5	4.8
2-Methylhexane	6.9	4.8	6.2
3-Methylhexane	8.3	6.9	8.5
2-Methylheptane	5.5	4.1	5.1
3-Methylheptane	7.1	5.6	6.7
4-Methylheptane	3.3	2.9	3.9
2-Methylundecane	4.8	3.9	4.9
3-Methylundecane	5.7	4.4	5.8
4-Methylundecane	5.6	3.9	5.7
5-Methylundecane	5.1	3.8	5.4
6-Methylundecane	2.5	1.4	2.4

[a] Adapted from Pichler *et al.* (*106*).

$$\phi'_{n+1}/\phi_n = \beta/(\alpha + \gamma) = af, \tag{7}$$

where α is the parameter for growth at the end carbon, β the constant for the penultimate addition, γ the parameter for desorption, and ϕ_i the rate of production in moles per unit time of a molecule with a carbon number i. If it is assumed that the parameters α, β, and γ have constant values or that certain ratios of these quantities are constant, the ratio of a branched isomer to the straight-chain molecule of the same carbon number is f or $2f$, and for dimethyls, f^2 or $2f^2$, and the carbon-number distribution is given by

Carbon number	*Relative rate of appearance*
2	$\frac{1}{2}$
3	a
4	$a^2(1 + f)$
5	$a^3(1 + 2f)$
6	$a^4(1 + 3f + f^2)$
7	$a^5(1 + 4f + 3f^2)$
8	$a^6(1 + 5f + 6f^2 + f^3)$
n	$a^{n-2}[1 + (n - 3)f + \frac{(n-4)(n-5)f^2}{2} + \cdots]$

(8)

Table 6
Carbon Number of the Maximum of Eq. (10)

	a					
f	*0.7*	*0.75*	*0.80*	*0.85*	*0.9*	*0.95*
0	(3.00)	3.48	4.48	6.15	9.49	19.50
0.04	3.12	3.97	5.32	7.81	13.69	37.35
0.08	3.36	4.36	5.99	9.10	16.54	43.59
0.12	3.48	4.55	6.33	9.74	17.76	45.68
0.16	3.50	4.60	6.45	10.00	18.33	46.67
0.20	3.46	4.57	6.45	10.10	18.62	47.24
0.24	3.38	4.49	6.40	10.11	18.78	47.61
0.28	3.30	4.41	6.32	10.90	18.88	47.86
0.32	3.22	4.31	6.24	10.06	18.94	48.05

The general term is valid for $n \geq 4$, and for $n \geq 8$, a term involving f^3 is required to account for trimethyl molecules. The $\frac{1}{2}$ for C_2 suggests that the C_2 intermediate grows twice as fast as other intermediates,

$$\phi_n = \phi_3 F(n) a^{n-3},$$

or

$$\ln \phi_n / F(n) = \ln \phi_3 + (n - 3) \ln a, \tag{9}$$

where $F(n) = [1 + (n - 3)f + (n - 4)(n - 5)f^2/2 + \cdots]$ with the value of f determined from the isomer distribution may be used to evaluate a. The maximum of the carbon-number-distribution equation plotted on a weight basis,

$$W_n = W_3 n F(n) a^{n-3}/3, \tag{10}$$

where W_n is the weight of the fraction with carbon number n, is given in Table 6 for selected values of a and f. The maximum occurs at larger carbon numbers as the value of f increases at a constant value of a. For lower values of a, the maximum occurs at $n = 3$.

Data of Achtsnit (*2*) for the synthesis at 1 atm on Co–ThO_2–kieselguhr will be used to illustrate the SCG equations. Detailed product distributions for hydrogenated hydrocarbons are presented, showing the effects of temperature and composition of feed gas. Monomethyl but not dimethyl molecules were separated. From the ratios of monomethyl to normal species, the values of f were calculated and average values taken, and the experimental ratios are compared with calculated values in Table 7. For tests 3 and 5, only a small amount of hydrocarbons were present in C_9 and larger carbon-number fractions. The agreement between calcu-

Table 7
Calculated and Experimental Isomer Distributions from Cobalt Catalysts

Experiment	1	2	3	4	5	6
Temperature, °C	187	187	187	171	208	190
H_2/CO Ratio in feed	1	2	4	4	2	2
Ratios of branched to normal molecules						
Calculated	0.042	0.082	0.132	0.021	0.222	0.051
2M3	0.020	0.063	0.139	0.010	0.079	0.127
3M5	0.048	0.118	0.181	0.033	0.300	0.071
4M7	0.061	0.105	0.128	0.028	0.318	0.071
Calculated	0.083	0.164	0.264	0.043	0.444	0.102
2M4	0.092	0.193	0.321	0.055	0.479	0.096
2M5	0.088	0.211	0.309	0.051	0.525	0.125
2M6	0.086	0.194	0.242	0.047	0.541	0.111
3M6	0.114	0.225	0.305	0.058	0.625	0.140
2M7	0.072	0.138	0.148	0.035	0.367	0.082
3M7	0.117	0.215	0.235	0.049	0.649	0.150
3M8	0.106	0.155	0.219	0.045	0.585	0.167
2M9	0.091	0.182	[a]	0.049	0.426	0.129
3M9	0.092	0.172		0.050	0.412	0.118
2M10	0.090	0.157		0.046	[a]	0.112
3M10	0.097	0.156		0.046		0.112
4M10	0.082	0.145		0.053		0.089
5M10	0.073	0.132		0.032		0.079
2M11	0.085	0.140		0.045		0.103
3M11	0.095	0.149		0.046		0.106
4M11	0.086	0.137		0.041		0.095
2M12	0.079	0.131		0.041		0.102
3M12	0.084	0.143		0.042		0.105
4M12	0.087	0.146		0.041		0.094
Calculated	0.125	0.245	0.396	0.064	0.666	0.154
4M + 5M9	0.105	0.205	[a]	0.049	0.405	0.126
5M + 6M11	0.110	0.189		0.048	[a]	0.119
Calculated	0.167	0.327	0.528	0.086	0.888	0.205
2M + 4M8	0.165	0.312	0.323	0.074	0.768	0.257
5M + 6M12	0.138	0.238	[a]	0.064	[a]	0.165

[a] The amount of hydrocarbons above C_8 was very small.

lated and experimental isomer distributions is only fair, and the SCG seems to be only a first approximation to predicting isomer distribution for these data.

The values of f from Table 7 were then used in Eq. (9) as shown by linear plots in Fig. 11. The points for CH_4 lie above the line, and those for C_2 below the line. Doubling the value for C_2 usually places it near the line. For $n > 2$, Eq. (9) represents the carbon-number distribution reasonably

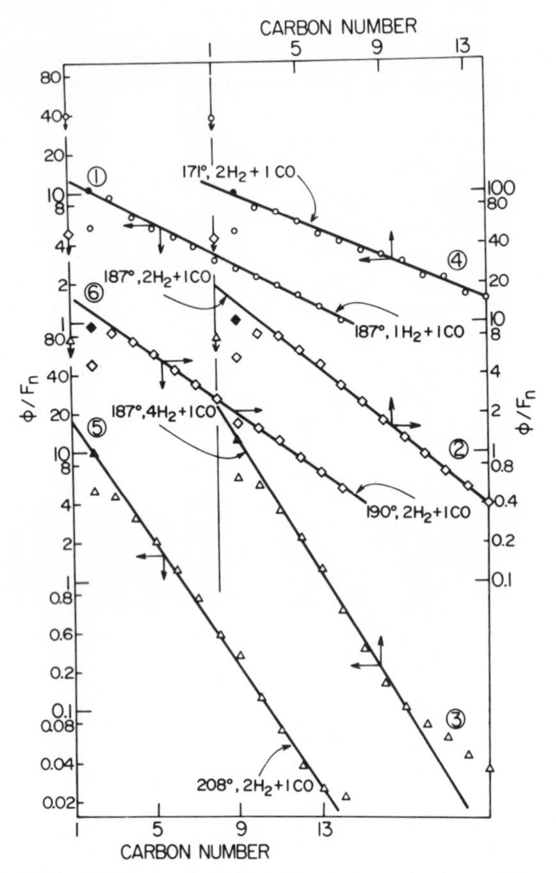

Fig. 11 Plots of Eq. (9) for products from Co at atmospheric pressure using values of *f* from Table 8. Temperatures in degrees Celsius.

well. Table 8 gives the values of $a, f,$ and the ratios β/α and γ/α from Eqs. (6) and (7). The ratios β/α and γ/α change in the same way as temperature or H_2/CO in the feed gas is changed, both ratios increasing with increasing temperature or increasing H_2/CO. Reasonable Arrhenius plots were obtained for γ/α and β/α, yielding activation energies of 20.8 and 15.7 kcal/mol, respectively. The principal constant for chain growth α may be expected to increase with increasing concentration of CO, and γ for termination, with increasing concentration of H_2; hence, γ/α may be a linear function of H_2/CO (Fig. 12). A similar pattern holds for β/α; in some schemes, branched molecules are produced from half-hydrogenated intermediates.

Table 8
Simple Chain-Growth Scheme Applied to Distributions from Cobalt[a]

Experiment	*Temp., °C*	*H_2/CO Ratio in feed*	*Hourly space velocity*	*Constants of simple chain-growth scheme*				*Sum of branched molecules divided by C_3*
				a	*f*	*β/α*	*γ/α*	
1	187	1	89	0.827	0.042	0.040	0.156	0.678
2	187	2	85	0.742	0.082	0.077	0.283	1.078
3	187	4	88	0.551	0.132	0.123	0.647	0.469
4	171	2	87	0.857	0.021	0.021	0.139	0.510
5	208	2	80	0.576	0.222	0.197	0.551	0.906
6	190	2	251	0.771	0.051	0.049	0.246	0.685

[a] From Achtsnit (*2*).

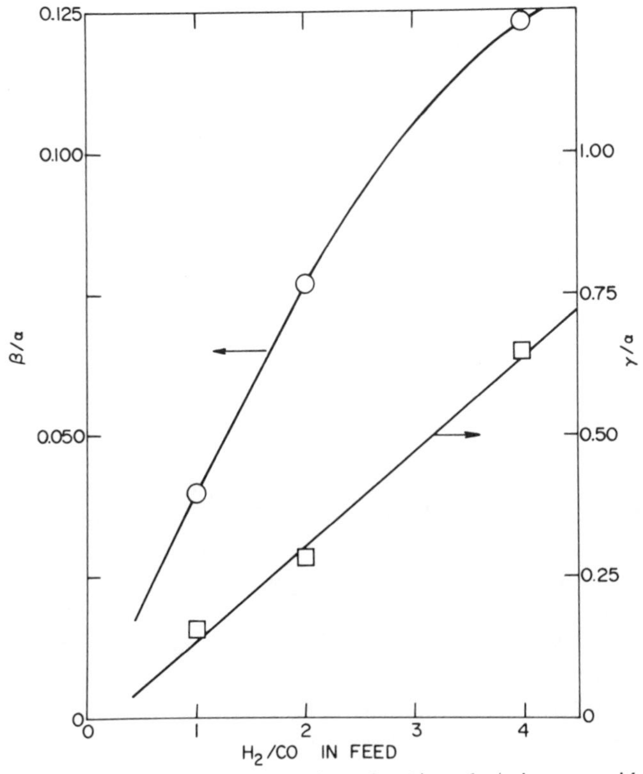

Fig. 12 For products from Co at 1 atm, the ratios β/α and γ/α increase with increasing H_2/CO of the feed.

Satterfield and Huff (*117*) obtained carbon-number distributions for products from a reduced, fused Fe in a stirred slurry reactor at 7.9 atm. The product was essentially straight chain ($f = 0.014$). In applying Eq. (9) to carbon-number distribution, the value of $F(n)$ was taken as 1, and good linear plots were obtained even in the C_1–C_3 range. In these plots the sum of moles of hydrocarbon plus oxygenates is used. The plots and the values of a are remarkably constant for temperatures from 234 to 269°C and wide changes in feed composition and conversion. At 234, 249, and 269°C, the values of a were 0.71, 0.68, and 0.67. This result for Fe should be compared with the data for Co in Table 8, where a varies from 0.58 to 0.83 and f from 0.020 to 0.222.

The SCG scheme has represented selectivity data for fixed-bed Co and fluidized and entrained Fe as good or better than the examples given in this section. The enolic growth scheme (Table IIIB) is appropriate for SCG. Two groups of data that are not properly predicted by the SCG equations are for fixed-bed Ni (*118*) in Fig. 5 and fixed-bed Fe (*107*) in Table 2; branching does not increase with carbon number, as required by the SCG equations. Schulz *et al.* (*121a*) have reported other isomer distributions from Fe, Co, and Ni catalysts that the SCG equations do not predict correctly. For products from Fe and Co, however, the isomer distribution can usually be represented accurately if the value of f is permitted to change, generally decreasing, as the point of growth moves down the growing chain, but with the value of a held constant (*5, 89, 121a*). The first branched species of a given type can be produced in only one way, but larger molecules of this type are made in two ways. For example, isobutane is produced by adding a carbon to the second carbon from the initial end of the chain, but 2-methylhexane is composed of two molecules with carbon addition at the second carbon in one molecule and at the fifth carbon in the other molecule. Individual values of f are assigned to each carbon.

The data of Achtsnit (*2*) have been used for considering further the postulate that branched molecules are produced by the incorporation of propylene; at least one propylene must be incorporated for each branched molecule produced. The moles of branched molecules produced were counted from the carbon-number and isomer distributions and divided by the moles of C_3 produced. This ratio, in the last column of Table 8, varies from 0.47 to 1.08. For distributions for which the values of a and f have been determined, Lee and Anderson (*89*) calculated the moles of propylene incorporated per mole of propylene produced based on all of the branched molecules being produced by incorporating propylene, using the summation

$$\sum_{4}^{\infty} a^{n-3}[(n-3)f + (n-4)(n-5)f^2]. \tag{11}$$

For products from Fe and Co, the ratios of C_3 incorporated to C_3 produced varied from 0.6 to 1.9. Thus, the postulate that methyl branches are produced largely or exclusively by the incorporation of propylene leads to absurd results. In addition, it should be noted that the formation of the 2,3-dimethylbutane chain requires the dimerization of two propylenes, and none of the carbon comes from CO.

More complicated chain-growth schemes have been devised by Friedel and Sharkey (*56*), Wojciechowski (*136a*), Taylor and Wojciechowski (*126a*), Anderson and Chan (*8*), and Anderson and Lee (*9*) to improve the agreement between calculated and observed isomer distributions. The first group (*56*) showed that their growth rules also predicted the composition of petroleum fractions. Wojciechowski (*126a, 136a*) developed equations for carbon-number and isomer distributions using an elegant mathematical procedure.

In the McMaster work (*8, 9*), schemes producing ethyl-substituted species were considered, first by one-carbon addition to the chain also at the third carbon from the growing end, and second, the addition of one- and two-carbon units at one end of the growing chain. The calculations were made by a large digital computer with the growing chains represented by a number or vector (11321) denoting the carbon chain of 2-methyl-3-ethylpentane. The computer was given the growth rules and trial values of the constants. The calculation involves developing the carbon chains, comparing calculated and observed distributions, and finally optimizing the values of the constants for the best agreement between predicted and experimental values. In addition to being an interesting academic problem, the calculations by computer are needed for isomer distributions to carbon numbers of 17 and more, because for growth schemes of even modest complexity, the hand calculations become large and tedious and the chances of mistakes are great.[1]

Lee and Anderson (*89*) also devised a chain-growth scheme with constant parameters for (a) addition of carbon atoms one at a time to one end of the growing chain, (b) addition of ethylene at one end of the growing chain, characterized by a parameter b and leading to straight-chain structures, and (c) addition to propylene to one end of the growing chain, characterized by a parameter f and allowing the methyl group in the added propylene to be placed adjacent or one carbon away from the growing chain with equal probability. This scheme was tested for predicting the isomer distribution from fixed-bed Co and entrained Fe processes for C_4–C_8 hydrocarbons. This mechanism predicts the isomer distribution from Co better than the SCG scheme, but it does not predict isomer distribution from Fe accurately.

[1] An improved SCG scheme that predicts ethyl-substituted isomers is described in the Addendum, p. 293.

Finally, to summarize the material presented on the incorporation of propylene as a source of branched molecules, the author's views are that

1. On Fe, propylene does not incorporate to an appreciable extent and cannot be the source of branched carbon chains.
2. On Co, propylene does incorporate, and the incorporated C_3 occurs to a large extent in branched molecules; however, incorporation and kinetic studies suggest that propylene is not an important source of branched species.
3. For both Co and Fe, accounting for all of the branched molecules as being produced from propylene requires the consumption of a ridiculously large amount of propylene.

5.5 Incorporation of Molecules Other than Olefins

5.5.1 Scavenger and Other Tests

In scavenger tests, small amounts of at least moderately reactive molecules that are not products of the FTS are introduced with the synthesis gas. It is hoped that the molecule added will react with the species on the catalyst surface to yield products that clearly delineate the nature of intermediates. The scavenger should not alter the course of the reaction significantly.

Eidus described numerous scavenger tests that suggest methylene as an intermediate. In a typical paper (*48*), cyclopentene was incorporated more rapidly in 2 H_2 + 1 CO on Co–ThO_2–kieselguhr at 1 atm than cyclohexene, which incorporated faster than benzene. Molecules suggesting that methylene groups had been added to the cyclic compounds introduced were found in the products.

Ekerdt and Bell (*13, 49*) added ~2% ethylene or cyclohexene to synthesis gas over Ru on silica. The cyclohexene addition led to production of methyl and ethylcyclohexane, propyl- and butylcyclohexane, and norcarane. Addition of ethylene increased the propylene yield substantially. These studies suggest methylene, alkyls, and alkylidenes as the intermediates in FTS. Baker and Bell (*12*) made similar but more detailed experiments using a precise gas chromatographic–mass spectrometric analytical method. Previous data for cyclohexene were confirmed. Benzene was largely hydrogenated to cyclohexane, but toluene and traces of methylcyclohexane and -hexenes and norcarane were found. Cyclopentene gave mostly cyclopentane but some methylcyclopentane and -pentene and traces of cyclohexane and -hexene. *cis*-2-Butene yielded largely butane and 1-butene with traces of *cis*- and *trans*-dimethylcyclopropane, 2-methyl-2-butene, and ethylcyclopropane. Hence, the intermediates suggested by these experiments include methylene, methyl, and higher alkyl groups. The chain-growth constant α of Eq. (2) decreased when scavengers were added to the feed. The scavenger reacts with the intermediates, removing them from the chain-growth process.

Another method for examining species on the surface, called "chemical trapping," involves addition of a reagent that will react with the adsorbed species on the catalyst surface by metathesis or double decomposition reactions. For example, when CO chemisorbed on Co and Ni catalysts was contacted with methyl iodide, *tert*-butylmethyl ether was produced, suggesting an adsorbed species M_3CO or M_3COM, where M is surface Ni or Co. Similarly, acetone was obtained from CO on Pd on silica, suggesting a bridged chemisorbed molecule (*40, 41, 71*). Ethyl iodide and diethyl or dimethyl sulfates have also been used. In a similar way, formyl was detected on Ni, and alkoxide and carboxylate species on Fe (*40*).

5.5.2 Incorporation of Ammonia and Amines

Ammonia and amines can be incorporated in the FTS to produce amines (*72*). Methylene and alkyls were added to the nitrogen by replacing a hydrogen. Similar reactions also occur in the Kölbel–Engelhardt synthesis.[2]

Nitrogenous bases, ammonia (*85*), primary (*72, 81*) and secondary (*1, 81*) aliphatic amines, and piperidine and pyrollidine (*36, 37*) added to the synthesis gas of the Kölbel–Engelhardt synthesis yielded terminal alkylamines in amounts up to 25 wt % of the total of amines plus hydrocarbons. The alkylation reactions follow the equations

$$3n\,CO + n\,H_2O + (CH_3)_xNH_{3-x} = C_nH_{2n+1}NH_{2-x}(CH_3)_x + 2n\,CO_2 \tag{12}$$

$$3n\,CO + n\,H_2O + HN\bigcirc = C_nH_{2n+1}N\bigcirc + 2n\,CO_2 \tag{13}$$

The direct synthesis of amines was made on Fe catalysts (100 Fe : 0.6 K_2CO_3 : 0.2 Cu) and on the Ruhrchemie Co catalyst in a fixed-bed reactor at conditions similar to those of the Kölbel–Engelhardt synthesis: 200–250°C, 11 atm, and 1.2–6 vol % of the base in the synthesis gas. Important results with Fe catalysts were the following:

1. Ammonia in the synthesis gas leads mainly to the formation of terminal primary *N*-alkylamines. Small quantities of secondary and ter-

[2] The following material was prepared by Professors H. Kölbel and M. Ralek.

tiary terminal amines result from alkylation of the primary *N*-alkylamines in a consecutive process.

2. Introduction of methylamine gives principally secondary terminal methylalkylamines. Adding dimethylamine leads entirely to the formation of dimethylalkylamines.
3. Trimethylamine is not alkylated.
4. The carbon number of the alkyl chains in the amines produced varies from 1 to 26. The average molecular weight of the alkylamines produced decreases considerably with the fraction of the base in the synthesis gas, because small-chain alkylamines form a larger part of the product.
5. Only a fraction of the base reacts to form amines, even in the presence of large amounts of the base, and the limited reaction of the base occurs for low as well as high conversions of CO. The amine portion of the product is small, and under optimum conditions reaches 45 g amines per normal cubic meter CO converted. The other products are hydrocarbons.
6. Alkylation of piperidine and pyrollidine gives *N*-alkyl species with alkyl chains from C_1 to C_{12}, with maxima for the ethyl derivatives. The products are not influenced by the concentration of the base in the feed gas.
7. Pyridine is not alkylated.
8. Adding 17 vol % ethylene to the synthesis gas containing amines leads to only a very small increase in the amount of propylamine. Piperidine and ethylene in the absence of CO do not react on the catalyst.

Thus, the alkylation occurs wholly by the substitution of a hydrogen on the nitrogen of the base by an alkyl group. The alkyl group is formed by the simultaneous FTS. The growth of the hydrocarbon chains terminates with the substitution, and the alkylamine desorbs. Presumably, this substitution could occur with oxygen-containing intermediates as well as with alkyls.

The mechanism of the alkylation reaction seems closely coupled with the mechanism of the FTS. The alkylation can best be explained with reference to the FTS, as is apparent from a comparison of hydrocarbons from the amine synthesis with those of the FTS. The hydrocarbon-product spectra for the amine synthesis and the FTS are similar at the same operation conditions.

These amines are useful as emulsifiers and as chemical intermediates. The alkylation reactions proceeded well in fixed-bed reactors, but not in a slurry reactor in which the conversion of NH_3 was noticeably smaller.

5.6 Mechanisms Based on Analogies to Organometallic Chemistry and the Quest for a Homogeneous Fischer–Tropsch Synthesis[3]

As early as 1940, Pichler (*104*) observed that the rate of the FTS increased with increasing pressure until carbonyls were produced. Incipient carbonyl formation was considered to be a desirable situation. Later, Wender *et al.* (*123, 132, 134*) proposed a sequence of reactions for the growth of straight carbon chains in the FTS, such as given in IVC of Table 1, except that details of the oxygen-removal reactions were not provided.

Analogous reactions involving the insertion of CO in carbon–metal bonds occur in hydroformylation and carbonylation reactions; however, these homogeneous processes never initiate a growing chain and usually only add one carbon to a reactive molecule.

Pichler and Schulz (*105*) and Schulz (*118, 120*) have presented a more detailed version of the CO-insertion scheme (IVC, Table 1), which is also based on reactions from homogeneous organometallic chemistry. In the main chain-growth scheme, oxygen from CO is removed before chain growth and the intermediate is a hydrocarbon radical. Methyl branches are produced in several ways, including by the incorporation of propylene, migration of a methyl radical, and by migration of the point of attachment of a hydrocarbon radical to the metal via a π-bonded intermediate. In producing oxygenated molecules, the oxygen in the final growth step is not eliminated.

The development of a homogeneous liquid-phase FTS has been of moderate interest; possible advantages of a homogeneous process are (a) the ability to adjust the selectivity delicately by changing ligands, etc., (b) sulfur tolerance of homogeneous catalysts, (c) relative ease of removal of heat of reaction, and (d) relatively simple construction of reactors. Disadvantages include (i) difficulty of separating the catalyst from the products and (ii) the fact that homogeneous FTS catalysts have not been found.

Muetterties (*96a*) suggested the use of multinuclear metal clusters as catalysts, these clusters being molecular approximations of a metal surface. Adjacent metal centers could multiply coordinate CO, through both carbon and oxygen, favoring hydrogen transfer to the now-weakened triple bond; the multiple coordination of CO might overcome the apparent instability of metal–formyl complexes. Another idea involved hydrogen

[3] The author is grateful to Dr. Raymond C. McLane for preparing a substantial part of this section.

transfer from a metal hydrido–carbonyl to a coordinated CO, and multiple coordination would not be needed (*52*).

Proving that a given system is homogeneous is not a trivial experimental problem. The metal cluster may decompose to metal under reaction conditions, and proving homogeneity requires tedious experimentation involving filtering, reactor cleaning, and blank experiments. One early "homogeneous" FTS was refuted (*46*). The metal clusters are often unstable at temperatures used for reaction studies, and they decompose to simpler organometallic compounds or to the metal.

Table 9 reviews more recent work on the homogeneous hydrogenation of CO, which can be summarized as follows:

1. True homogeneous CO hydrogenations require severe conditions, usually more severe than the heterogeneous FTS or the methanol synthesis.
2. Unless a Lewis acid cocatalyst is present, the products are oxygenates and not hydrocarbons (*127*).
3. Selectivity is almost exclusively to oxygenates, of which methanol, ethylene glycol, methyl formate, and methyl acetate usually predominate.
4. Selectivity and reactivity can be modified by the presence of other ligands and modifiers, such as phosphines, chelating ethers, and amine additives.

If severe conditions are necessary, the active species will likely be the carbonyl complexes or clusters, regardless of which metal precursor is used. Furthermore, if more recent papers on the stabilities of metal clusters under CO–H_2 atmospheres are representative of typical cluster behavior (*45, 54, 74, 131*), only simple clusters or mononuclear carbonyls will be stable under reaction conditions. Because complicated clusters are usually prepared by CO elimination from adjacent metal centers (*58*), it would appear that the reverse action, cluster decomposition, is inevitable.

The almost total selectivity to oxygenates is striking. Only when a Lewis acid cocatalyst is present is the C—O bond cleaved and hydrocarbons formed. Although the field is too new to make sweeping generalizations, one conclusion seems to be generally accepted: homogeneous systems produce oxygenates whereas heterogeneous systems produce hydrocarbons (*24, 46*). In other words, an unambiguous test for solution homogeneity is the type of product.

Methanol and ethylene glycol are the primary products of homogeneous CO hydrogenations (*50, 53*); secondary products arise mainly from known methanol-homologation reactions (*133*) (mainly to acetic acid, ethanol, and higher alcohols and acids) and subsequent esterification with

Table 9
Homogeneous CO Hydrogenations

Main author	*Ref.*	*Conditions*	*Metal*	*Products*[a]
J. S. Bradley	*24*	225–275°C, 1300 atm	Ru carbonyls[b]	Oxygenates
R. L. Pruett	*110*	200°C, 1000 atm	Rh carbonyls[c]	Oxygenates, glycols
B. D. Dombeck	*45*	200°C, 400 atm	$Ru_3(CO)_{12}$	Oxygenates, glycols
H. M. Feder J. M. Rathke	*52*	200°C, 300 atm	$Co_2(CO)_8$, $Mn_2(CO)_{10}$	Oxygenates
H. M. Feder, J. M. Rathke	*53*	200°C, 200–300 atm	Co carbonyls	Oxygenates, glycols
G. Jenner	*39*	1000–4000 atm	Rh carbonyls	Glycols, oxygenates
W. F. Gresham	*59*	200°C, 3000 atm	$Co(OAc)_2$	Oxygenates, glycols
D. R. Fahey	*50*	225°C, 2600 atm	$Re_2(CO)_{10}$, $Ir_4(CO)_{12}$, $V(CO)_6$, $Cr(CO)_6$, $Mn_2(CO)_6$, $Mo(CO)_6$	Glycols, oxygenates
W. Keim	*74*	230°C, 2000 atm	$Fe_3(CO)_{12}$, $Co_2(CO)_8$, $Ni(acac)_2$, $Ru_3(CO)_{12}$, $Pd(acac)_2$, $Os_3(CO)_{12}$, $Ir_4(CO)_{12}$, H_2PtCl_6	Oxygenates, glycols
G. Wilkinson	*34*	200°C, 200 atm	$Fe(CO)_5$, $Ru_3(CO)_{12}$, $Os_3(CO)_{12}$	Oxygenates
G. Wilkinson	*33*	200°C, 300 atm	$Co_2(CO)_8$[d]	Oxygenates
E. L. Muetterties	*127*	180°C, 1–2 atm	$Ir_4(CO)_{12}$[e]	Hydrocarbons
E. L. Muetterties	*44*	140°C, 2 atm	$Os_3(CO)_{12}$,$Ir_4(CO)_{12}$[e]	Hydrocarbons
S. Olivé G. Henrici-Olivé	*68*	200°C, 120 atm	$W(CO)_6/AlCl_3$ in Benzene[f]	Alkyl benzenes
J. A. Labinger	*137*	25–45°C, 1 atm	Cp_2NbH_3 + $M(CO)_6$[g]	Ethane

[a] Oxygenates = methanol, methyl esters, ethers, and other alcohols or esters; Glycols = ethylene glycol and glycerine.
[b] Ruthenium cluster breakdown; $Ru(CO)_5$ only detectable carbonyl.
[c] Catalyst includes amine additives.
[d] Catalyst includes chelating ethers.
[e] Solvent is molten $NaCl \cdot 2AlCl_3$
[f] Reaction stops as a result of $AlCl_3$ consumption.
[g] M = Cr, Mo, W; others are $Mn_2(CO)_{10}$, $Fe(CO)_5$, $Co_2(CO)_8$; product derives from metal hydride and coordinated CO; reaction is not catalytic; Cp = η^5-C_5H_5 (cyclopentadienyl).

methanol or glycol. Because the primary reaction requires severe conditions, it appears inevitable that secondary products will be formed.

Mechanisms for homogeneous CO hydrogenations indicate that a key intermediate is formaldehyde (*50, 53*). Although formaldehyde is not a thermodynamically favored product, enough may exist bound to a metal center to be an intermediate.

A controversial aspect of the mechanism is the initial hydrogen insertion into a carbonyl ligand. All other steps can either be justified or have precedents in organometallic chemistry, but the initial insertion, to form a metal–formyl complex, has never been observed. No metal–formyl compound has been prepared directly from CO and H_2 (*29, 94*); those that have been synthesized by other means are unstable with respect to decomposition to the hydrido–carbonyl complex, suggesting that the initial step is thermodynamically unfavorable. It was with this view that simultaneous coordination to both carbon and oxygen, presumably achieved by polynuclear metal clusters, was considered necessary (*127*). Later papers suggested that the difficulties with the apparent instability of formyl complexes can be avoided (e.g., side-on coordination using only one metal center, bridging coordination between two mononuclear metal centers, solvent stabilization or activation of a complex, or kinetic variations that have more than one step occurring simultaneously with the initial hydrogen transfer). At any rate, the initiation step may not be as unfavorable as once thought (*53*).

A homogeneous FTS that produces hydrocarbons does not exist, despite extensive research. Possibly, energetic or mechanistic limitations make the development of this process unlikely. Homogeneous reactions producing oxygenates from H_2 + CO are known, but the severe operating conditions will limit the use of these reactions.

A paper by Kiennemann *et al.* (*76*) emphasizes some of the previous conclusions. At 230°C and 2000–3000 atm of 2 H_2 + 1 CO, substantial yields of primary alcohols up to at least C_{10} were obtained on a homogeneous Ru catalyst. The distribution of the alcohols followed the Bureau of Mines or Flory equation [Eq. (2)]. Ruthenium acetyl acetonate was less active than $Ru_3(CO)_{12}$ and produced more gaseous products and alcohols, but only up to C_3, despite the fact that both of these compounds should be transformed to $Ru(CO)_5$ during the reaction. The Ru compounds would not be expected to reduce to the metal during the reaction under the high-pressure conditions (i.e., the process may be assumed to be homogeneous). A formyl species, produced by inserting CO in a hydride, was postulated. This species is hydrogenated to methanol, and higher alcohols are produced by homologation.

5.7 Mechanisms with Intermediates Containing Oxygen

These mechanisms are enolic-2 (IIIB, Table 1), CO insertion–2 (VD), and alkoxy (VIE).

5.7.1 The Enolic Mechanism

This reaction scheme dates from the 1951 work of Anderson (*4, 124*). The production of straight-chain molecules is described in Table 1, and the formation of methyl branches involve a half-hydrogenated intermediate:

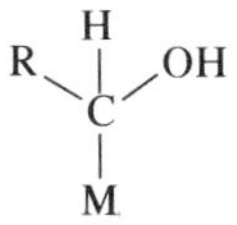

```
   H
R  |  OH
 \ | /
   C
   |
   M
```

Reasonable results are obtained for carbon-number and isomer distributions for alcohols and hydrocarbons, and many aspects of the FTS are described correctly. The elegant investigations of Emmett *et al.* (*63, 79, 87, 88*) gave moral support to the enolic scheme. In these experiments, tagged alcohols were incorporated in the synthesis on Fe. Ethyl, *n*-propyl, and isobutyl alcohols initiated growing chains, but isopropyl alcohol was incorporated to a much smaller extent than the primary alcohols. Methanol both initiated chains and built in repeatedly. The position of the hydroxyl seemed to be the place where the next carbon was added in the chain growth. Intermediates containing hydroxyl groups or in the form of hydrocarbonyls suggest an explanation of the promoter action of alkali, particularly on Fe catalysts.

However, since the mid-1950s no substantial evidence has been reported that supports the enolic scheme; for example, enolic structures have not been found in spectroscopic studies. Blyholder *et al.* (*19, 22*) studied the interaction of alcohols on Fe, Co, and Ni by infrared spectroscopy. No enolic species were found; alcohols do form alkoxides, as described in the next section.

Evidence for an oxygenated intermediate was obtained by Takeuchi and Katzer (*126*) in a double-tracer experiment on the synthesis of methanol on Rh on TiO_2 at 150°C and ~1 atm. A 25 H_2 + 1 CO feed was used in a recycle reactor, with about half of the CO as $^{13}C^{16}O$ and the other half as $^{12}C^{18}O$. Products obtained (in mole percentages) were 4.25 methanol, 11.3 ethanol, 74.3 CH_4 plus C_2–C_4 olefins and paraffins, and 1.7 C_{5+}. The methanol contained in one test, 54% $^{13}CH_3{}^{16}OH$ and 44% $^{12}CH_3{}^{18}OH$ compared with analyses of methanes of 55% $^{13}CH_4$ and 45% $^{12}CH_4$. In the formation of methanol, the CO does *not* dissociate on Rh; however, the nature of the intermediate is not defined by these experiments.

Analyses were also made of the ethanol obtained in a similar double-tracer experiment (*125*) using $^{13}C^{16}O$ and $^{12}O^{18}O$ on Rh on TiO_2. The isotopic distribution of the ethanol is a statistical mixture suggesting that all of the CO had dissociated in the reaction. Mechanisms IIIB, IVC, and VD of Table 1 require that one of the two CO groups forming ethanol dissociate and the other remain intact in the process. The data on CH_3OH indicate that the CO does not dissociate. The authors suggested that the initial intermediate is a hydroxy carbene (CHOH) that hydrogenates to methanol or to carbene (CH_2). The CH_2 reacts with CO to produce ketene, which is in equilibrium with oxirene and causes the scrambling of isotopic species. The ketene is subsequently hydrogenated to ethanol.

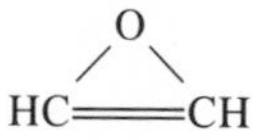

O
HC═CH

Oxirene

5.7.2 Alkoxy Intermediates

Blyholder *et al.* (*19–22*) studied the interactions of methyl and *n*-propyl alcohols and H_2 + CO on supported Co and Ni by infrared spectroscopy. On Co at 25–125°C, the alcohols form an alkoxide surface species. This intermediate reacts with chemisorbed CO to yield a surface carboxylate with one additional carbon atom than the alkoxide:

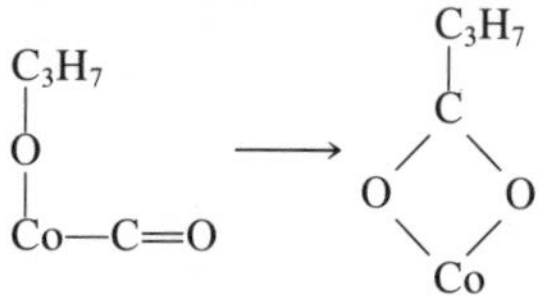

$$C_3H_7\text{—O—Co—C=O} \longrightarrow C_3H_7\text{—C}(\text{O})_2\text{Co} \quad (14)$$

This reaction might be a step in the chain-growth process. The Ni catalyst also produced an alkoxide that degraded, losing first H_2 and then CO, rather than following reaction (14).

Benziger and Madix (*14*) studied the reactions of alcohols on an Fe(100) surface. Methanol and ethanol formed alkoxides at room or lower temperatures, losing the hydroxyl hydrogen in the process. Above 400 K the alkoxides decompose to H_2, CO, hydrocarbons, and the original alcohol; the process was first order and activation energies were 25.1 and 26.5 kcal/mol for the methoxide and ethoxide. Isopropyl alcohol did not produce a stable alkoxide; the principal reaction was dehydrogenation to acetone. Other molecules studied were methyl chloride, olefins, and formic and acetic acids.

Deluzarche *et al.* (*38–43*) proposed the CO-insertion mechanism (VE of Table 1). The CO is inserted between the oxygen and the next group, H or alkyl, and the resulting species is successively reduced to an alkyl. Some of the intermediates have the structure of metal salts, formates, acetates, etc., and formyl, formate, methoxy, acetate, and ethoxy species have been identified by chemical trapping. This growth scheme is of the "away from the surface'' type and leads to straight carbon chains. No steps are given for producing branched chains. The initial step is the reaction of CO with a hydroxyl group. The same intermediate can also be produced from the reaction of CO_2 with an adsorbed hydrogen atom.

Sapienza *et al.* (*62, 115*) proposed mechanisms VIE of Table 1 and called it the ''oxide mechanism.'' Alcohols and hydrocarbons, all with straight carbon chains, are products of this reaction. Sapienza *et al.* claimed many unique results for their postulated scheme, including new explanations for (a) CO_2 formation in the FTS and (b) oxidation of Fe and Co catalysts during the FTS. On point (a), work before 1955 had demonstrated convincingly that virtually all CO_2 is produced by a subsequent water–gas shift. The relative rates of the water–gas shift to the FTS determine the amount of CO_2 produced. For Fe and Mo, the rates of the water–gas shift and the FTS are about equal, but for Co, Ni, and Ru, the water–gas shift is very much slower than the FTS.

Argument (b) suggests that the oxygen left on the catalyst in step D (Table 1) leads to oxidation of Co and Fe catalysts. To the author's knowledge, bulk cobalt oxide is never produced during the FTS, and this result would be expected because the ratio $[H_2O]/[H_2]$ required for producing CoO exceeds 100 at pertinent temperatures. For Fe the ratio $[H_2O]/[H_2]$ required for producing magnetite is less than 0.1; the actual ratio is a function of conversion of $H_2 + CO$ and particle size. Diffusion-in-pore limitations may cause the ratio at the center of a particle to be substantially larger than the ratio in the gas phase surrounding the particle, and the particle may oxidize from the inside.

5.8 Surface Carbides and/or CH_{1-3} Radicals as Intermediates

On Fe, Co, Ni, and Ru at temperatures above 200°C, small pulses of CO react quantitatively according to $2\ CO = C + CO_2$ to deposit carbon on the surface. Pulse-type experiments using gas chromatographic apparatus provides a convenient way of introducing known amounts of surface carbon and of following its reactions with H_2. We note at this point that the formation of surface carbon followed by its reaction with a pulse of H_2 does not necessarily provide even a first approximation of the FTS, because in the synthesis the catalyst is normally strongly inhibited by CO.

Rabo *et al.* (*111*) found that surface carbon could be produced conveniently at 300°C on Ni, Co, and Ru and that this carbon reacted rapidly with H_2 at 300°C to produce CH_4. This carbon also reacted with H_2 at lower temperatures, yielding C_1–C_4 paraffins. For example, at 23°C the product contained 76 mol % CH_4, 21 C_2H_6, 2.4 C_3H_8, and 0.2 C_4H_{10}; at 80°C, 93 CH_4, 5.3 C_2, 1.1 C_3, and 0.6 C_4; and at higher temperatures, all CH_4. At room temperature, CO chemisorbed nondissociatively, and this adsorbate did not react with H_2 at 25°C and only slowly at 200°C. On Pd the chemisorption of CO was nondissociative even at 300°C, and the adsorbed CO reacted very slowly at 200°C. Thus, on Ni, Co, and Ru, the surface carbon was very reactive. Similar results were obtained earlier by Wentrcek *et al.* (*135*).

McCarty and Wise (*95*) prepared four types of carbon on an alumina-supported Ni catalyst: chemisorbed carbon (α), bulk Ni_3C, amorphous carbon (β), and crystalline elemental carbon. Small amounts of carbon were deposited by CO, but because the rate of deposition decreased rapidly with coverage, large coverages were obtained using ethylene. At ~500 K most of the deposit was of the α type, and at 600 K mostly β. The α carbon and the first few layers of Ni_3C were very much more active to hydrogenation than the β type and the crystalline carbon. The reactivities of the carbons decreased rapidly with the amount deposited. Heating the carbons to 668 K or higher temperatures for 6 or 7 min converted the α to β carbon and part of the β type was transformed to crystalline carbon. These and other studies suggest that the surface carbons "age," that is, become less reactive to H_2 with increasing time, increasing temperature, and increasing coverage by the carbon. At 550 K the hydrogenation of the α carbon was rapid enough to be a possible intermediate in methanation.

Martin *et al.* (*93*) studied the chemisorption of CO on reduced Ni on silica by infrared spectroscopy and magnetic methods; the sample was completely reduced and the Ni-particle diameters were ~60 Å. On adsorbing CO in less than monolayer amounts in the temperature range 300–400 K, the CO was in the form of linear ($NiCO_S$) and bridge (Ni_2CO_S) species, as shown by infrared bands at 2070–2040 and 1935 cm^{-1}, respectively, plus an unknown band at 1800 cm^{-1}. Here the subscript S denotes a species on the surface.

Magnetic data (Fig. 13) show plots of α_{CO} and the amount of CO_2 removed on pumping at various temperatures, where α_{CO} is the decrease in saturation magnetization in Bohr magnetons at room temperature per molecule adsorbed. The plateau from 300 to 400 K corresponded to an average bond number of 1.85. Another plateau for α_{CO} was attained above 530 K after part of the adsorbate had been removed as CO_2. The new value of α_{CO} corresponds to a bond number of 3.3. Evacuating the sample at 600 K eliminated all infrared bands. On heating in a closed system, bands for linear CO decreased and those for bridge CO disappeared above

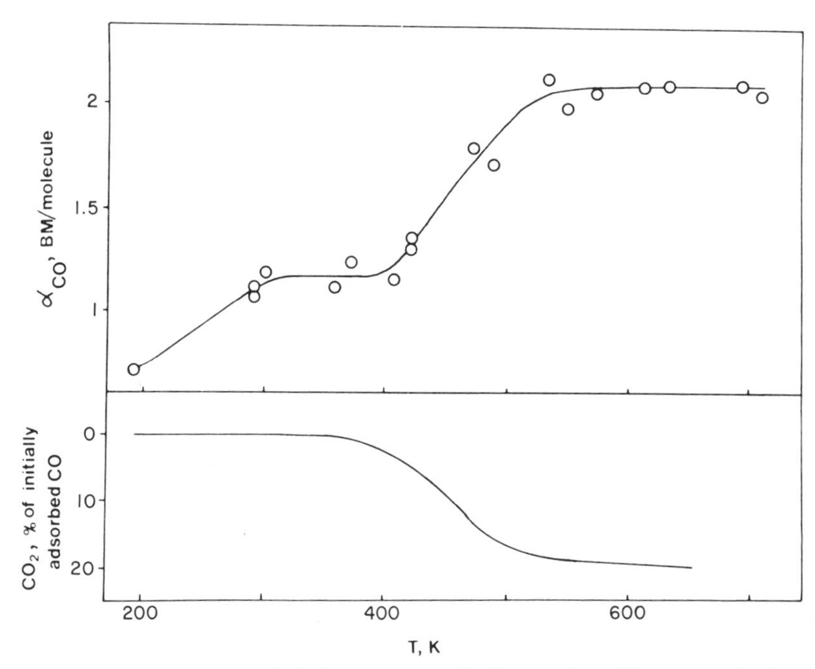

Fig. 13 Variations of α_{CO} in Bohr magnetons (BM) and of the CO_2 content in the gas phase (expressed as a percentage of the initial adsorbed CO) against the holding temperature. Reproduced with permission from Martin *et al.* (*93*).

400 K and were replaced by a band at 1830 cm^{-1} that persisted to at least 600 K. This species had a bond order of ~4 (i.e., Ni_4CO_S) and was in equilibrium with the CO_2 in the gas phase by 2 CO = CO_2 + C occurring through rupture of the C—O bond to give $Ni_3C_S + O_S$. This carbon also reversibly dissolved in the bulk Ni.

Hinderman *et al.* (*71*) and Martin *et al.* (*93*) have interpreted the bonding as a mixture of

O
C
|
Ni

and

O
C
/\
NiNi

for the low-temperature forms with a bond number of 1.84,

for the species with a bond number of ~3, and for the adsorbed material with a bond number of 4,

```
 C———O
/|\    |
NiNiNi  Ni
```

which may dissociate to

```
 C     O
/|\    |
NiNiNi  Ni
```

Ollis and Vannice (*102*) derived a power rate law for the FTS based on a scheme similar to IIA of Table 1; the derivation involves reasonable approximations, such as representing the Langmuir equation by the Freundlich expression, that is, $\theta_x = KP_x/(1 + Kp_x) \propto (Kp_x)^n$, where θ_x is the fraction of the surface covered by component x, K the adsorption equilibrium constant, and n the exponent of the Freundlich equation. The rate-determining step was postulated to be

$$CHOH_{ads} + (y/2)\ H_{2\ ads} = CH_{y\ ads} + H_2O \tag{15}$$

and the rate was proportional to $\theta_C \theta_H^{y/2}$, where C denotes CHOH and H hydrogen. Then, for the power law $r = kp_{H_2}^X p_{CO}^Y$, $y = 2(X - Y)$ and $n = X/(1 + y/2)$. The values of y and n obtained by applying these equations to exponents of the power rate law for the FTS on supported metals are reasonable (Table 10). Except for Ru, y was about 1, 2, or 3 and n about 0.5. In Table 10B the equations were applied to data of Bond and Turnham (*23*) for Ru–Cu alloys on silica. As the amount of Cu increased, the number of hydrogen atoms in the intermediate y decreased from 3.2 and 1.2.

Happel and co-workers (*64, 65, 102a*) studied methanation on Ni on kieselguhr at 1 atm and 210°C by transient-isotope tracing in a circulating glass system. After steady-state operation was attained with $^{12}CO + H_2$, ^{13}CO was substituted for part of the ^{12}CO to give temporal plots of the appearance of ^{13}C in CO, CO_2, and CH_4 (Fig. 14). $^{13}CO_2$ appeared almost at once, but $^{13}CH_4$ had a time delay. Similar experiments were made with the catalyst "equilibrated" with H_2 + CO, and the feed changed to D_2 + CO. Analysis of the response curves led the following conclusions:

1. Chemsorption of CO on Ni is rapid, but CO does not exchange readily with surface carbidic carbon. Thus, the dissociation of CO to carbon is unidirectional.
2. The surface carbon that is reactive in methanation is in equilibrium with an inactive surface carbon. Both forms are converted to CH_4 in pure H_2.
3. The active surface carbon reacts stepwise and unidirectionally with adsorbed hydrogen atoms to produce adsorbed CH, CH_2, CH_3, and finally gaseous CH_4. Adsorbed CH is the principal species of this

Table 10
Interpretation of Power Rate Law Exponents Data for Supported Metals

Metal	*Exponent for* H_2, *X*	*Exponent for* *CO*, *Y*	*Hydrogens in complex*, *y*	*Freundlich*, *n*
A. Data of Ollis and Vannice (*102*)				
Pd	0.7	0.2	1.0	0.467
	0.8	0.3	1.0	0.533
Ni	0.8	−0.3	2.2	0.38
Ir	1.0	0.1	1.8	0.526
Pd	1.0	0	2.0	0.5
Pt	0.8	0	1.6	0.445
Rh	1.0	−0.2	2.4	0.455
Fe	1.1	−0.1	2.4	0.5
Co	1.2	−0.5	3.4	0.445
Ru	1.6	−0.6	4.4	0.5
Percentage Cu				
B. Data of Bond and Turnham (*23*), 1% Ru–Cu on silica				
0	1.2	−0.43	3.3	0.46
0.016	0.66	−0.19	1.7	0.36
0.063	0.89	−0.36	2.5	0.4
0.32	0.64	−0.29	1.9	0.33
0.63	0.75	+0.14	1.2	0.47

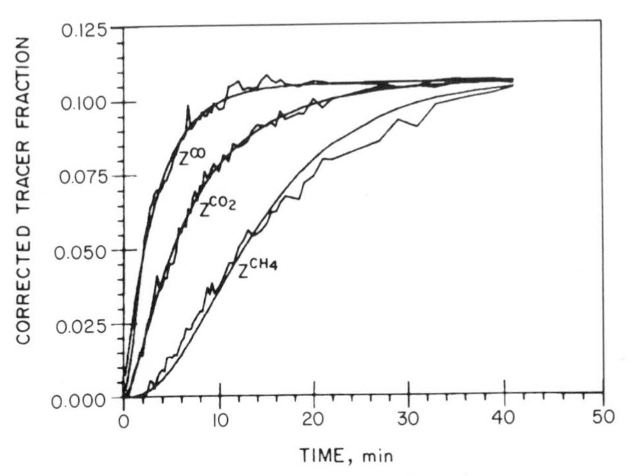

Fig. 14 Carbon-13 monoxide tracing (Ni catalyst). Smooth curve represents theoretical results; uneven curve represents data. T = 258°C; feed molal ratio H_2/CO = 4/1. Reproduced with permission from Happel *et al.* (*65*).

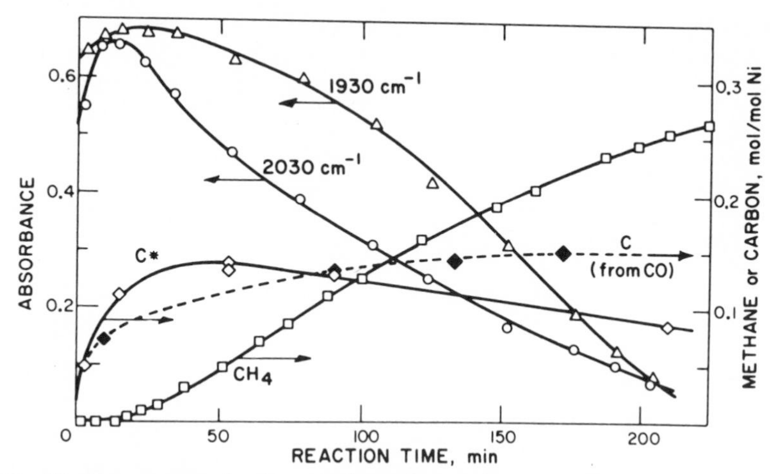

Fig. 15 Time course of carbon on the surface during methanation. Reproduced with permission from Galuszka *et al.* (*57*).

type, the concentrations of adsorbed CH_2 and CH_3 being always less than 40% that of CH. The concentration of the active surface carbon is at most less than 10% of the concentration of CH.

4. Approximate surface coverages during methanation are active surface carbon, 1%; inactive surface carbon, 14%; CH_x, 10%; and CO, 46%. The surface also has a substantial coverage by dissociatively adsorbed hydrogen.

Biloen *et al.* (*17a*) performed similar transient-isotope experiments by changing stepwise from $^{12}CO + H_2$ to $^{13}CO + H_2$ on a Co powder, Ni on SiO_2, and Ru on γ-Al_2O_3 at 210–215°C and 3 atm. The catalysts produced higher hydrocarbons as well as CH_4. Analysis of the response curves led to the following estimates of the upper limit of the fraction of the surface covered by the intermediates yielding CH_4 and propane: Ni, 8–13% for CH_4 and 1.4% for propane; Co, 4–5% for CH_4 and 1.0% for propane; and Ru, 13–14% for CH_4 and 6% for propane.

The transient-response analyses by both groups indicate that only a small fraction of the surface is effective in the synthesis reactions. These results probably explain the difficulty in observing surface species other than adsorbed CO by any spectroscopic method. The small concentration of the intermediate makes the determination of its composition less precise, and questions such as, Does the intermediate contain oxygen? may be hard to answer categorially.

Galuszka *et al.* (*57*) studied methanation on Ni on alumina in a batch-recycle system at 100°C. Pertinent data are shown in Fig. 15. Methane production passed thru an induction period and then increased. Infrared bands for CO on Ni at 1930 and 2030 cm^{-1} increased to a maximum in the first few minutes; however, the rate did not seem related to the magnitude of these bands. Surface carbon (C*), calculated from a carbon balance, seemed proportional to methane production. Experiments with D_2 indicated that the surface species was probably CH, although no infrared bands were observed for this species. CH_2 and CH_3 bands were observed in special cases, but these seem unrelated to the methanation rate.

Araki and Ponec (*11*) and Sachtler *et al.* (*114*) observed that elemental ^{13}C from CO disproportionation was incorporated in methanation with 5 H_2 + 1 ^{12}CO at ~1 torr and 250–300°C in an ultrahigh-vacuum apparatus on evaporated films of Ni, Co, and Ru. With Ni the initial rate of formation of $^{13}CH_4$ was very much larger than the rate of $^{12}CH_4$, for Co the initial rates of formation of $^{13}CH_4$ and $^{12}CH_4$ were about equal, and for Ru, $^{12}CH_4$ was produced at the larger rate from the start. For the Ni film, H_2O was the main oxygenated product, but on Co and Ru, CO_2 was obtained.

Sachtler *et al.* (*17, 114a*) studied the incorporation of surface "carbide" in the FTS on catalysts of Ni, Co, and Ru supported on silica gel. Carbon-13 was deposited on the reduced catalysts from ^{13}CO at $\frac{1}{2}$ atm; the carbon atoms on the catalyst were taken to be equal to the $^{13}CO_2$ produced. Various reactions that could exchange ^{13}C and ^{12}C were shown to be too slow to complicate a simple interpretation of the data. The FTS at 1 atm with 3 H_2 + 1 ^{12}CO gas was carried out on catalysts covered with known amounts of surface carbide as ^{13}C. By measuring the ^{13}C in synthesis products during the course of the test, the coverage of the surface by ^{13}C could be determined. Plots of the percentage of ^{13}C in the CH_4 produced were linear functions of the fraction of the surface covered by ^{13}C, as shown in Fig. 16 for Ni catalysts. Higher hydrocarbons were produced, and these followed approximately the Flory distribution. Carbon-13 was found in the higher hydrocarbons in the same atom fraction as in CH_4. Within the accuracy of the isotopic measurements, these relative abundances approximated a binomial distribution. It was concluded that surface carbon or CH_x groups that could be produced by adding hydrogen atoms were both the initiating and building species in the FTS. Dissociation of CO was not the rate-determining step.

Mori *et al.* (*96*) obtained pertinent information on methanation on supported Ni using pulsed tests in which a pulse of CO was introduced into flowing H_2 at 180–210°C. Methane and H_2O were evolved in equal amounts at the same time, but relatively slowly, half in ~80 s, whereas H_2O from a pulse of O_2 was desorbed rapidly, half in 10 s. Pulses of CO and CO_2 reacted at identical rates, but H_2O was produced very rapidly at the start from CO_2. The pulses decayed according to first-order kinetics,

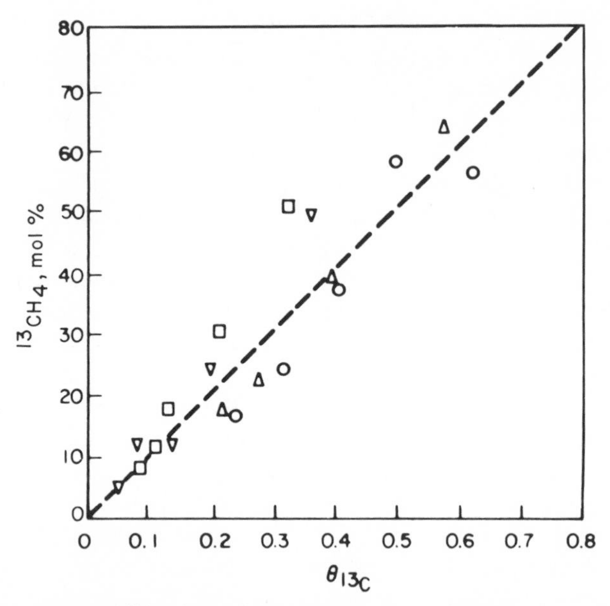

Fig. 16 Percentage of $^{13}CH_4$ in total CH_4 versus $\theta_{^{13}C}$; $\theta_{^{13}C}$ is defined as the number of ^{13}C atoms present per surface Ni atom. Each symbol represents a different experiment. Reproduced with permission from Sachtler (*114a*).

and an inverse isotope effect was observed when D_2 was used instead of H_2. The authors postulated that the slow rate of desorption of CH_4 and H_2O from the pulse was the result of the formation and reaction (by dissociation) of a hydroxycarbene species. Transition-state-theory calculations confirmed this postulate. Both steady-state and pulse tests suggested that the decomposition of the M—COH_m species was the rate-determining step, with other steps at equilibrium. This mechanism corresponds to either the enolic-1 or -2 species in Table 1. Nothing can be said about the chain-growth process because it did not occur significantly. The slow formation of CH_4 is also shown in the curves of Happel *et al.* (*65*) (Fig. 14).

In regular-flow experiments the rate equation had H_2 and CO exponents of 0.90 to 1.03 and −0.39 to −0.47, respectively, and the activation energy was 21 kcal/mol. An identical inverse isotope effect was also found in the flow experiments, $k_H/k_D = 0.77$. The inverse isotope effect was postulated to be the result of an equilibrium situation involving adsorbed hydrogen or deuterium and CO in forming the hydroxycarbene, that is, hydrogen was not involved in the rate-determining step.

At this point in the discussion of mechanisms, we may pause to say that negative and positive isotope effects have been observed as well as no effect at all. As noted by Wilson (*136*) and Ozaki (*102b*), an isotope effect on the rate may result from a thermodynamic isotope effect on the concentration of an intermediate, as well as from a kinetic effect in the rate-determining step. In H_2 + CO reactions, a positive effect was found on Ru powder (*126*), no effect for Ru on alumina (*133*), and a negative effect for Ru on silica or alumina (*14*); no effect was observed for Ni on zirconia (*133*), a negative effect for Ni on silica, and no effect on Pt on alumina (*133*).

Cant and Bell (*28*) studied transient responses using infrared, mass spectrometry, and tagged molecules on Ru on SiO_2. Principal conclusions were the following:

1. Chemisorbed CO covered more than 85% of the surface, and exchange between adsorbed and gaseous CO was more rapid than methanation in the steady state.
2. Forming surface carbon requires surface vacancies and is produced by a reversible dissociation of CO as indicated by isotopic scrambling of $^{13}C^{16}O$ and $^{12}C^{18}O$.
3. Rapid scrambling of H_2 + D_2 indicates that dissociative adsorption of H_2 is at equilibrium under synthesis conditions.
4. Although there is a substantial inventory of surface carbon during synthesis, the catalyst does not contain oxygen.
5. Surface carbon hydrogenates more rapidly than CO.
6. The rate-determining step in methanation is the reaction of CH_3 with atomic hydrogen.

The scrambling of CO isotopes, point 2, actually proceeded very much slower than expected, and the rate was slower in H_2 than in its absence. The slow rate of scrambling was attributed to slow surface diffusion of surface carbon and chemisorbed oxygen.

Blyholder and Emmett (*18*) incorporated tagged ketenes (CH_2CO) in the FTS over a reduced Co–ThO_2–MgO–kieselguhr catalyst and a reduced alkali-free fused-Fe catalyst at atmospheric pressure and 185°C for Co and 240°C for Fe. Addition of ketene increased the yields of C_1–C_3 hydrocarbons. Plots of radioactivity as a function of carbon number are given in Figs. 17 and 18 for carbonyl-labeled ketene and in Fig. 19 for methylene-labeled ketene. These data suggest that the ketene dissociates to CH_2 and CO and the tagged CO mixes with ordinary CO and reacts repeatedly to give the linear curves passing through the origin. The tagged CH_2 initiates hydrocarbon chains. In this capacity, Fe is very effective with most hydrocarbons containing only one tagged carbon; however, the slope of the curve for Fe in Fig. 19 suggests that some molecules contain

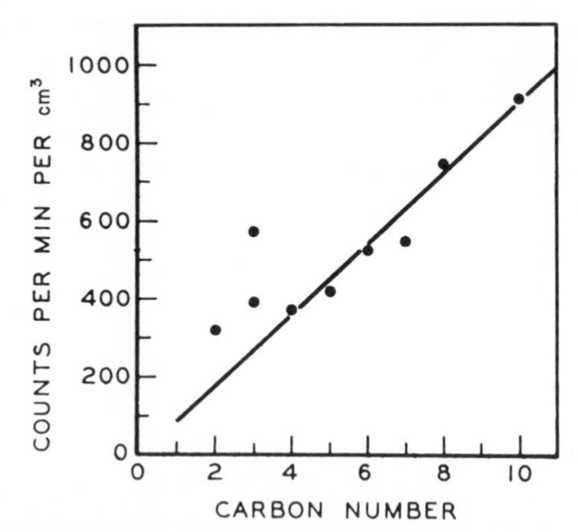

Fig. 17 Radioactivity of hydrocarbons formed on adding 0.7% carbonyl-tagged ketene to 1 H_2 + 1 CO feed to a singly promoted Fe catalyst. Reproduced with permission from Blyholder and Emmett (*18*). Copyright 1960 American Chemical Society.

more than one tagged carbon. For Co, about one-third of the hydrocarbons are initiated by CH_2.

Brady and Pettit (*25*) placed methylene groups on the surface of metal catalysts by the dissociative chemisorption of diazomethane (CH_2N_2). When CH_2N_2 in He or N_2 was passed over supported Ni, Pd, Fe, Co, Ru, and Cu at 1 atm and 25–200°C, the product was predominantly ethylene plus N_2. When the CH_2N_2 in H_2 was introduced, the products changed drastically. On Co, Fe, and Ru, C_1–C_{18} paraffins and olefins were pro-

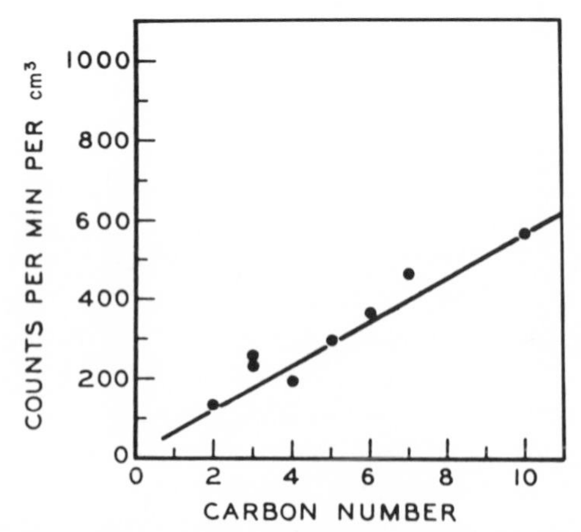

Fig. 18 Radioactivity of hydrocarbons formed on adding 0.5% carbonyl-tagged ketene to 2 H_2 + 1 CO over a Co catalyst. Reproduced with permission from Blyholder and Emmett (*18*). Copyright 1960 American Chemical Society.

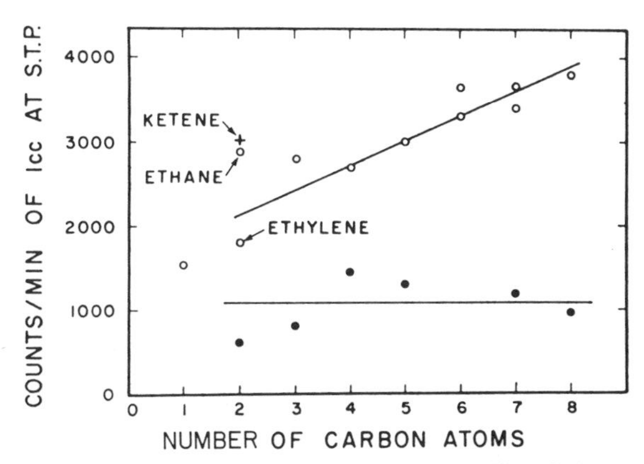

Fig. 19 Radioactivity of hydrocarbons formed on adding 2% methylene-tagged ketene to 1 H_2 + 1 CO over an Fe catalyst at 240°C (○), and for 0.25% ketene in 2 H_2 + 1 CO over a Co catalyst at 185°C (●). Reproduced with permission from Blyholder and Emmett (*18*). Copyright 1960 American Chemical Society.

duced, resembling the products of the FTS on both types of molecules and amounts. Ruthenium also produced a white solid identified as polymethylene. Nickel and Pd produced some higher hydrocarbons, but mostly CH_4, from CH_2N_2 + H_2, but Cu still produced ethylene. These data suggest that the methylene plus H_2 reacted in the same way as postulated in the surface-carbide mechanisms.

In a second communication (*26*), Brady and Pettit performed two tests to discriminate between the CO-insertion (IVC, Table 1), the enolic-2 (IIIB), and the surface-carbide (IA) schemes. In the first, synthesis on Co was performed using a diluted H_2 + CO feed in the early portion of the test, and then CH_2N_2 was added. The conversion of CO was 1% in the first part, and with the addition of CH_2N_2 the hydrocarbon yield increased 90%. Plots of Eq. (2), the Bureau of Mines or Flory equation, for products from the initial period gave a growth constant $\alpha = 0.24$, and with addition of CH_2N_2, $\alpha = 0.51$. According to the authors, in the enolic-2 scheme the enolic groups should not react with methylenes, and each intermediate will produce its own distribution, resulting possibly in a bimodal carbon-number distribution. In the CO-insertion mechanism, the added methylenes should increase the number of chains but not their growth rate. For the surface-carbide hypothesis, the added methylenes are the same intermediates as found in the normal synthesis, and the net result of CH_2N_2 addition is to increase the value of α.

In the second test, a feed containing ^{13}CO, $^{12}CH_2N_2$, and H_2 in different proportions was passed over Co on Cab-O-sil at 250°C. The propene produced was analyzed with a mass spectrometer to give the number of ^{13}C atoms per molecule. According to the authors, the enolic-2 scheme should give none or three ^{13}C atoms per propene; the CO insertion, two or three; and the surface carbide, all possible combinations. The results of tests in which the ratio $^{13}CO/^{12}CH_2N_2$ was changed corresponded within experimental error to calculated predictions of the surface carbide scheme. Deviations were moderate for the CO-insertion hypothesis and large for enolic-2.

The experiments of Brady and Pettit (*25, 26*) provide convincing arguments favoring the surface-carbide mechanism. A word of caution seems pertinent. Highly reactive or strongly chemisorbed "intermediates" added to the system may dominate and control the catalytic process and provide a distorted picture of the reaction scheme.[4]

Deluzarche *et al.* (*41*) presented a comprehensive review of the mechanism of the FTS, including a critique of the role of surface carbon. Their work will serve as the basis for a summary of this section, in which arguments favoring surface carbon intermediates are presented first followed by opposing information. The author has added the comments in brackets.

1. Surface carbon may be hydrogenated to CH_4 on Ni, Co, Ru, Fe, and Rh (*11, 17, 47, 111, 122, 128, 135*), but higher hydrocarbons are not produced on Ni, Ru, Fe, and Rh (*47, 98, 122, 128*). [However, the hydrogenation of surface carbon to higher hydrocarbons may not be favorable thermodynamically.]

2. Surface carbon on Ni can be hydrogenated at room temperature (*111*), but not CO. This result does not prove that CO cannot be hydrogenated directly at higher temperatures. At 200°C on Ni, about one-third of a pulse of CO is dissociated to surface carbon and two-thirds is chemisorbed. With a pulse of H_2, CH_4 corresponding to 60% of the total carbon on the surface is produced. Thus, both the surface carbon and the chemisorbed CO must react to produce CH_4, unless the chemisorbed CO dissociates rapidly in forming CH_4.

3. CO_2 precedes CH_4 in the reaction on Ni and Fe (*11*); however, these data are not a proof that CO must dissociate to surface carbon initially, if the water–gas shift is faster than CH_4 formation. [These data (*11*) seem atypical; H_2O is almost always the primary product.] However, the turnover number (TON) for the formation of surface carbon was smaller than the TON for CH_4 on Ni, Fe, Ru, and Co (*17, 129*). On the other hand, the presence of H_2 may accelerate the rate of formation of surface carbon (*17, 129*).

4. Surface ^{13}C is incorporated into growing hydrocarbon chains in $^{12}CO + H_2$ on Ni, Co, and Ru (*17, 114a*), but the presence of CO seems

[4] An improved SCG scheme involving stepwise addition of methylene is described in the Addendum, p. 293.

necessary for the formation of higher hydrocarbons. When CO is removed abruptly from the synthesis gas, the production of higher hydrocarbons stops, but the formation of CH_4 continues (*128*). [See also an earlier paper (*73*).]

5. On Ni in the presence of $^{12}CO + H_2$, surface ^{13}C is converted to $^{13}CH_4$ more rapidly than $^{12}CH_4$ is produced (*11, 17*). If the rupture of the bond in the hypothetical oxygenated intermediate is the slow step, and if this step is more rapid than the dissociation of CO, the same result is obtained. However, the rates of conversion to CH_4 of surface ^{13}C compared with ^{12}CO are equal on Co, and for Ru the hydrogenation of surface ^{13}C is slower than methanation of ^{12}CO. Two explanations have been given: a transient of nondissociated CO or a very rapid deactivation of the surface carbon on Ru (*17*).

6. Magnetic data suggest the sequence $Ni_4CO_S \rightarrow Ni_3C_S + NiO_S$ for the interaction of CO on Ni at 120°C (*93*).

7. Alloying Ni with Cu decreases the TON for CH_4, which has been attributed to the disappearance of multisites (*128*). This idea is supported by data on the breaking of the C—O bond in alcohols, which also requires multisites. However, H_2 plays no role in the reaction $2\ CO \rightarrow C + CO_2$, whereas H_2 has a positive effect on the overall methanation (*129*). H_2 may increase the rate of reactions forming surface carbon by removing oxygen produced in $CO \rightarrow C_S + O_S$. [In sulfur-poisoning studies on Co and Ni catalysts (*3*), the relative activity decreased linearly with the square of the fraction of unpoisoned sites remaining (see Chapter 6).]

8. The absence of a kinetic isotope effect of H_2/D_2 in the synthesis on Ni, Ru, and Pt was reported (*32*), but another group (*75*) found an isotopic effect on Ru. [Additional isotope effects are listed in this chapter. Generally, isotope effects have no large diagnostic value.]

9. Methylene groups from CH_2Cl_2 are incorporated into metal–alkyl bonds in organometallic compounds (*109, 138*). In $H_2 + CO$, the CH_2 from ketene builds into as well as initiates growing alkyl chains (*109*), and in H_2 or synthesis gas, alkyl chains are produced from azomethane on Ni, Pd, Fe, Co, and Ru (*25*). [However, the original ketene study (*18*) indicated that the CH_2 mostly initiated rather than built into chains on Co, whereas on Fe almost all chains were initiated by the CH_2.]

10. No oxygenated species were observed on Fe by *in situ* infrared, but bands for CH_2 and CH_3 were found (*78*). Possibly the oxygen-containing species are too short-lived to be observed (*78, 129*). On the other hand, formaldehyde and other oxygenated primary products have been observed by mass spectrometry (*10, 60, 80*). [Transient-isotope tracing (*17a, 64, 65, 102a*) suggests that the primary surface intermediate occupies only a small fraction of the surface, a few percent or less. The surface of operating catalysts contains substantial amounts of inactive carbon,

which can, however, be converted to CH_4 in pure H_2. Coverages by inactive carbon varies from 10% to several monolayers. The small concentrations of the primary intermediate may preclude its observation by most spectrocopies and makes the determination of the nature of the intermediate difficult.]

11. On Rh there is evidence for ethylidene intermediates, but none containing oxygen (*86*); however, it is probably not possible to observe an absorbed species covering less than 20% of the surface. Formates were observed on heating catalysts in the absence of H_2, but this species was not considered to be an intermediate. On the other hand, on Ni species containing carbon, hydrogen, and oxygen [called "oxyhydrocarbonated" species, with M—CHOH or M—CHO given as examples] cover the carbon on the Ni, as shown by LEED and HREL (*15, 16*) [HREL is vibrational electron energy loss spectroscopy]. Thermogravimetry indicated complexes containing oxygen on Ni at 200°C (*51*). [A time delay in the formation of CH_4 and H_2O on Ni has been attributed to the presence of a hydroxycarbene intermediate (*96*), i.e., step II or III and not step I of Table 1.]

12. Alcohols can initiate chains but are not involved in chain-growth reactions (*108*); however, other workers found that chain growth also occurs (*112*).

13. On Ni the activation energy of 2 CO → C + CO_2 is 33 kcal/mol; of surface carbon + 2 H_2 → CH_4, 17 kcal/mol; and CO + H_2 → CH_4, 24 kcal/mol. Neither of the first two reactions seems to control the rate of the third (*129*).

14. For Fe the clean surface does not produce alcohols and does not promote the growth of the carbon chain (*47*). On Rh and oxidized Rh, the activation energies and preexponential factors are different, suggesting two mechanisms. The oxidized Rh yields oxygenates and growth of the carbon chain (*30*).

15. A large amount of oxygenated products from Fe cannot come from the dissociation of CO. One may imagine initiation by surface carbon, propagation by CH_x, and insertion of CO to stop the chain (*17, 109*). [With Rh on TiO_2, double-tracer experiments (*126*) show that CO does not dissociate in forming methanol. A hydroxycarbene intermediate was postulated. The intermediate can hydrogenate to methanol or to carbene. To produce ethanol, the carbene plus CO forms ketene, which hydrogenates to ethanol (*125*).]

16. Auger electron spectroscopy shows the presence of carbon on the surface during the synthesis on Ni (*135*). On the other hand, Fe catalysts are mostly Fe_3O_4 with very little metallic Fe (*47*). [These results may be atypical of the FTS on Fe.] Because the results obtained by the majority

of workers are experiments on the metals, the presence of an oxide phase may be very important. [Older work has shown that Fe is extensively oxidized, but the active sites seem to be metal or carbide (see Chapter 3).]

Thus, a variety of interesting studies suggest that surface carbon and/or CH, CH_2, and CH_3 radicals on the surface may be intermediates in the FTS. However, it seems premature to conclude that these postulates provide even a correct general picture of the reaction mechanism.

References

1. Abdulahad, I., Dissertation, Technical University, Berlin (1974).
2. Achtsnit, H. D., Dissertation, University of Karlsruhe (1973).
3. Agrawal, P. K., Fitzharris, W. D., and Katzer, J. R., *in* "Catalyst Deactivation" (B. Delmon and G. F. Froment, eds.), p. 179. Am. Elsevier, New York, 1980.
4. Anderson, R. B., *in* "Catalysis" (P. H. Emmett, ed.), Vol. 4, pp. 346–367. Van Nostrand-Reinhold, Princeton, New Jersey, 1956.
5. Anderson, R. B., to be published.
6. Anderson, R. B., *J. Catal.* **55,** 114 (1978).
7. Anderson, R. B., *J. Catal.* **60,** 484 (1979).
8. Anderson, R. B., and Chan, Y.-C., *Adv. Chem. Ser.* **178,** 113 (1979).
9. Anderson, R. B., and Lee, C. B., unpublished work.
10. Ansorg, J., and Förster, H., *React. Kinet. Catal. Lett.* **10,** 99 (1979).
11. Araki, M., and Ponec, V., *J. Catal.* **44,** 439 (1976).
12. Baker, J. A., and Bell, A. T., *J. Catal.* **78,** 165 (1982).
13. Bell, A. T., *Catal. Rev.—Sci. Eng.* **23,** 203 (1981).
14. Benziger, J. B., and Madix, R. J., *J. Catal.* **65,** 36, 49 (1980).
15. Bertolini, J. C., and Imelik, B., *Ned. Tijsdschr. Vacuumtech.* **16,** 24 (1978).
16. Bertolini, J. C., and Imelik, B., *Surf. Sci.* **80,** 586 (1979).
17. Biloen, P., Helle, J. N., and Sachtler, W. M. H., *J. Catal.* **58,** 95 (1979).

17a. Biloen, P., Helle, J. N., van den Berg, F. G. A., and Sachtler, W. M. H., *J. Catal.* **81,** 450 (1983).

18. Blyholder, G., and Emmett, P. H., *J. Phys. Chem.* **63,** 962 (1959); **65,** 470 (1960).
19. Blyholder, G., and Neff, L. D., *J. Phys. Chem.* **70,** 893, 1738 (1966).
20. Blyholder, G., and Shihabi, D., *Proc. Int. Congr. Catal., 6th, 1976* Vol. 1, p. 440 (1977).
21. Blyholder, G., Shihabi, D., Wyatt, W. V., and Bartlett, R., *J. Catal.* **43,** 122 (1976).
22. Blyholder, G., and Wyatt, W. V., *J. Phys. Chem.* **70,** 1745 (1966).
23. Bond, G. C., and Turnham, B. D., *J. Catal.* **45,** 128 (1976).
24. Bradley, J. S., *J. Am. Chem. Soc.* **101,** 7419 (1979).
25. Brady, R. C., III, and Pettit, R., *J. Am. Chem. Soc.* **102,** 6182 (1980).
26. Brady, R. C., III, and Pettit, R., *J. Am. Chem. Soc.* **103,** 1287 (1981).
27. Bridge, M. E., Comrie, C. M., and Lambert, R. M., *J. Catal.* **58,** 28 (1979).
28. Cant, N. W., and Bell, A. T., *J. Catal.* **73,** 257 (1982).
29. Casey, C. P., Neumann, S. M., Andrews, M. A., and McAlister, D. R., *Pure Appl. Chem.* **52,** 625 (1980).
30. Castner, D. G., Blackadar, R. L., and Somorjai, G. A., *J. Catal.* **66,** 257 (1980).

31. Collis, F. P., and Schwarz, J. A., *J. Comput. Chem.* **3,** 135 (1982).
32. Dalla Betta, R. A., and Shelef, M., *J. Catal.* **49,** 383 (1977).
33. Daroda, R. J., Blackborrow, J. R., and Wilkinson, G., *J. Chem. Soc., Chem. Commun.* p. 1098 (1980).
34. Daroda, R. J., Blackborrow, J. R., and Wilkinson, G., *J. Chem. Soc., Chem. Commun.* p. 1101 (1980).
35. Dautzenberg, F. M., Helle, J. N., van Santen, R. A., and Verbeek, H., *J. Catal.* **50,** 8 (1977).
36. Deckert, H., Dissertation, Technical University, Berlin (1977).
37. Deckert, H., Kölbel, H., and Ralek, M., *Chem.-Ing.-Tech.* **47,** 1022 (1975).
38. Deluzarche, A., Cressly, J., and Kieffer, R., *J. Chem. Res., Synop.* p. 136 (1979); *J. Chem. Res., Miniprint* p. 1657 (1979).
39. Deluzarche, A., Fonseca, R., Jenner, G., and Kiennemann, A., *Erdoel Kohle, Erdgas, Petrochem.* **32,** 313 (1979).
40. Deluzarche, A., Hindermann, J. P., and Kieffer, R., *J. Chem. Res., Synop.* p. 72 (1981); *J. Chem. Res., Miniprint* p. 934 (1981).
41. Deluzarche, A., Hindermann, J. P., Kieffer, R., Cressely, J., and Kiennemann, A., *Bull. Soc. Chim. Fr.* **2,** II329 (1982).
42. Deluzarche, A., Hindermann, J. P., Kieffer, R., Muth, A., Papadopoulous, M., and Tanielian, C., *Tetrahedron Lett.* p. 797 (1977).
43. Deluzarche, A., Kieffer, R., and Muth, A., *Tetrahedron Lett.* p. 3357 (1977).
44. Demitras, G. C., and Muetterties, E. L., *J. Am. Chem. Soc.* **99,** 2796 (1977).
45. Dombek, B. D., *J. Am. Chem. Soc.* **102,** 6855 (1980).
46. Doyle, M. J., Kouwenhoven, A. P., Schaap, C. A., and van Oort, B., *J. Organomet. Chem.* **174,** C55 (1979).
47. Dwyer, D. J., and Somorjai, G. A., *J. Catal.* **52,** 291 (1978).
48. Eidus, Ya. T., Nefedov, B. K., and Labzova, A. V., *Izv. Akad. Nauk SSSR, Ser. Khim.* No. 4, p. 726 (1964).
49. Ekerdt, J. G., and Bell, A. T., *J. Catal.* **62,** 19 (1980).
50. Fahey, D. R., *J. Am. Chem. Soc.* **103,** 136 (1981).
51. Farrauto, R. J., *J. Catal.* **41,** 482 (1976).
52. Feder, H. M., and Rathke, J. W., *J. Am. Chem. Soc.* **100,** 3623 (1978).
53. Feder, H. M., and Rathke, J. W., *Ann. N.Y. Acad. Sci.* **333,** 45 (1980).
54. Fox, J. R., Gladfelter, W. L., and Geoffrey, G. L., *Inorg. Chem.* **19,** 2574 (1980).
55. Friedel, R. A., and Anderson, R. B., *J. Am. Chem. Soc.* **72,** 1212, 2307 (1950).
56. Friedel, R. A., and Sharkey, A. G., Jr., *Rep. Invest.—U.S., Bur. Mines* **7122,** 1 (1968).
57. Galuszka, J., Chang, J. R., and Amenomiya, Y., *Proc. Int. Congr. Catal., 7th, 1980* (T. Seiyama and K. Tanabe, eds.), Part A, p. 529. Elsevier, Amsterdam, 1981.
58. Geoffrey, G. L., *Acc. Chem. Res.* p. 469 (1980).
59. Gresham, W. F., British Patent 655,237 (1951); *Chem. Abstr.* **46,** 7115h (1951).
60. Griogorev, V. V., Alekseev, A. M., Golosman, E. Z., Sobolevskii, V. S., and Yakerson, V. I., *Kinet. Catal.* **16,** 975 (1975).
61. Gupta, R. B., Viswanthan, B., and Sastri, M. V. C., *J. Catal.* **26,** 212 (1972).
62. Haggin, J., *Chem. Eng. News* **59**(43), 24 (1981).
63. Hall, W. K., Kokes, R. J., and Emmett, P. H., *J. Am. Chem. Soc.* **79,** 2983 (1957).
64. Happel, J., Cheh, H. Y., Otarod, M., Ozawa, S., Severdia, A. J., Yoshida, T., and Fthenakis, V., *J. Catal.* **75,** 314 (1982).
65. Happel, J., Suzuki, I., Kokayeff, P., and Fthenakis, V., *J. Catal.* **65,** 59 (1980).
66. Henrici-Olivé, G., and Olivé, S., *Angew. Chem., Int. Ed. Engl.* **15,** 136 (1976).
67. Henrici-Olivé, G., and Olivé, S., *J. Catal.* **60,** 481 (1979).
68. Henrici-Olivé, G., and Olivé, S., *Angew. Chem., Int. Ed. Engl.* **18,** 77 (1979).

69. Henrici-Olivé, G., and Olivé, S., *J. Catal.* **60,** 482 (1979).
70. Herington, E. F. G., *Chem. Ind.* (*London*) p. 347 (1946).
71. Hinderman, J. P., Deluzarche, A., Kieffer, R., and Kiennemann, A., *Can. J. Chem. Eng.* **61,** 21 (1983).
72. Kanowski, S., Dissertation, Technical University, Berlin (1973).
73. Karn, F. S., Shultz, J. F., and Anderson, R. B., *Ind. Eng. Chem. Prod. Res. Dev.* **4,** 265 (1965).
74. Keim, W., Berger, M., and Schlupp, J., *J. Catal.* **61,** 359 (1980).
75. Kellner, C. T., and Bell, A. T., *J. Catal.* **67,** 175 (1978).
76. Kiennemann, A., Jenner, G., Bagherzadah, E., and Deluzarche, A., *Ind. Eng. Chem. Prod. Res. Dev.* **21,** 418 (1982).
77. King, D. L., *J. Catal.* **51,** 386 (1978).
78. King, D. L., *J. Catal.* **61,** 77 (1980).
79. Kokes, R. J., Hall, W. K., and Emmett, P. H., *J. Am. Chem. Soc.* **79,** 2989 (1957).
80. Kölbel, H., and Hanus, D., *Chem.-Ing.-Tech.* **46,** 1042 (1974).
81. Kölbel, H., Kanowski, S., Abdulahad, I., and Ralek, M., *React. Kinet. Catal. Lett.* **1,** 267 (1972).
82. Kölbel, H., Ludwig, H.-B., and Hammer, H., *J. Catal.* **1,** 156 (1962).
83. Kölbel, H., Patzschke, G., and Hammer, H., *Brennst.-Chem.* **47,** 4 (1966).
84. Kölbel, H., and Tillmetz, K. D., *Ber. Bunsenges. Phys. Chem.* **76,** 1156 (1972).
85. Kölbel, H., and Trapper, J., *Angew. Chem.* **78,** 908 (1966).
86. Kroeker, R. M., Kaska, W. C., and Hansma, P. K., *J. Catal.* **61,** 87 (1980).
87. Kummer, J. T., and Emmett, P. H., *J. Am. Chem. Soc.* **75,** 5177 (1953).
88. Kummer, J. T., Podgurski, H. H., Spencer, W. D., and Emmett, P. H., *J. Am. Chem. Soc.* **73,** 564 (1951).
89. Lee, C. B., and Anderson, R. B., *Int. Congr. Catal., 8th, 1984* (preprint).
90. Madon, R. J., *J. Catal.* **57,** 183 (1979).
91. Madon, R. J., *J. Catal.* **60,** 485 (1979).
92. Manes, M., *J. Am. Chem. Soc.* **74,** 3148 (1952).
93. Martin, G. A., Primet, M., and Dalmon, J. A., *J. Catal.* **53,** 321 (1978).
94. Masters, C., *Adv. Organomet. Chem.* **17,** 61 (1978).
95. McCarty, J. G., and Wise, H., *J. Catal.* **57,** 406 (1979).
96. Mori, T., Masuda, H., Imai, H., Miyamoto, A., Baba, S., and Murakami, Y., *J. Phys. Chem.* **86,** 2753 (1982).
96a. Muetterties, E. L., *Bull. Soc. Chim. Belg.* **84,** 959 (1975).
97. Nijs, H. H., and Jacobs, P. A., *J. Catal.* **65,** 328 (1980).
98. Nijs, H. H., and Jacobs, P. A., *J. Catal.* **66,** 401 (1980).
99. Nijs, H. H., Jacobs, P. A., and Uytterhoeven, J. B., *J. Chem. Soc., Chem. Commun.* p. 180 (1979).
100. Nijs, H. H., Jacobs, P. A., and Uytterhoeven, J. B., *J. Chem. Soc., Chem. Commun.* p. 1095 (1979).
101. Novak, S., Madon, R. J., and Suhl, H., *J. Catal.* **77,** 141 (1982).
102. Ollis, D. F., and Vannice, M. A., *J. Catal.* **38,** 514 (1975).
102a. Otarod, M., Ozawa, S., Yin, F., Chew, M., Cheh, H. Y., and Happel, J., *J. Catal.* **84,** 156 (1983).
102b. Ozaki, A., "Isotopic Studies of Heterogeneous Catalysis." Academic Press, New York, 1977.
103. Pannell, R. B., Kibby, C. L., and Kobylinski, T. P., *Proc. Int. Congr. Catal., 7th, 1980* (T. Seiyama and K. Tanabe, eds.), Part A, p. 447. Elsevier, Amsterdam, 1981.
104. Pichler, H., *Adv. Catal.* **4,** 336 (1952).
105. Pichler, H., and Schulz, H., *Chem.-Ing.-Tech.* **42,** 1162 (1970).

106. Pichler, H., Schulz, H., and Elstner, M., *Brennst.-Chem.* **48,** 3 (1967).
107. Pichler, H., Schulz, H., and Kühne, D., *Brennst.-Chem.* **49,** 344 (1968).
108. Ponec, V., *Catal. Rev.—Sci. Eng.* **18,** 15 (1978).
109. Ponec, V., and van Barneveld, W. A., *Ind. Eng. Chem. Prod. Res. Dev.* **18,** 268 (1979).
110. Pruett, R. L., *Ann. N.Y. Acad. Sci.* **295,** 239 (1977); *Chem. Eng. News* **58,** 44 (1980); also Union Carbide patents.
111. Rabo, J. A., Risch, A. P., and Poutsma, M. L., *J. Catal.* **53,** 295 (1978).
112. Riegert-Kamel, S., Diploma of Advance Studies, Strasbourg (1979).
113. Rofer-DePoorter, C. K., *Chem. Rev.* **81,** 447 (1981).
114. Sachtler, J. W. A., Kool, J. M., and Ponec, V., *J. Catal.* **56,** 284 (1979).
114a. Sachtler, W. M. H., *Chem.-Ing.-Tech.* **54,** 901 (1982).
115. Sapienza, R. S., Sansome, M. J., Spaulding, L. D., and Lynch, J. F., *in* "Fundamental Research in Homogeneous Catalysis" (M. Tsutsui, ed.), Vol. 3, p. 179. Plenum, New York, 1979.
116. Sastri, M. V. C., Gupta, R. B., and Viswanthan, B., *J. Catal.* **32,** 325 (1974).
117. Satterfield, C. N., and Huff, G. A., Jr., *J. Catal.* **73,** 187 (1982).
118. Schulz, H., *Erdoel Kohle, Erdgas, Petrochem.* **30,** 123 (1977).
119. Schulz, H., *in* "Chemierohstoffe aus Kohle" (J. Falbe, ed.), pp. 344–354. Thieme, Stuttgart, 1977.
120. Schulz, H., *in* "Chemrawn I" (L. E. St. Pierre and G. R. Brown, eds.), p. 175. Pergamon, Oxford, 1980.
121. Schulz, H., Rao, B. R., and Elstner, M., *Erdoel Kohle, Erdgas, Petrochem. Brennst.-Chem.* **23,** 651 (1970).
121a. Schulz, H., Rösch, S., and Gökcebay, H., *Proc. CIC Coal Symp., 64th, 1982* p. 486 (1982).
122. Sexton, B. A., and Somorjai, G. A., *J. Catal.* **46,** 167 (1977).
123. Sternberg, H. W., and Wender, I., *Spec. Publ.—Chem. Soc.* **13,** 53 (1959).
124. Storch, H. H., Golumbic, N., and Anderson, R. B., "The Fischer–Tropsch and Related Syntheses," pp. 585–595. Wiley, New York, 1951.
125. Takeuchi, A., and Katzer, J. R., *J. Phys. Chem.* **86,** 2438 (1982).
126. Takeuchi, A., and Katzer, J. R., *J. Phys. Chem.* **85,** 937 (1981).
126a. Taylor, P. D., and Wojciechowski, B. W., *Can. J. Chem. Eng.* **61,** 98 (1983).
127. Thomas, M. G., Beier, B. F., and Muetterties, E. L., *J. Am. Chem. Soc.* **98,** 1296 (1976).
128. van Barneveld, W. A., and Ponec, V., *J. Catal.* **51,** 426 (1978).
129. van Ho, S., and Harriot, P., *J. Catal.* **64,** 272 (1980).
130. Vanhove, D., Makambo, P., and Blanchard, M., *J. Chem. Soc., Chem. Commun.* p. 605 (1979).
131. Vidal, J. L., Schoening, R. C., and Walker, W. E., *ACS Symp. Ser.* **155,** 61 (1981).
132. Wender, I., Friedman, S., Steiner, W. A., and Anderson, R. B., *Chem. Ind. (London)* p. 1694 (1958).
133. Wender, I., Levine, R., and Orchin, M., *J. Am. Chem. Soc.* **71,** 4160 (1949).
134. Wender, I., and Sternberg, H. W., *Adv. Catal.* **9,** 594 (1957).
135. Wentrcek, P. R., Wood, B. J., and Wise, H., *J. Catal.* **43,** 363 (1976).
135a. White, J. M., *J. Phys. Chem.* **87,** 915 (1983).
136. Wilson, T. P., *J. Catal.* **60,** 167 (1979).
136a. Wojciechowski, B. W., personal communication.
137. Wong, K. S., and Labinger, J. A., *J. Am. Chem. Soc.* **102,** 3652 (1980).
138. Young, C. B., and Whiteside, G. M., *J. Am. Chem. Soc.* **100,** 5808 (1978).

CHAPTER 6

Sulfur Poisoning of Fischer–Tropsch Catalysts

6.1 Introduction

Fischer (*16*) in 1935 cited a practical upper limit for sulfur concentration of 1–2 mg/m^3 for feed gas for the Fischer–Tropsch synthesis (FTS). In 1955, Rapoport (*37*) cited the same limit. Sulfur poisoning of iron catalysts in the ammonia synthesis had been encountered a quarter of a century earlier (*29*), and poisoning data from the ammonia synthesis are useful for the FTS, because catalysts are often similar.

Poisoning experiments are difficult to perform and interpret. Hence, the experimenter should carefully consider the kind of information that is required before designing the test. Many poisons are effective in trace amounts; therefore, special procedures may be required for analyses and for introducing known concentrations of the poison. Reaction systems of materials that will not adsorb or react with the poison are needed. For example, special gas cylinders are required for storing mixtures of synthesis gas and H_2S in which the concentration of H_2S will remain constant for 1 or 2 months (*42*). The reaction system for sulfur-poisoning tests should have stainless steel in contact with the gas stream throughout.

Interpretation of poisoning data is generally difficult because the poison is normally nonuniformly adsorbed with respect to position within the particle and, in a fixed bed of catalyst, also with respect to position within the bed, particularly bed length. In a fixed bed of catalyst the poison may be selectively adsorbed near the inlet of the bed, which usually decreases the effectiveness of the poison. Here the inlet portion of the bed serves as a purification unit. Within a catalyst particle the pores may act as long reactors, with the poison selectively adsorbed near the periphery of the particle. This factor increases the effectiveness of the poison; the most accessible portion of the particle is preferentially poisoned.

The poison may be merely chemisorbed on the surface, or the chemisorption may be followed by diffusion and reaction with the bulk of the active material. Poisoning data may be conveniently expressed as relative activity *F*, the ratio of the activity of the poisoned catalyst to the activity of the fresh catalyst. The activity should, if possible, be expressed in a rational way, such as the rate constant of a kinetic equation (*4*). In the FTS, sulfur poisoning is almost always permanent, that is, the activity is not restored after the poison is removed from the feed.

For a complete understanding of the poisoning process the following information seems pertinent: (a) the amount of poison required to poison the catalyst completely, (b) the number of active sites on the catalyst as measured by chemisorption, (c) the relationship between average relative activity of a fixed bed and the average poison concentration in the catalyst, (d) the relationship of amount of poison on the catalyst as a function of bed length, the concentration profile, (e) the poison distribution within the catalyst particles, (f) the effect of particle size on poisoning, (g) the intrinsic relationship between relative activity, selectivity, and amount of poison as measured on a single particle, (h) the effect of poison concentration in the feed on poisoning rate, and (i) the nature of the bonding of the poison to the active component. Obviously, few poisoning experiments consider all of these aspects of the problem, although most of them can be determined directly. In some cases, item (g) the intrinsic relationship between relative activity and amount of poison on the catalyst, may be obtained by prepoisoning the catalyst and using it in a fixed bed.

Wheeler considered effects of diffusion in porous catalysts on single particles (*46*) and, subsequently, poisoning in a fixed bed (*47*). Anderson and Whitehouse (*7*) studied the effect of poison-concentration–bed-length profiles on the activity of fixed beds of catalyst.

Two reviews, one to 1965 (*5*) and the other to 1977 (*25*), are available. Agrawal *et al.* (*2*) summarized surface physics studies showing for Co, Ni, Pt, Fe, and Mo that a saturated surface sulfide consists of two metal atoms to one sulfur atom.

Although sulfur-resistant catalysts would be desirable for many reasons, most of the sulfur compounds in the raw synthesis gas would probably still have to be removed, because environmental standards may be expected to require the use of essentially sulfur-free fluid fuels.

6.2 Sulfur Poisoning of Iron Catalysts

A literature review to 1955 (*3*) revealed few important contributions to the sulfur poisoning of iron catalysts. For iron in the ammonia synthesis, resistance to sulfur poisoning increased in the order: pure Fe < $Fe\text{–}K_2O$

$<$ Fe–Al_2O_3–K_2O $<$ Fe–Al_2O_3–CaO–K_2O (*12*). In the ammonia synthesis at 100 atm, workers at the Tennessee Valley Authority (TVA) (*17*) found that ~20 mg H_2S per gram of catalyst were required to decrease the activity of a doubly promoted catalyst to a low value. Sulfur dioxide was a temporary poison similar to oxides of carbon and water vapor. Balatnikova *et al.* (*8*), in the atmospheric-pressure ammonia synthesis, found an order similar to that of Brill (*12*). The amount of sulfur required to decrease activity to zero was 1.7, 3.2, and 8.1 mg per gram catalyst for Fe–K_2O, Fe–Al_2O_3, and Fe–Al_2O_3–K_2O, respectively, and in milligrams per square meter of surface area the values were 2.5, 0.4, and 0.9.

Rapoport and Muzovkaya (*38*) investigated the poisoning of precipitated iron oxide–copper oxide by organic sulfur compounds in the FTS. Catalysts reduced only to magnetite or not reduced at all were remarkably resistant to poisoning, and the synthesis was continued at moderate to high yields until the sulfur in the catalyst increased to ~10 wt %. If the catalyst was completely reduced in H_2 at 500°C, it was inactive at a sulfur content of 0.9 wt %. The partly reduced iron oxide apparently is an effective purifier, protecting the remainder of the bed.

British workers studied sulfur poisoning of reduced catalysts prepared from mill scale (*19*). Prepoisoning by H_2S introduced to an evacuated bulb containing the catalyst was more effective than poisoning by H_2S in the feed gas. Sulfur poisoning increased methane production.

From 1955 to 1965 a substantial amount of research was performed at the U.S. Bureau of Mines in Bruceton, Pennsylvania, including prepoisoning, *in situ* poisoning, and catalyst characterization. The principal catalysts studied were fused iron catalyst D3001 (Fe, 67.4 wt %; Mg, 4.61; Cr, 0.65; and K_2O, 0.57), usually as 6- to 8-mesh particles, and activated SAE1018 steel lathe turnings L2201, approximate size 8 × 3 × 0.4 mm, oxidized 25% with steam at 600–700°C and impregnated with aqueous K_2CO_3 to give a concentration of 0.13 wt % K_2O.

The surface area and pore volume of D3001 increased linearly with extent of reduction, and the surface areas for completely reduced samples were ~23 m^2 per gram Fe for a reduction temperature of 400°C, 14 for 450°C, and 6 for 550°C. The pore volume was ~0.13 cm^3 per gram Fe for catalysts completely reduced at all of the temperatures. Converting the iron to Hägg carbide or nitride did not change the surface area significantly but increased the pore volume by 11–12%. Lathe turnings L2201 after reduction at 450°C had surface areas varying from 0.45 to 0.83 m^2 per gram Fe.

During synthesis, large changes occur in the fused catalysts and smaller changes in the lathe turnings. Changes in 9 days of synthesis with 1 H_2 + 1 CO gas at 21.4 atm at 260°C for D3001 and 286–266°C for L2201 are given in Table 1 (*45*); used catalysts were extracted in boiling heptane to re-

Table 1
Surface Area and Composition of Catalysts[a]

Catalyst and state	*Surface area, m^2 per gram Fe*	*Composition, atom ratios*		*Phases from X-ray diffraction*[b]
		C/Fe	*O/Fe*	
Turnings (L2201)				
Reduced	0.45	~0	0	α
Used	0.54	0.55[c]	0.15[c]	χ, C(?)[d]
Fused iron oxide (D3001)				
Reduced particles	14.8	~0	0	α
Used				
Inlet	1.00	0.31	0.92	M, α
Middle	0.77	0.18	1.01	M, α
Outlet	0.74	0.15	1.06	M, α

[a] Reproduced with permission from Shultz *et al.* (*45*).
[b] Phases listed in order of intensity of diffraction pattern. α, Metallic iron; χ, Hägg carbide; C, cementite; M, magnetite. Phases with faint diffraction patterns are followed by (?).
[c] These analyses are given for activated layer. Composition can be expressed on the basis of entire turnings by dividing these values by 4.
[d] Diffraction analysis for activated layer.

move wax before examination. The turnings were carburized and D3001 was oxidized. Tables 19 and 20 in Chapter 3 present similar data for D3001, including carbided and nitrided samples.

To provide a measure of metallic iron and alkali on the catalyst surface, the chemisorption of CO at $-195°C$ and CO_2 at $-78°C$ as well as physical adsorption of N_2 at $-195°C$, were determined on some catalysts used in poisoning test (*9*). In Table 2, the amount of chemisorbed CO divided by V_m, the monolayer value for N_2, is given as a measure of the fraction of metallic iron surface, and similarly V_{CO_2}/V_m for surface alkali. Catalysts with MgO as a structural promoter have a larger fraction of surface metallic iron than samples promoted with Al_2O_3. Magnesia seems to chemisorb a small amount of CO_2. In other experiments the effect of chemisorbed sulfur compounds on the chemisorptions of CO and CO_2 was measured. H_2S was chemisorbed at 450°C and the catalyst treated with H_2 at 450°C for 1 h and then evacuated before the chemisorption of CO and CO_2. Figures 1 and 2 show the results for four of the fused catalysts. The sulfur content can be expressed as sulfur atoms per molecule of N_2 in the monolayer by multiplying the present horizontal scale by 3.09. Thus 0.5 mg S per square meter is equivalent to 1.55 S atoms per molecule of N_2. At this sulfur coverage, V_{CO}/V_m and in some cases V_{CO_2}/V_m have substantial values. Apparently the sulfur is not uniformly dispersed over the

Table 2
Chemisorption of CO and CO_2 on Iron Catalyst[a]

Catalyst	*Composition, weight per 100 Fe*	*Surface area, m² per gram Fe*	*Chemisorption on reduced catalysts*	
			V_{CO}/V_m[b]	V_{CO_2}/V_m[c]
L1022.1	0.2 Al_2O_3, 0.6 SiO_2	8.4	0.58	0
P2007.6[d]	3.3 Al_2O_3, 0.3 SiO_2	19.8	0.41	0.02
L3119	3.2 Al_2O_3, 0.3 SiO_2, 0.37 K_2O	13.4	0.30	0.18
L2021	2.8 MgO	15.2	0.50	0.08
D3001	6.8 MgO, 1.1 SiO_2, 1.0 Cr_2O_3, 0.85 K_2O	14.7	0.39	0.26
P1039.2[e]	0.01 K_2O	0.9	0.65	0

[a] Except where noted all are fused catalysts reduced at 450°C. From Bayer *et al.* (*9*).
[b] Chemisorbed CO at −195°C divided by the V_m value for physically adsorbed nitrogen at −195°C.
[c] Chemisorbed CO_2 at −78°C.
[d] Final part of reduction at 500°C.
[e] Oxidized lathe turnings without alkali, reduced at 400°C.

surface and chemisorbs in some places several layers deep. For D3001 the chemisorption of CO_2 decreased more rapidly than that of CO as the sulfur increased, suggesting that the sulfur may be preferentially adsorbed by the alkali.

Reduced lathe turnings L-2201 and reduced and reduced-and-nitrided D3001 were prepoisoned by immersion in a 1% solution of the sulfur compound in heptane for 24 h at room temperature, and the poisoned catalyst tested with 1 H_2 + 1 CO gas at 21.4 atm (*41*). For H_2S and ethyl mercaptan on reduced D3001, the relative activity F decreased linearly down to 20%; SO_2 poisoning was similar but more severe. Tests of COS on reduced and H_2S on nitrided D3001 were erratic. Nevertheless, for all of the sulfur compounds, 8–10 mg S per gram Fe decreased F to less than 2%. For the steel turnings, 0.18 mg S (as H_2S) per gram Fe decreased F to 8.1% compared with 7.8 mg S per gram Fe for D3001. Sulfur concentrations in milligrams S per square meter of surface needed to decrease F to 8.1% were 0.22 and 0.58 for turnings and D3001, respectively.

Catalyst D3001 was rapidly oxidized in the synthesis, as shown in Table 1. Samples of D3001 were used in the synthesis at 21.4 atm for 5–9 weeks, removed from the reactor, extracted with heptane to remove wax, prepoisoned with H_2S, and tested. For equal amounts of sulfur, the F's of the used catalyst were only 20% of those of the fresh prepoisoned catalyst. The F's of the fresh and used catalyst were more nearly equal when

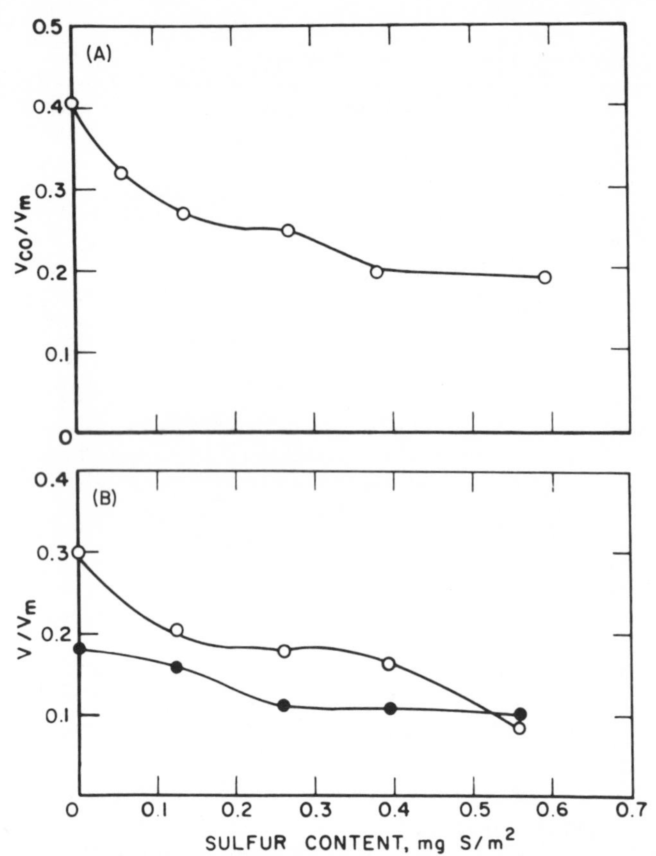

Fig. 1 Effect of H_2S on the chemisorption of CO (○) and CO_2 (●) on reduced, fused iron oxide catalysts containing alumina: (A) P2007.6 (Fe_3O_4–Al_2O_3), $V_{CO_2}/V_m < 0.02$; (B) L3119 (Fe_3O_4–Al_2O_3–K_2O). Reproduced with permission from Bayer *et al.* (*9*).

compared on the basis of milligrams S per square meter, the fresh catalyst with an area of 14.8 m^2/g and the used sample 0.75.

In these tests higher temperatures were used with the poisoned catalyst, and the yield of gaseous hydrocarbons increased with increasing amounts of sulfur on the catalyst. This change in selectivity also was observed for several poisoned catalysts that were operated at about the same temperature as the fresh catalyst. At about the same conversion, the usage ratio $H_2:CO$ increased from 0.70 for the fresh catalyst to ~0.77.

In the Bureau of Mines research, tests were also made with H_2S in the feed gas (*5, 20, 21*). Tests were started on a sulfur-free 1 H_2 + 1 CO gas, total sulfur content less than 1 mg/m^3, at an hourly space velocity of 300 and at 200°C, and the temperature was increased slowly so that ~48 h were required for the apparent CO_2-free contraction to attain 65%. Usu-

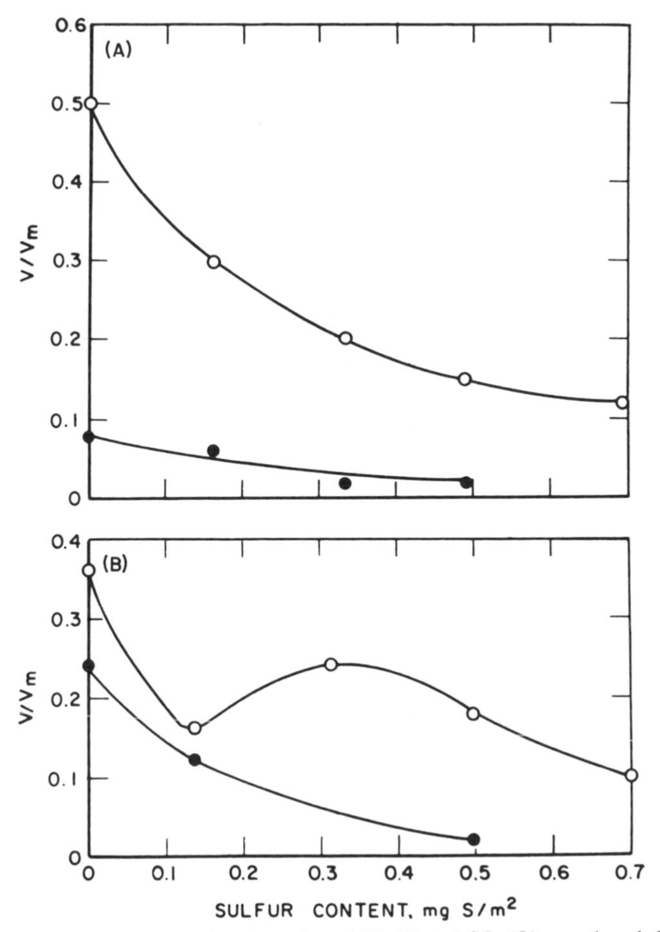

Fig. 2 Effect of H_2S on the chemisorption of CO (○) and CO_2 (●) on reduced, fused iron catalysts promoted with magnesia: (A) L2021 (Fe_3O_4–MgO); (B) D3001 (Fe_3O_4–MgO–K_2O). Reproduced with permission from Bayer *et al.* (*9*).

ally a constant activity was reached at 260°C, and synthesis was continued for 9–20 days to establish a starting point for the poisoning tests. Synthesis gas, 1 H_2 + 1 CO, containing concentrations of H_2S of 6.9, 23, and 69, mg S per cubic meter, was then introduced by a separate feed system without depressurizing the reactor.

Two modes of operation were followed after the change to feed gas containing sulfur:

1. Constant temperature, where the temperature was maintained constant at the temperature of the previous period of constant activity and the conversion permitted to decrease with time

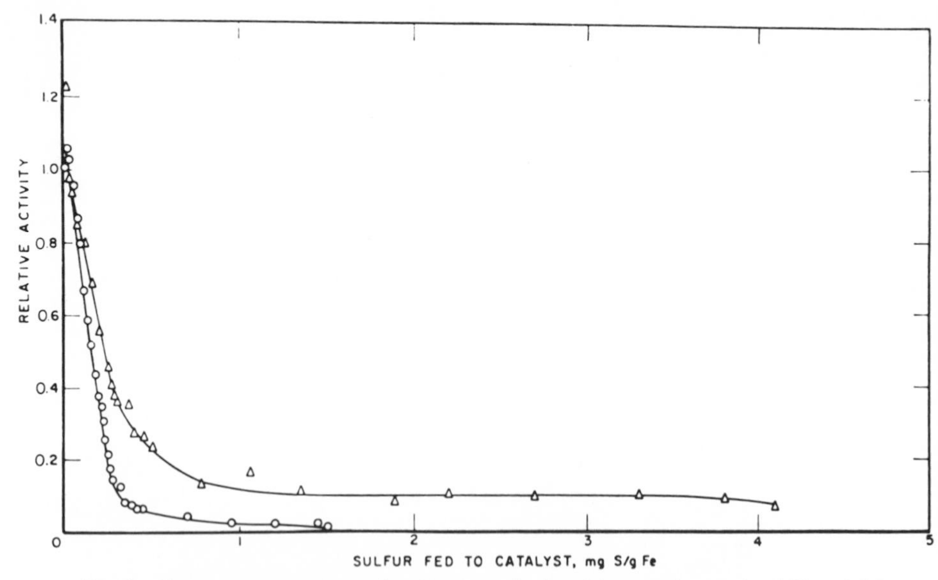

Fig. 3 Constant-temperature poisoning tests of reduced iron catalysts using 1 H_2 + 1 CO gas containing 69 mg S (as H_2S) per cubic meter: △, catalyst D3001; ○, turnings L2201. From Anderson *et al.* (*5*).

2. Constant productivity, where the temperature was increased as required to maintain the CO_2-free contraction at ~60%.

Tests of the first type were usually terminated voluntarily after the conversion had decreased to a low constant value. Tests of the second type were usually terminated by clogging of the catalyst bed. Products were recovered weekly. Sulfur balances were made; 80–100% of the sulfur introduced was found in the catalyst. The exit gas and synthesis products did not contain significant concentrations of sulfur (*5, 21*).

Constant-temperature poisoning tests are shown in Fig. 3 for the poisoning of reduced catalysts D3001 and L2201 with 1 H_2 + 1 CO gas containing 69 mg S (as H_2S) per cubic meter. Similar data are presented in Fig. 4 for a sulfur concentration of 23 mg/m^3. For both carbided catalysts and reduced catalysts, following the initial short period of high activity, the relative activity decreased linearly with increasing sulfur fed to the catalyst until 50–75% of the activity was lost; subsequently, the activity decreased less rapidly. For nitrided catalysts, the linear portion of the poisoning curves did not extend beyond a relative activity of 0.7 for 69 mg S per cubic meter, and for 6.9 mg S per cubic meter the linear portion was absent. These data are given in Table 3 but are not shown as poisoning curves. For catalyst L2201, the relative activity F decreased steadily to

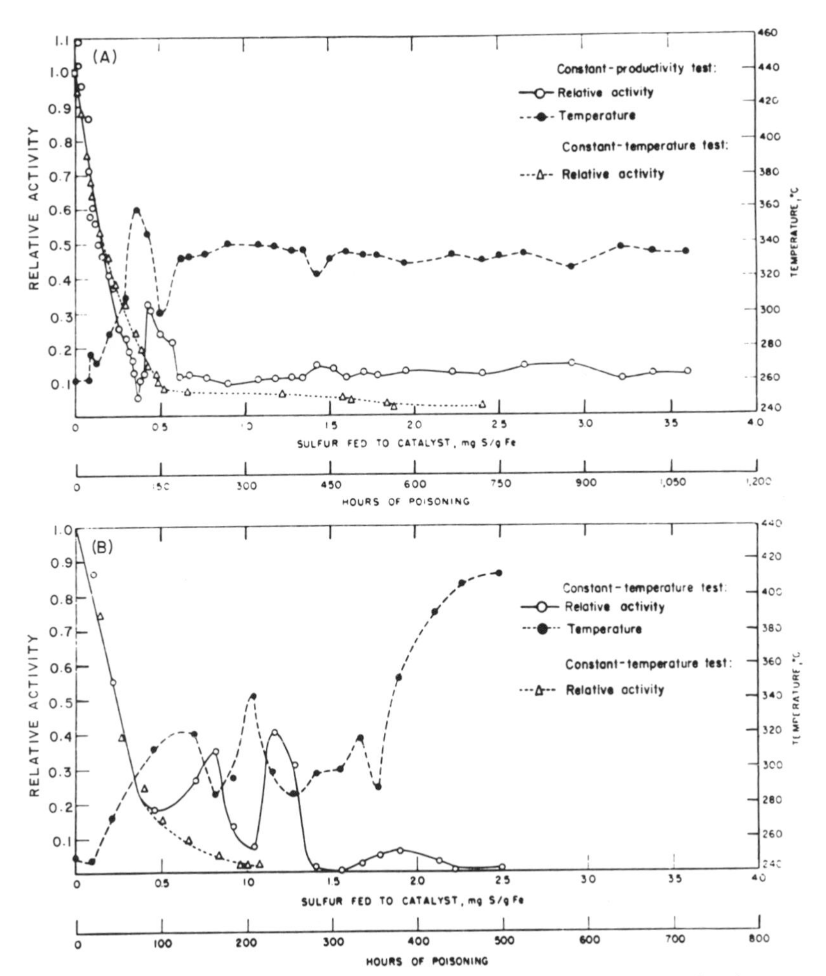

Fig. 4 Comparison of constant-productivity and constant-temperature poisoning tests of catalysts D3001 (A) and L2201 (B) with 1 H_2 + 1 CO gas containing 23 mg S (as H_2S) per cubic meter. From Anderson *et al.* (*5*).

zero when 1–2 mg S per gram Fe had been introduced, but the relative activity of reduced and carbided catalyst D3001 decreased to a constant value of 0.03–0.10. For nitrided catalyst D3001, *F* decreased to a constant value of 0.1–0.2. Table 3 presents data for all tests, read from experimental curves at equal intervals of relative activity.

In the constant-temperature tests the *F*'s of the lathe turnings approached zero at ~1.5 mg S per gram Fe in approximate agreement with the prepoisoning experiments. The reduced and the carbided D3001 lost 90% of their activity when 0.5–1.75 mg S per gram Fe had been fed to the

Table 3
Summary of Constant-Temperature Poisoning Tests: Sulfur Required to Decrease Relative Activity to a Given Value[a]

	Catalyst D3001						*Catalyst L2201, reduced*	
	Reduced, synthesis at 260°C			*Carbided, synthesis at 242°C*	*Nitrided, synthesis at 250°C*		*Synthesis at 260°C*	*Synthesis at 262°C*
Relative activity	*6.9*[b]	*23*[b]	*69*[b]	*69*[b]	*6.9*[b]	*69*[b]	*23*[b]	*69*[b]
0.9	0.031	0.031	0.046	0.092	0.015	0.090	0.082	0.062
0.8	0.060	0.056	0.092	0.180	0.050	0.180	0.113	0.086
0.7	0.090	0.090	0.135	0.242	0.103	0.260	0.147	0.112
0.6	0.120	0.125	0.175	0.352	0.155	0.410	0.185	0.137
0.5	0.155	0.165	0.218	0.455	0.283	0.740	0.215	0.160
0.4	0.194	0.216	0.275	0.552	0.450	1.020	0.250	0.185
0.3	0.242	0.300	0.353	0.628	0.600	1.460	0.350	0.220
0.2	0.300	0.379	0.600	0.980	0.970	>3.0[c]	0.400	0.253
0.1	0.950	0.440	1.750	1.750	>1.3[c]	—	0.600	0.340
0.05	>1.5[c]	1.600	>4.2[c]	[d]	—	—	0.800	0.700

[a] In milligrams S per gram Fe. All tests with 1 H_2 + 1 CO gas at 21.4 atm. From Anderson *et al.* (*5*).
[b] H_2S in feed gas, mg S per cubic meter.
[c] Tests were voluntarily terminated when this amount of sulfur had been introduced.
[d] Temperature was increased after 1.9 mg S per gram Fe was fed to the catalyst.

catalyst; these values are far smaller than 8–10 mg S per gram Fe found in prepoisoning tests. In *in situ* prepoisoning, D3001 is largely oxidized at the end of the initial period on pure gas and has a surface area of 0.8 m^2 per gram Fe compared with an area of 14.8 m^2 for the fresh catalyst that is prepoisoned.

Poisoning curves for constant productivity tests with 23 mg of S (as H_2S) per cubic meter are shown in Fig. 4. The relative activity decreased initially in about the same manner as in the constant-temperature tests. However, after the temperature had been increased above 300°C, the activity often increased sharply and usually decreased again rapidly. These periods of sudden transient activation probably result from the generation of fresh active surface by processes related to carbon deposition that cause disintegration of the catalyst particles. The constant-productivity tests also exhibited periods of sustained higher activity, compared with the constant-temperature tests, probably for the same reason. This type of activation of iron catalysts at moderately high temperatures has been described (*44, 45*). Carbon deposition was more rapid at the higher temperatures used in the constant-productivity tests, and the catalyst bed eventually clogged, presumably owing to carbon deposition. For

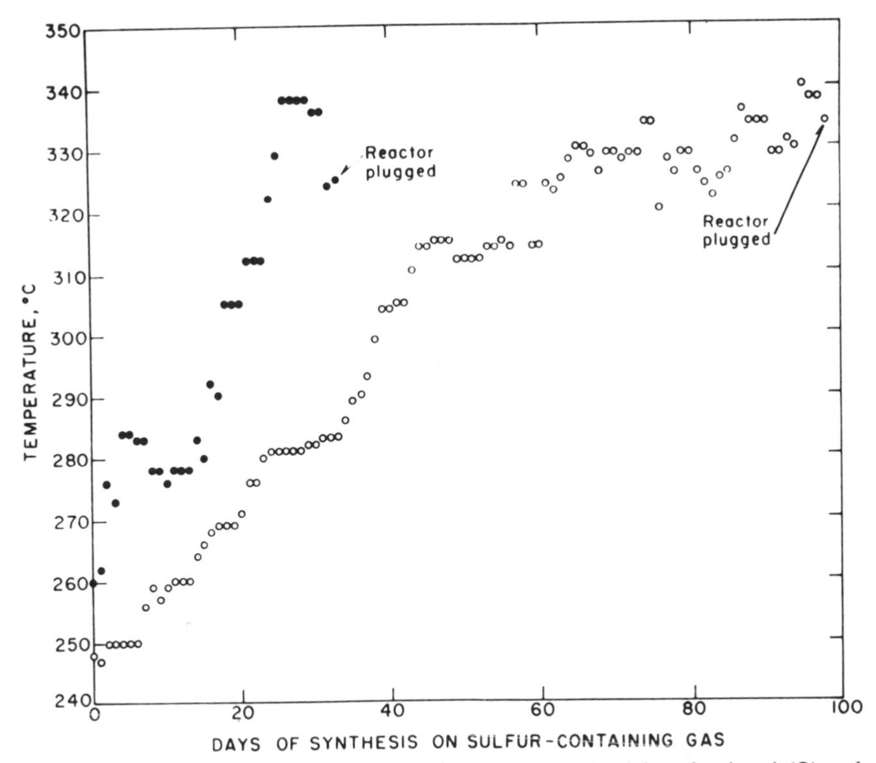

Fig. 5 Temperatures required to maintain constant productivity of reduced (●) and nitrided (○) catalyst D3001 using 1 H_2 + 1 CO gas containing 6.9 mg S (as H_2S) per cubic meter. From Anderson *et al.* (*5*).

tests with catalyst D3001, productivity could be maintained constant until the test was terminated by clogging of the reactor, but for catalyst L2201 the productivity decreased substantially before the reactor clogged. Figure 5 is a plot of temperatures of synthesis as a function of time for constant-productivity tests of reduced and nitrided catalyst D3001 using gas containing 6.9 mg S (as H_2S) per cubic meter. With the lower concentration of H_2S, the activity changed less rapidly and large fluctuations in activity usually were not observed.

Table 4 lists the life of catalysts in constant-productivity tests before clogging occurred for catalyst D3001, or before productivity decreased to a low value for catalyst L2201. Because conditions leading to clogging are diverse, the time before clogging occurs can be regarded as only an approximate value. Changes in reactor type or size may lead to widely differing values of time before clogging occurs.

Table 5 gives product distributions for constant-productivity experiments. As the temperature was increased substantially during poisoning,

Table 4
Duration and Maximum Temperature in Constant-Productivity Poisoning Experiments[a]

Catalyst and pretreatment	*Sulfur concentration, mg S per cubic meter*	*Duration, days*	*Reason for terminating*[b]	*Temperature, °C*		
				Start of poisoning	*Maximum*	*End of test*
Fused iron oxide (D3001)						
Reduced	6.9	33	C	260	338	325
	23	41	C	260	360	330
	69	18	C	256	355	355
Nitrided	6.9	98	C	248	338	338
	23	56	C	256	350	319
	69	28	C	249	349	346
Carbided	69	55	C	242	344	344
Turnings (L2201)						
Reduced	23	17.5[c]	I	250	410	410
	69	11.7[c]	I	260	396	396

[a] From Anderson *et al.* (*5*).
[b] C, Bed clogged; I, constant productivity could not be maintained.
[c] Time for which constant productivity was maintained. Test continued for 50–80 h at low productivity before bed clogged.

Table 5
Product Distribution in Constant-Productivity Poisoning Experiments

Catalyst and test	*Pretreatment*	*Sulfur concentration, mg S per cubic meter*	*Sulfur fed to catalyst, mg S per gram Fe*	*Temperature, °C*	*Distribution of hydrocarbons,[a] wt %*			
					C_1	C_2	$C_3 + C_4$	C_{5+}
Turnings (L2201) Z210	Reduced	0	0[b]	262	6	6	13	75
		23	0–0.8	257–320	10	6	13	71
		23	0.8–1.7	288–350	17	10	24	49
		23	1.7–2.5	322–357	26	12	23	39
Fused iron oxide (D3001)								
Z132	Reduced	23	0–1.3	261–364	10	3	11	76
			1.3–2.6	331	16	9	13	62
			2.6–2.8	330–349	16	10	17	57
Z245	Carbided	69	0[b]	236	10	19	12	59
			0–1.8	242	8	6	18	68
			3.6–7.1	298	25	16	24	39
			8.9–10.6	303	17	11	25	47
			10.6–12.2	316	18	10	45	27
Z198	Nitrided	6.9	0[b]	248	28	12	18	41
			0.2–0.6	274	23	13	21	43
			2.2–2.5	328	26	12	19	43
			2.5–2.9	327	17	10	22	51
Z184	Nitrided	23	0[b]	256	18	8	16	58
			1.3–5.6	327	20	10	18	52

[a] Hydrocarbons include oxygenated molecules dissolved in liquid hydrocarbons.
[b] Products collected for first week of operation of fresh catalyst. Products from this period usually contain a greater fraction of gaseous hydrocarbons than products from subsequent weekly periods.

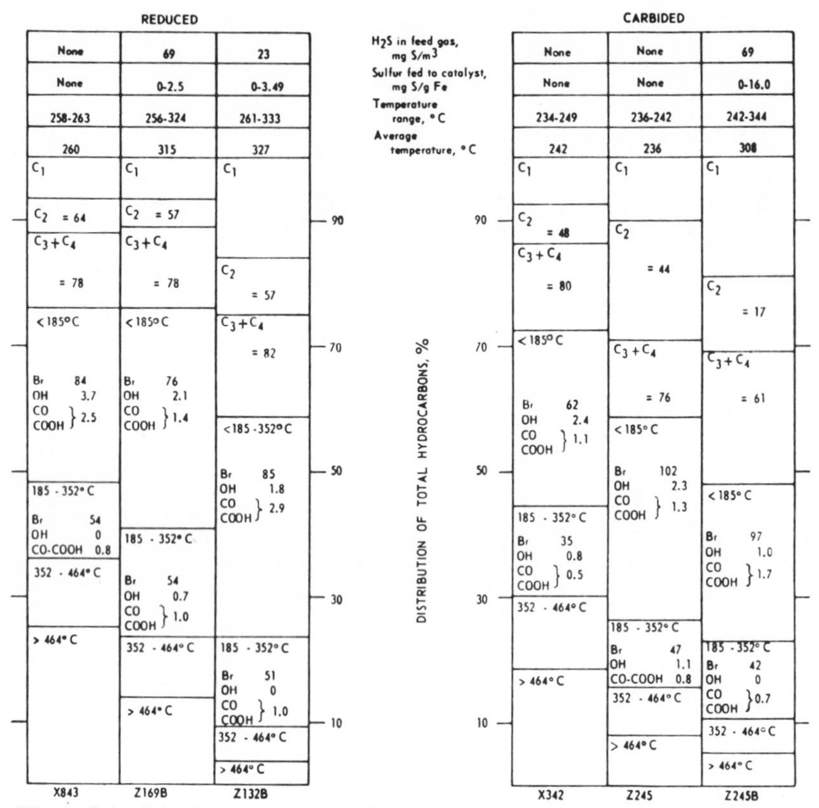

Fig. 6 Selectivity in constant-productivity poisoning tests of reduced and carbided catalyst D3001. Total hydrocarbons includes oxygenated material dissolved in liquid phase. From Anderson *et al.* (*5*).

the selectivity data reflect, in addition to the effect of poisoning, the effects of high temperatures and of changes in the catalyst that occur at these temperatures.

For reduced catalyst L2201 and reduced or carbided catalyst D3001, yields of C_{5+} hydrocarbons decrease and yields of gaseous hydrocarbons increase during poisoning in constant-productivity tests. For nitrided catalyst D3001 the distribution of products into C_{5+} and C_1–C_4 fractions did not change significantly during poisoning.

Liquids plus solid products from the tests in Table 4, which usually comprised 2 or more weeks of operation, were combined and fractionated by a simple distillation. These data and infrared analyses of the two lower boiling fractions are presented in Figs. 6 and 7. During poisoning with reduced or carbided catalysts, the yields of higher boiling fractions de-

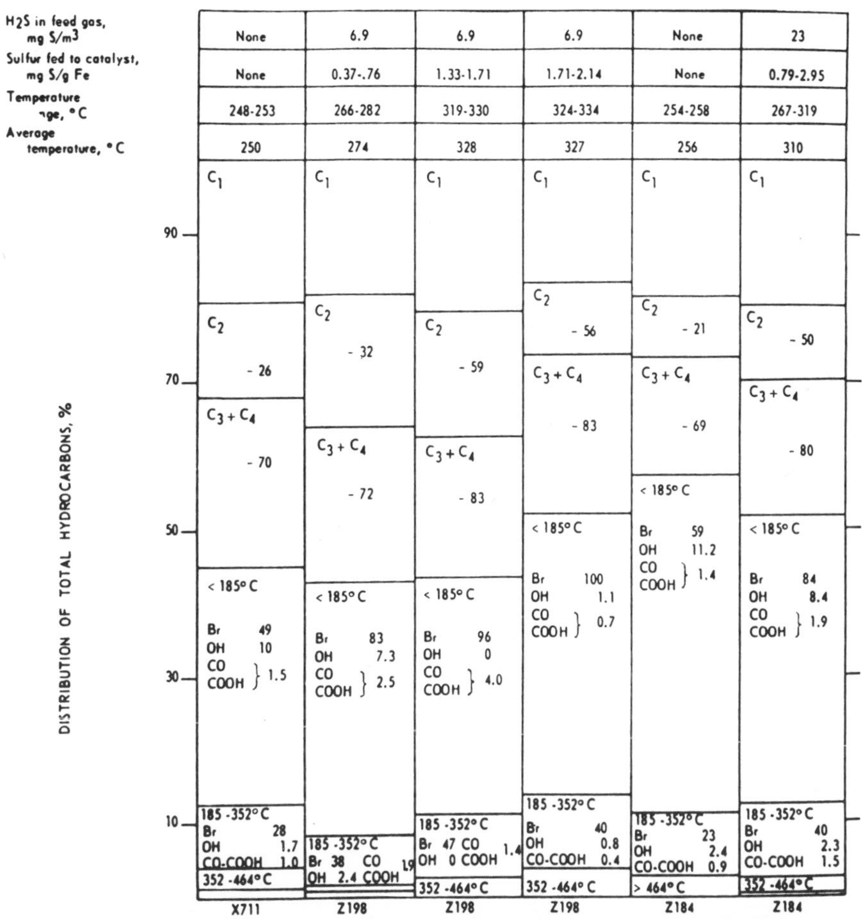

Fig. 7 Selectivity in constant-productivity poisoning tests of nitrided catalyst D3001. Total hydrocarbons includes oxygenated molecules dissolved in liquid phase. From Anderson *et al.* (5).

creased substantially and the yields of gaseous hydrocarbons increased. The fraction of oxygenated molecules in distillation fractions decreased with poisoning, but the gasoline fraction and the olefin content remained essentially constant.

For nitrided catalysts (Fig. 7), the distribution based on boiling point did not change greatly during poisoning. The yield of C_3 to 185°C remained ~55%, and C_1 + C_2 ~35%. The olefin content of the fractions increased with poisoning, and the alcohols (OH group) decreased. The effects are produced by the higher temperatures required to maintain constant productivity; the higher temperatures in turn cause the catalyst to lose its nitrogen rapidly. At the end of the test, the catalyst in Z198 had the following atom ratios to iron: carbon, 0.94; nitrogen, 0.03; and oxygen, 0.53.

An objective of these *in situ* poisoning tests (*5, 20, 38*) was to determine if gas containing 6.9 mg S per cubic meter, in 1960 the lowest concentration attainable using hot potassium carbonate scrubbing (*10, 11*), could be used. The following conclusions were drawn:

1. For the present catalysts operated at constant temperatures, or at increasing temperatures no higher than 290°C, catalyst life seems too short to be of practical interest.
2. For maximum temperatures of 350–400°C, lives of reduced, nitrided, or carbided D3001 at least approach those desired for practical synthesis processes. Because these tests are terminated by clogging, the length of time before the reactor becomes inoperable is probably strongly dependent on the configuration of the catalyst bed. Difficulties resulting from carbon deposition, particle disintegration, and other processes leading to clogging of the catalyst bed would probably become apparent earlier in life tests in large fixed-bed reactors than in the present laboratory units.
3. In constant-productivity operation the ability to control selectivity by adjusting temperature is lost. Gasoline production is not too seriously decreased by operation at higher temperatures. However, for the reduced and the carbided catalyst the yield of gaseous hydrocarbons increases substantially at the expense of fractions boiling above gasoline. For nitrided catalysts, the yield of products did not change significantly in the poisoning tests, but the yields of alcohols decreased and olefins increased.
4. On the basis of life, selectivity, and possible severe operating difficulties, it was concluded that additional gas purification would be desirable to reduce the sulfur content to less than 0.1 mg/m^3.

In a later Bureau of Mines paper (*6*), a number of factors in the poisoning of iron catalysts were examined. Table 6 shows data from poisoning tests with D3001 at 21.4 atm. Most of the sulfur was found in the inlet portion, and we may assume that this part had lost nearly all of its activity. The small concentrations in the rest of the bed, 0.2–0.9 mg S per gram Fe, were sufficient to decrease the relative activity in the last two-thirds or one-half of the bed to 0.15–0.3. Other tests were made on the effect of particle size and pore geometry on poisoning. It was observed that the average relative activity of a bed of catalyst, $\bar{F}$, decreased linearly initially with the average poison concentrations of the bed, $\bar{S}$, according to

Table 6
Sulfur Concentrations in Beds of Catalyst D3001 as a Function of Bed Length

Test number	X920	X999	Z295	
Mesh size	6–8	6–8	28–32	
H_2S in feed gas, mg S per cubic meter	69	188	69	
Relative activity at end of test	0.19	0.15	0.13	
Position in bed from inlet, cm	*Sulfur concentration, mg S per gram Fe*			*Position in bed from inlet, cm*
0–10	2.3	6.1	31.6	0–14
10–20	0.6	0.9	0.9	14–21
20–30	0.6	0.2	0.6	21–30

$\bar{F} = 1 - \alpha\bar{S}$ or $1 - \bar{F} = \alpha\bar{S}$, where α is a constant. The last form plotted on logarithmic scales is a useful way of expressing data for long poisoning tests. An initial linear part appears as a 45° straight line on these plots. The reciprocal of the constant, $1/\alpha$, is a useful criterion for the effectiveness of the poison. Figure 8 shows several tests of D3001 plotted in this way. The base test, Z181, used 6- to 8-mesh particles, a reduction temperature of 450°C, and was operated on pure gas for a week or more before introducing the feed containing H_2S. The surface area and average pore radius of the catalyst reduced at 450°C were 15 m^2 per gram Fe and 371 Å, compared with the catalyst in Z243 reduced at 400°C, 23 m^2 per gram Fe and 242 Å. Despite its larger initial surface area, the activity of the catalyst reduced at 400°C was the same as that reduced at 450°C. However, in the poisoning tests the catalyst reduced at 400°C lost activity more rapidly; values of $1/\alpha$ were 0.45 and 0.32 mg S per gram Fe for Z181 and Z243, respectively. Decreasing particle size increased resistance to poisoning substantially, as shown by Z295 with a value of $1/\alpha$ of 1.50 mg S per gram Fe. Decreasing the mesh size from 6–8 to 28–32 increases the external area per gram by a factor of 5.16, increases the activity in pure gas by a factor of 3.22 (*40*), and increases the poisoning constant by 3.34. These data suggest that the active portion of the catalyst and the part involved in the early part of poisoning tests is confined to a depth of ~0.1 mm from the outside of the particle. Additional data supporting this conclusion are that the surface areas at the start of the poisoning (after extraction of hydrocarbons) were 0.75 and 3.9 m^2 per gram Fe for the 6- to 8- and 28- to 32-mesh samples, respectively, and in Table 6 the concentration of sulfur in the initial portion of the bed of fine-mesh particles was very much larger than for the 6- to 8-mesh sample.

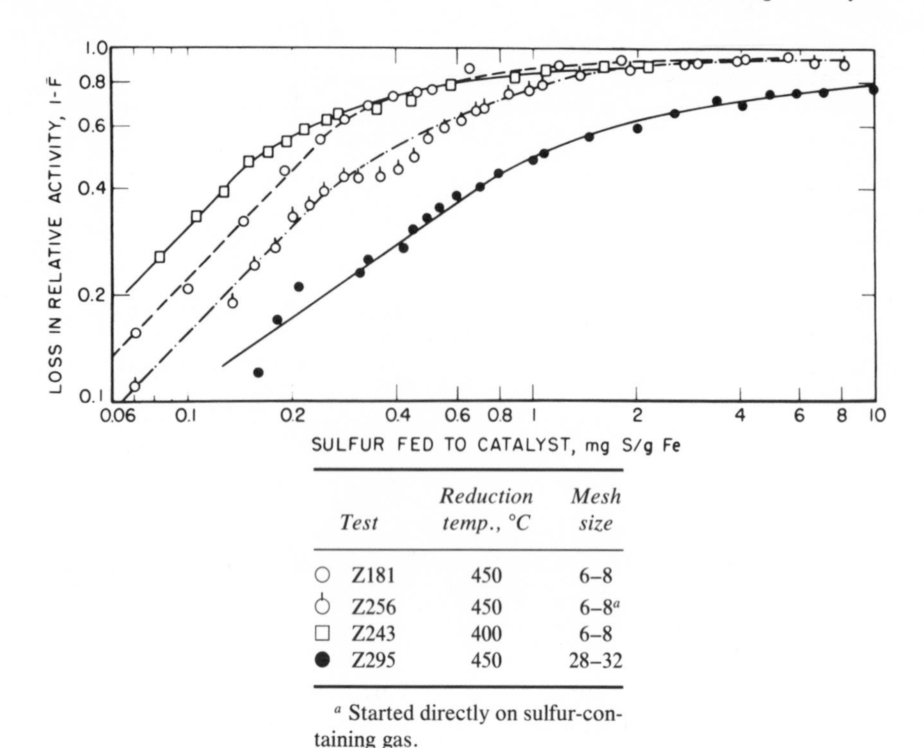

	Test	*Reduction temp., °C*	*Mesh size*
○	Z181	450	6–8
♁	Z256	450	6–8[a]
□	Z243	400	6–8
●	Z295	450	28–32

[a] Started directly on sulfur-containing gas.

Fig. 8 Effect of catalyst geometry on H_2S poisoning. Catalyst D3001 with 1 H_2 + 1 CO gas containing 69 mg S per cubic meter. Reproduced with permission from Anderson *et al.* (*6*).

K_2O or K_2CO_3 is the important promoter for iron catalysts. Table 7 shows the effect of changing alkali content on activity and resistance to poisoning. For test Z262, the alkali content was decreased by extracting the raw catalyst, and for Z276 the alkali was increased by impregnating the reduced catalyst with alcoholic KOH. L3140 and L3137 were prepared by impregnating raw P2007 with aqueous K_2CO_3 and aqueous $KMnO_4$, respectively. The resistance to poisoning as depicted by $1/\alpha$ values and poisoning curves of the type of Fig. 8 increased with increasing alkali content. The best results were obtained with $KMnO_4$ as a promoter.

Madon, Bucker, and Taylor (*24, 27*) studied the poisoning of a precipitated 60- to 120-mesh 100 Fe : 22 Cu : 1 K_2CO_3 catalysts. The catalyst, charged in seven identical reactor tubes, was pretreated at atmospheric pressure with 1 H_2 + 1 CO gas for 24 h at 225°C, 24 h at 230°C, and 24 h at 235°C. Four of the tubes were then treated at 1 atm and 235°C with 2 H_2 + 1 CO containing 150 ppm H_2S at an hourly space velocity of 4000 for

Table 7
Effect of Alkali on Poisoning[a]

Test no.	Catalyst no.	*Promoters per 100 Fe, by weight*				Mesh size	Initial activity on pure gas	*Poisoning test*	
		K_2O	MgO	Al_2O_3	*Other*			*Temp., °C*	*$1/\alpha$, mg S per gram Fe*
Z181	D3001	0.85	6.8	—		6–8	55	254	0.45
Z262	D3001X	0.25	6.8	—		6–8	62	248	0.35
Z276	D3001K	1.32	6.8	—		6–8	127	242	1.01
Z292	P2007	0.04	—	3.5		8–10	25	274	0.25
Z311	L3140	0.51	—	3.5		8–10	146	232	1.13
Z297	L3137	0.79	—	3.5	0.9 MnO_2	8–10	159	230	2.08

[a] All tests at 21.4 atm with 1 H_2 + 1 CO gas containing 69 mg S (as H_2S) per cubic meter. Adapted from Anderson *et al.* (*6*).

Table 8
Effect of H_2/CO Ratio on Poisoned, Precipitated-Iron Catalyst[a]

Conditions and results	H_2/CO 1.61				H_2/CO 1.03			
Nominal sulfur, wt %	0	0.08	0.24	0.40	0	0.08	0.24	0.40
CO_2-free contraction, %	76.6	74.1	76.2	75.8	84.4	84.3	84.2	85.4
H_2 + CO converted, %	81.3	78.5	81.8	80.0	87.2	87.4	87.1	88.1
Selectivity, % CO converted to								
CO_2	32.0	36.7	33.7	35.2	44.4	45.8	39.3	40.9
CH_4	6.7	7.1	6.9	6.2	4.5	4.7	5.1	4.7
C_{5+}	39.5	31.7	37.7	39.4	38.3	35.4	43.9	43.0
$C_{5+}/C_1–C_4$, wt basis	1.40	1.00	1.31	1.55	2.20	1.88	2.62	2.69
Olefins								
Ethylene/ethane	0.12	0.14	0.16	0.19	0.27	0.33	0.25	0.35
Propylene/propane	2.69	3.11	2.83	3.23	4.62	4.89	3.84	4.33
1-Butene/*n*-butane	2.05	2.39	2.17	2.50	3.80	4.16	3.16	3.58

[a] Temperature, 250°C; pressure, 10 atm; hourly space velocity, 195. Adapted from Madon *et al.* (*24*).

varying times. H_2S was not found in the exit gas, and the sulfur content of the four samples varied from 0.08 to 0.40 wt % (1.4 to 7.2 mg S per gram Fe). Concentration–bed-length profiles were determined; except for the largest concentration, nearly all of the sulfur was found in the first 7% of the bed. For the tube with 0.40 wt % S, most of the sulfur was found in the first fourth of the bed.

The pretreated catalysts in their reactor tubes were then tested in pure synthesis gas under different conditions. The presulfided catalysts were as active as the unpoisoned catalyst, as shown by representative data in Table 8. This work confirms the observations of Rapoport and Muzovkaya (*38*), that precipitated iron catalysts that have not been reduced to the metal in H_2 at substantial temperatures have remarkable resistance to sulfur poisoning. The differences of activity and selectivity data in Table 8 are about the uncertainties expected in catalyst tests. Although the conversions in Table 8 are relatively large for testing for poisoning, other data at lower conversions also show that the fresh and the prepoisoned catalysts have the same activity.

At SASOL, synthesis gas from the Lurgi Rectisol purification units routinely contains ~0.03 mg S per cubic meter (*15*). In a fixed-bed unit with iron catalyst, "notably steadier" conversion was obtained when the sulfur content of the feed was lowered from 0.2 to 0.02 mg S per cubic

Table 9
Loss in Activity of an Iron Catalyst in Fluid-Bed Reactor at 320°C[a]

Sulfur content of synthesis gas, mg S per cubic meter	*Decrease in percentage conversion per day*
0.1	Very low
0.4	0.25
2.8	2.0
28	33

[a] Reproduced with permission from Dry (*15*).

meter. When a fixed-bed reactor is poisoned by sulfur compounds, the sulfur is predominantly adsorbed near the inlet of the bed. Activity was determined as a function of distance from the inlet. The inlet portion was inactive, and for subsequent parts the activity increased progressively with increasing distance from the inlet. In the fixed-bed units ethyl mercaptan was found to be a more severe poison than H_2S. Apparently the H_2S was adsorbed more rapidly than ethyl mercaptan, and for this reason the H_2S was adsorbed in a shorter zone near the inlet than the mercaptan (*15*).

In the fluid- and entrained-bed reactors the catalyst is continually mixed and the inlet portion does not act as a purifying unit. Table 9 presents data for the FTS in a fluid-bed reactor at 320°C. With the Rectisol gas purification, sulfur poisoning is not a problem (*15*).

In SASOL tests (*15*), increasing the alkali promoter concentration was not effective in preventing sulfur poisoning.

6.3 Sulfur Poisoning of Cobalt, Nickel, Ruthenium, and Rhenium Catalysts

Poisoning studies on cobalt catalysts have been made on the standard catalyst of Fischer, 100 Co : 18 ThO_2 : 100 kieselguhr. King (*22*) reported the catalyst could be used effectively in synthesis gas containing CS_2. Catalysts containing up to 1.5 wt % S continued to produce liquid hydrocarbons if the temperature was increased from 185 to 210°C. At 210°C the unpoisoned catalyst would produce large yields of light hydrocarbons. Herington and Woodward (*18*) observed that addition of small quantities of H_2S or CS_2 increased the yield of liquids and decreased the gaseous hydrocarbons. A patent by Myddleton (*32*) claimed that the Fischer catalyst alkalized with K_2CO_3 could be used in gas containing up to 57 mg of organic S per cubic meter without substantial loss of activity, and that the sulfided catalyst produced more olefins and more high boiling liquids than the fresh catalyst. Pichler (*36*) gave a sulfur–bed-length profile for a Ruhrchemie cobalt catalyst after 1 year of synthesis at 10 atm. The sulfur concentration in the catalyst decreased linearly from 0.8% at the inlet to nearly zero at the middle of the bed. The first 20% of the bed contained sulfate produced by traces of oxygen in the synthesis gas.

Madon and Taylor (*26*) checked the work of Myddleton using a 60- to 120-mesh catalyst containing 100 Co : 16 ThO_2 : 93 kieselguhr : 2 K_2CO_3 in a unique metal-block reactor with 0.77-mm internal diameter (I.D.) tubes containing a packed bed about a meter long. Care was taken to avoid diffusional problems and overheating. Activation energies, pressure dependence, and turnover numbers were determined and agreed well with literature values.

The reduced catalyst was sulfided at 180°C with 2 H_2 + 1 CO gas containing 250 ppm H_2S at atmospheric pressure for different times to introduce the desired amount of sulfur. No sulfur compounds were found in the outlet stream. The catalyst bed was divided into 17 sections; most of the sulfur was found in the first 3 parts.

The poisoned catalysts were tested at 6–16 atm at 197°C, as shown in Table 10. Most of the poisoned catalysts were as active as the fresh catalyst. The selectivity for producing C_{5+} hydrocarbons was 75–85% and seemed independent of sulfur content. For 2 H_2 + 1 CO feed the olefin content of the C_3 and C_4 hydrocarbons increased with sulfur content. Carbon-number distributions were reported; however, these data are not typical of products from cobalt at atmospheric pressure (*13*); having maxima or shoulders in the ranges C_9–C_{12} and C_{20}–C_{25}. Possibly the unusual distributions result from the mode of operation, 5 or 6 h of synthesis followed by an overnight period in H_2. At the lower pressures, 6 and 11 atm, more high molecular weight products were obtained with increasing sulfur content.

Similar tests were made with a 1% ruthenium on alumina in the form of a 60- to 140-mesh powder (*24*). The catalyst was tested after reduction in H_2 and after presulfiding at 200°C with a space velocity of 300 h^{-1} of 2 H_2 + 1 CO containing 260 ppm H_2S. Another portion of the catalyst was impregnated with aqueous KOH to give a K_2O content of 10 wt % in the final catalyst and reduced and sulfided as described before. Sulfur concentration profiles were essentially independent of bed length, and H_2S was found in the exit gas from the nonalkalized catalysts.

Although alkalizing and sulfiding the catalyst decreased activity, the effects were remarkably smaller at an operating pressure of 31 atm than at 22 atm, as shown in Table 11. At 240°C the alkalized catalyst in the

Table 10
Sulfur Poisoning of a Standard Cobalt Catalyst[a]

Sulfur, wt %	*H_2/CO ratio*	*Pressure, atm*	*Conversion of H_2 + CO, %*	*Selectivity, %*[b]	*Ratio of olefins to paraffins*	
					C_3	*C_4*[c]
0	1.9	6	91	84	1.25	0.64
	1.5	6	80	83	2.37	1.40
	1.9	11	96	78	0.61	0.23
	1.5	11	81	80	2.22	1.36
	2.0	16	87	75	0.82	0.39
0.29	1.9	6	86	81	1.45	0.79
	1.5	6	82	83	2.33	1.36
	1.9	11	98	79	0.62	0.19
	1.5	11	83	82	1.80	1.08
	2.0	16	97	84	0.41	0.13
0.50	1.9	6	78	85	2.25	1.39
	1.5	6	78	85	2.43	1.44
	1.9	11	98	75	—	—
	1.5	11	79	78	2.00	1.06
	2.0	16	88	83	1.29	0.69

[a] Temperature, 197°C; hourly space velocity, 200–210. Adapted from data in Madon and Tayor (*26*).
[b] Percentage CO converted to C_{5+} hydrocarbons.
[c] 1-Butene to *n*-butane.

reduced and in the sulfided form had about equal activity. Alkalizing and/or sulfiding decreased the conversion of CO to gaseous hydrocarbons and CO_2.

Agrawal *et al.* (*2*) studied H_2S poisoning of cobalt on fused α-alumina plates in an internal recycle reactor at 390°C and 1 atm with H_2 containing 1–4% CO. The methanation activity decreased by more than a factor of 1000 when exposed to feed containing 13 ppb and by 10,000 in 90 ppb. These concentrations are too low for the formation of bulk sulfides. Auger electron spectroscopy showed the formation of a surface sulfide with one sulfur to two cobalt atoms. The relative activity decreased with the square of the sulfur-free metal surface, as shown in Fig. 9. Poisoning rates suggest that H_2S molecules adsorb with a sticking coefficient of 1.0 up to a surface coverage of 70%, and lower values at larger coverages. Random adsorption of H_2S rather than formation of islands was postulated.

Catalysts poisoned by carburization and the fresh preparations were compared in sulfur poisoning. Both ended at about the same low activity.

Table 11
Tests of Alkalized and/or Sulfided Ruthenium on Alumina Catalysts[a]

Pressure, atm	*Pretreatment*[b]	*Conversion of $H_2 + CO$*	*Average sulfur content,*[c] *wt %*	*Percentage conversion of CO to*		
				C_{5+}	CH_4	CO_2
22	R	87.7	0	85.3	4.1	2.7
	R + S	9.4	0.23	99.6	0.1	0
	K + R	7.4	0	97.2	1.6	0.4
	K + R + S	4.8	0.34	97.3	0.8	0.9
31	R	95.5	0	99.6	0.2	0
	R + S	52.9	0.23	99.9	0.02	0.1
	K + R	49.1	0	97.2	0.1	2.5
	K + R + S	35.8	0.34	99.8	0.05	0.1

[a] 2 H_2 + 1 CO gas at an hourly space velocity of 200; Temperature, 240°C. Adapted from Madon *et al.* (*24*).
[b] R, Reduced; S, sulfided; K, alkalized with KOH.
[c] Average sulfur concentration calculated from longitudinal sulfur distribution.

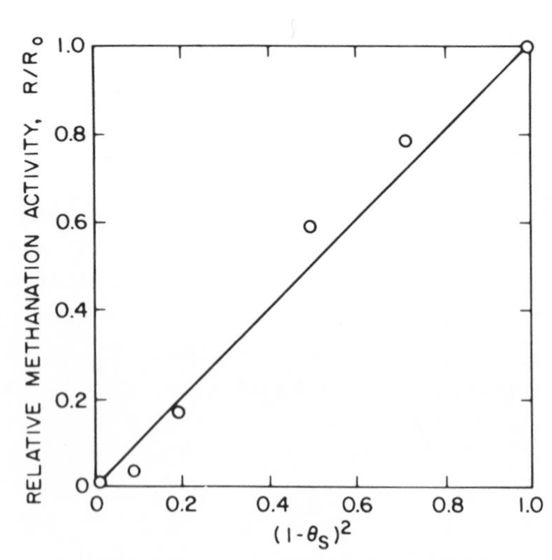

Fig. 9 Relative steady-state methanation activity of Co/Al_2O_3 as a function of the square of the sulfur-free surface fraction $(1 - \theta_S)^2$. Reproduced with permission from Agrawal *et al.* (*2*).

Table 12
Poisoning of Supported Nickel, Ruthenium, and Rhenium[a]

	Turnover numbers N, $s^{-1} \times 10^4$					
	Steady state on pure gas		*After 24 h on 10 ppm H_2S*		*On pure gas again*	
Catalyst	*CH_4*	*C_{2+}*	*CH_4*	*C_{2+}*	*CH_4*	*C_{2+}*
At 400°C						
1.5% Ru/Al_2O_3	550	63	8.2	6.6	58	30
0.5% Ru/Al_2O_3	348	21	7.0	5.8	81	46
5% Ru/Al_2O_3 (heat treated)	987	0	2.0	1.9	14	7.4
5% Ni/ZrO_2	156	0	30	32	30.5	45
2% Ni/Al_2O_3	519	0	10.6	7.7	52	24
Raney nickel	629	9	0.5	0.8	5.1	3.9
1% Re/Al_2O_3[a]	400	100	26	44	40	10
At 250°C						
1.5% Ru/Al_2O_3	110	220	0.005	0.48	0.07	0.36
5% Ni/ZrO_2	373	807	1.42	3.42	0.20	0.28
2% Ni/Al_2O_3	173	495	0.14	0.14	0.12	0.08
Raney nickel	149	410	0.84	2.3	0.44	1.7

[a] Tests at 0.75 atm; 3.8 H_2 + 1 CO. Adapted from Dalla Betta *et al.* (*13*).
[b] Data were read from a graph. The C_{2+} values have large uncertainties.

The catalysts after sulfur poisoning, carbon poisoning and carbon-plus-sulfur poisoning had the same activation energy for methanation, 16 kcal/mol, compared with 28 for the fresh catalyst.

Dalla Betta *et al.* (*13*) examined a number of supported nickel, ruthenium, and rhenium catalysts and Raney nickel for resistance to poisoning by 10 ppm H_2S. The tests were made at 250 and 400°C in the following way. The catalysts were reduced in H_2 and operated in 3.8 H_2 + 1 CO until a constant activity was attained, then 10 ppm H_2S was introduced into the feed for 24 h and the activity determined; finally, pure gas was passed over the catalyst for 24 h and the activity again measured as given in Table 12. The synthesis data were reported in terms of turnover numbers for CH_4 and C_{2+} hydrocarbons, as moles produced per site · second; the data for C_{2+} are ambiguous because the molecular weight is not defined. At both temperatures the 24-h exposure to 10 ppm H_2S decreased both turnover numbers drastically, and 24 h on pure gas at 400°C caused modest, but far from complete, reactivation. At 400°C, nickel on zirconia was most resistant to poisoning, and Raney nickel the least. The produc-

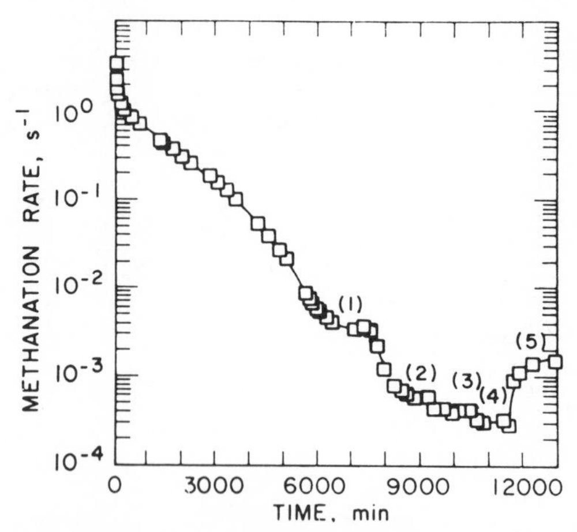

Fig. 10 *In situ* poisoning of $Ni-Al_2O_3$. Reaction conditions: 100 kpa, 661 K, 4% CO, 96% H_2; H_2S (ppb): (1), 14; (2), 52; (3), 73; (4), 93; (5), 15. Reproduced with permission from Agrawal *et al.* (*1*).

tion of C_{2+} was substantially larger for the poisoned catalysts, compared with the fresh samples, even at 400°C. The rhenium catalyst was not very active before poisoning; it was severely poisoned in the period in H_2S, and the activity remained low in the final period on pure gas.

Agrawal *et al.* (*1*) studied the poisoning of nickel dispersed on nonporous alumina in a recycle reactor without concentration gradients in the gas phase at low levels of H_2S. As shown in Fig. 10, the methanation decreased and attained a steady state corresponding to equilibrium between the catalyst and the gas phase. When the concentration was increased, the activity either decreased further and attained a new steady state or remained constant. Decreasing the concentration of H_2S increased the activity. Figure 11 shows the steady-state relative activity of iron, cobalt, ruthenium, and nickel in methanation as a function of concentration of H_2S. For the nickel on alumina, the methanation activity varied linearly with the square of the fraction of the unpoisoned surface (Fig. 12). Apparent activation energies were reported for supported cobalt, nickel, and ruthenium; the values were not changed by sulfur poisoning. The authors concluded that the catalysts studied were more sensitive to sulfur poisoning than previously considered. Concentrations of H_2S of 13 ppb decreased the activity 5000-fold. Chemisorbed sulfur is more stable than bulk sulfides. The primary effect of sulfur is to block the metal surface.

Den Besten and Selwood (*14*) and Ng and Martin (*33*) made magnetic measurements on nickel-on-silica catalysts as H_2S was added at room temperature. Initially the decrease in magnetization corresponded to the

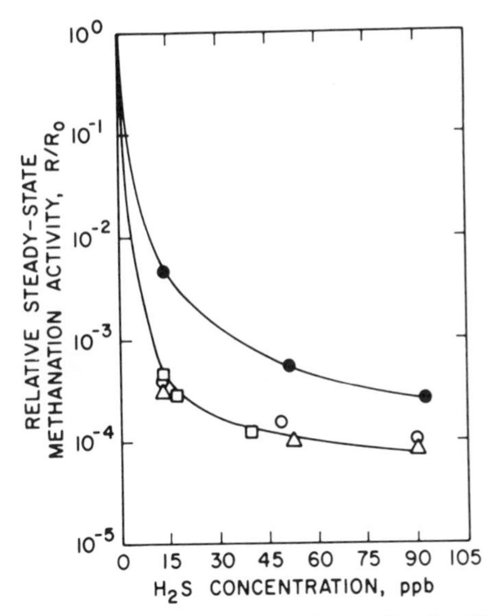

Fig. 11 Relative *steady-state* methanation activity profiles for Ni (●), Co (△), Fe (□), and Ru (○) as a function of gas-phase H_2S concentration. Reaction conditions: 100 kpa, 663 K, 1% CO, 99% H_2, for Co, Fe, and Ru; 100 kpa, 661 K, 4% CO, 96% H_2, for Ni. Reproduced with permission from Agrawal *et al.* (*1*).

formation of four bonds to the Ni, 2 H, and 1 S. As the amount of H_2S adsorbed increased, the decrease became smaller corresponding to about two bonds and H_2 was desorbed. The chemisorption of CO (at 1 torr) and H_2 at room temperature decreased linearly with the amount of H_2S adsorbed. The H_2S interacted only with the surface nickel, and H_2 and CO

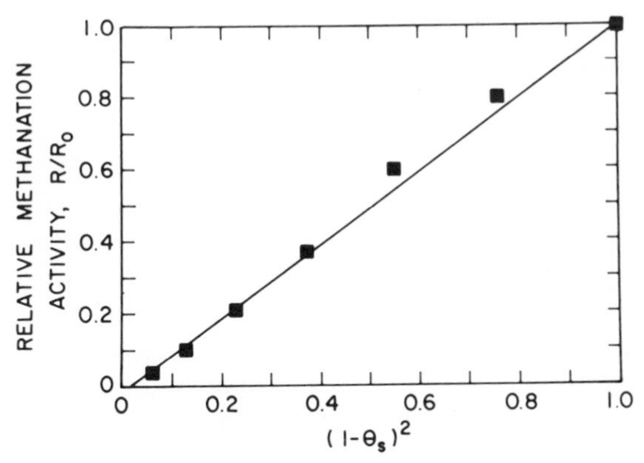

Fig. 12 Methanation turnover number of $Ni–Al_2O_3$ as a function of $(1 - \theta_S)^2$. Reaction conditions: 100 kpa, 661 K, 4% CO, 86% H_2. Reproduced with permission from Agrawal *et al.* (*1*).

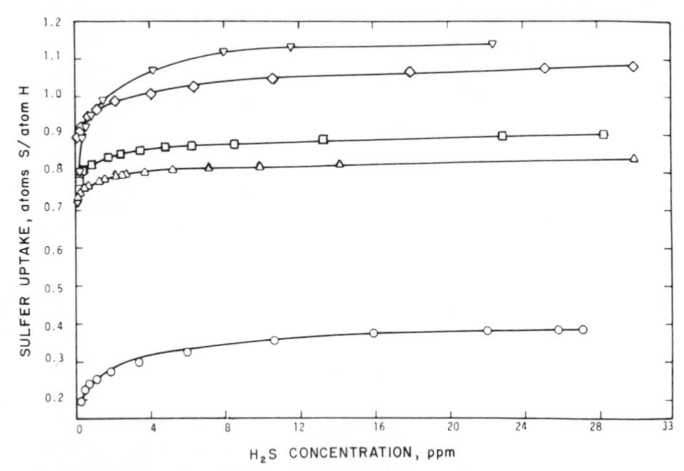

Fig. 13 H_2S desorption isotherms at 450°C normalized to H_2 uptake: ◇, 3% Ni; ▽, Ni powder; △, Ni–Pt; □, Ni–Co; ○, Ru. Reproduced with permission from Oliphant *et al.* (*34*).

were chemisorbed on surface nickel and not on sulfur-poisoned nickel (*33*).

Klostermann and Hobert (*23*) studied the reactions of organic sulfur molecules on reduced nickel on silica, using infrared and thermal desorption; these compounds included four mercaptans, three sulfides, two disulfides, and thiophene. At room temperature, the sulfur compounds dissociatively chemisorbed, involving rupture of S—H, C—S, and S—S bonds. Hydrocarbons were found in the gas phase. Subsequently, the sample was evacuated and then subjected to thermal desorption. Additional hydrocarbons were evolved in TD up to ~500 K. In other thermal-desorption tests, chemisorbed CO was evolved in two peaks, one at 400–500 K, and a second at 800–1100 K. Preadsorption of organic sulfur molecules nearly eliminated the higher temperature peak and decreased the temperature at which the lower peak occurred.

The reversible adsorption of sulfur on nickel (*28, 34, 35, 39*), nickel alloys (*34*), and ruthenium (*34*), has been reported. The sulfur coverage was determined as a function of the ratio of H_2S/H_2 being flowed or circulated over the sample at equilibrium. These experiments yield isotherms of sulfur coverage as a function of ratio H_2S/H_2, or as shown in Fig. 13, the concentration of H_2S in H_2, or plots of log $[H_2S]/[H_2]$ as a function of reciprocal temperature at constant surface coverage, as shown in Fig. 14. The ratios H_2S/H_2 for nickel in Fig. 14 are smaller by a factor of 10^5 than those required for forming the most stable bulk sulfide Ni_3S_2. McCarty and Wise (*28*) found the heat of adsorption of H_2S increases with coverage; at 75% coverage is −37 kcal/mol. The chemisorption of CO at

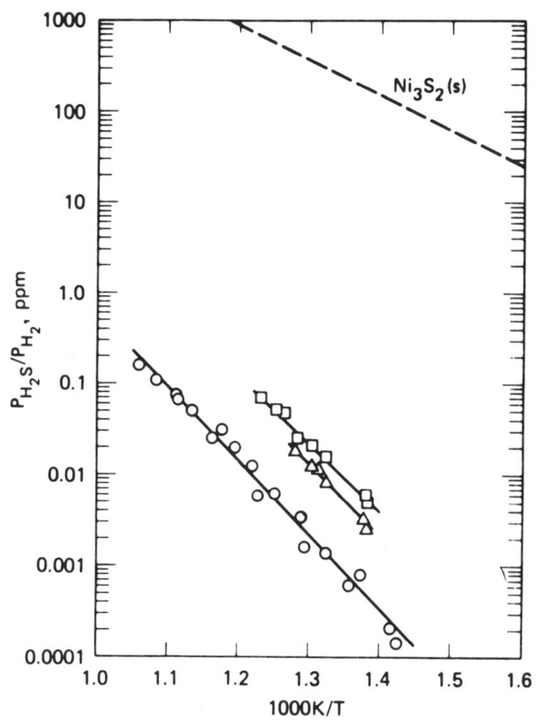

Fig. 14 Thermodynamic activity of sulfur chemisorbed on supported nickel. ○, 5 wt % Ni/α-Al_2O_3, $\xi = 0.75$; □, 5 wt % Ni/γ-Al_2O_3, $\xi = 0.674$; △, Ni sponge, $\xi = 0.5$, where ξ is the coverage normalized to CO uptake at 300 K. Reproduced with permission from McCarty and Wise (*28*).

300 K was used to estimate the nickel area and the adsorption ratio S/CO = 1.11 was used in their final calculations of surface coverage.

6.4 Sulfur Poisoning of Molybdenum and Tungsten Catalysts

Shultz *et al.* (*43*) studied sulfur poisoning of coprecipitated molybdena–alumina catalysts. The catalyst in one test, Z335, contained MoO_3, and another preparation in Z329 was precipitated in the presence of H_2S and contained molybdenum sulfide. After reduction in H_2 at 400°C, both catalysts were moderately active at 400°C and 21.4 atm using pure 3 H_2 + 1 CO gas at an hourly space velocity of ~300. As shown in Fig. 15, the contraction decreased sharply when a feed containing 6.9 mg S (as H_2S)

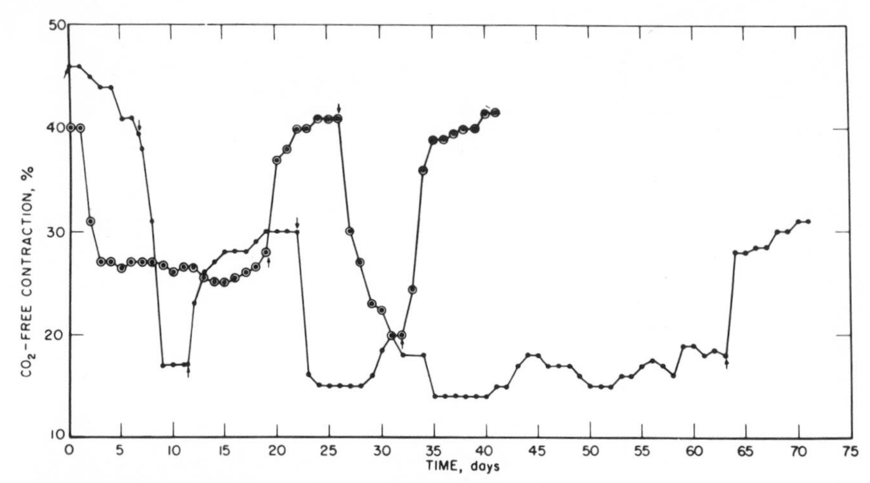

Fig. 15 Poisoning of precipitated Al_2O_3–MoO_3 and Al_2O_3–MoS_3 catalysts. ↑, Pure 3 H_2 + 1 CO in; ↓, 3 H_2 + 1 CO + H_2S in; •, test Z335 MoO_3; ⊙, test Z329 MoS_3. From Shultz *et al.* (*43*).

per cubic meter was used. For the presulfided catalyst in Z329, the activity was completely restored when H_2S was removed from the feed; in test Z335 the activity was only partly regained. In most operating periods the hydrocarbon product was largely methane, 84–93 wt % of the total.

Murchison and Murdick (*31*) found that a molybdenum catalyst in the oxide form was not adversely affected by using 1 H_2 + 1 CO gas containing 7 ppm H_2S at 400°C and 62 atm. More than half of the product was C_{2+}. In similar tests with 70 ppm H_2S, the catalyst rapidly lost activity and the ability to produce C_{2+}. Reduction in H_2 at 600°C restored both activity and selectivity. Good performance in the presence of sulfur compounds was obtained with presulfided catalysts. Results with preparations on alkalized carbon were better than for those on alkalized alumina. A presulfided molybdenum on alkalized carbon had a constant activity in 0.8 H_2 + 1 CO feed containing 24 ppm H_2S at 360°C and 25 atm for at least 500 h, and C_2–C_5 amounts in the products were 70–75%. A similar catalyst maintained nearly the same activity and selectivity on 100 ppm COS (for 170 h) as on pure gas. In tests with molybdenum on alkalized alumina, both the activity and the yield of LPG decreased in synthesis with 100 ppm COS, and the original performance of the catalyst was not fully restored in subsequent operation on pure gas. Further studies (*30*) showed that alkalized catalysts were resistant to poisoning by 100 ppm of COS (Table 13). Carbon was a better support than alumina, and the activity and selectivity of the carbon-supported catalyst were virtually un-

Table 13
Effect of Support on Sulfur Tolerance of Alkalized Molybdenum Catalyst[a]

	Supported on			
	Alumina		*Carbon*	
Time	*Conversion of CO, %*	*Selectivity to LPG, %*	*Conversion of CO, %*	*Selectivity to LPG, %*
Before sulfur addition	63	53	58	61
After exposure to 100 ppm COS	(90 h) 48	37	(170 h) 56	56
After reduction in H_2 at 500°C	57	40	61	60

[a] Reproduced with permission from Murchison (*30*).

changed in 170 h of use with the poisoned feed gas. Treatment in H_2 at 500°C improved activity and selectivity. Alkalized molybdenum on a low-area alumina for producing C_2–C_4 hydrocarbons maintained its selectivity for olefins in 120 h with a feed containing 20 ppm H_2S (Table 14).

Madon, Bucker, and Taylor (*24*) studied the FTS on a presulfided fine-mesh cobalt molybdate catalyst (5% CoO, 11% MoO_3 on alumina) in a fixed-bed reactor. A portion of the catalyst was impregnated with K_2CO_3 to give 3.45% K_2O in the material. Both were reduced in H_2 at 450°C. Some of the reduced samples were treated with 0.1% H_2S in H_2 at 300°C

Table 14
Sulfur Tolerance of Molybdenum Supported on a Low-Area Alumina[a]

		Selectivity for gaseous hydrocarbons, %	
Hydrocarbons		*370 h on pure synthesis gas*	*+120 h on synthesis gas with 20 ppm H_2S*
Olefins	C_2	23	20
	C_3	17	20
	C_4	4	7
Paraffins	C_1	40	43
	C_2	7	6
	C_3	8	2
	C_4	1	1
Conversion of CO, %		42	37

[a] 4 m^2/g; 10% MoO_3–2% K_2O–alumina. Reproduced with permission from Murchison (*30*).

Table 15
Tests of Cobalt Molybdate on Alumina and MoS_2[a]

Catalyst	*Pretreatment*[b]	*Conversion of H_2 + CO*	*Percentage conversion of CO to*				
			CO_2	*CH_4*	*C_2*	*$C_3 + C_4$*	*C_{5+}*
A. At 15 atm and 302°C							
Cobalt molybdate	R	43	50.3	26.2	17.5	10.4	0
on alumina	R + S1	48	38.2	22.8	16.9	3.2	18.9
	R + S2	42	42.9	21.9	23.4	5.7	5.9
	K + R	45.5	41.2	14.9	15.2	9.5	19.2
	K + R + S1	45	29.4	12.9	10.9	5.5	41.2
	K + R + S2	40.5	29.8	10.6	15.1	6.3	38.2
B. At 20 atm and ~350°C							
MoS_2	K + R	51	5.1	2.7	1.1	0.7	90.4
CoMo on alumina	K + R + S2	54	39.5	18.4	21.1	13.8	13.8

[a] 2 H_2 + 1 CO gas and hourly space velocity of 200. Adapted from Madon *et al.* (*24*).
[b] R, Reduced in H_2 at 450°C for 24 h; K, alkalized; S1, presulfided with H_2S in H_2 to an average S concentration of 1%; S2, presulfided in H_2S in H_2 to ~5% S.

to introduce ~1% sulfur, and others were treated with 10% H_2S in H_2 to completely sulfide the catalyst. The latter group of catalysts contained ~5% sulfur throughout the bed, and the former had most of the sulfur in the first two-thirds of the bed. All of the samples had modest activity in FTS at 300°C, as shown in Table 15. Here, the prepoisoned alkalized catalysts produced moderate yields of C_{5+}. Alkalized and/or sulfided catalysts produced less CO_2. The hydrocarbons were principally normal paraffins with a moderate amount of isoparaffins and only a very small amount of olefins. At 350°C for the sulfided catalysts the ratio of isobutane to *n*-butane was larger than 0.22.

A finely divided MoS_2, surface area 2.2 m^2/g, was impregnated with 3% KOH and tested in FTS to give results shown in part B of Table 15. The activity was low, requiring temperatures of 350°C or higher, but the yields of C_{5+} as percentage of CO converted were 90.4 at 350°C and 82.3 at 400°C. The liquid product contained only a small amount of *n*-paraffins. The C_4 fraction contained isobutane (16%), isobutene (11%), *n*-butane (31%), 1-butene (28%), and cis-2-butene (14%).

Madon, Bucker, and Taylor (*24*) also tested tungsten (10% WO_3) on alumina and a nickel–tungsten (3% NiO, 10% WO_3) on alumina in the reduced and sulfided forms, with and without K_2CO_3 addition. Short tests of 4- to 8-h duration are shown in Table 16. The activity of all of these catalysts was low, but the selectivity for producing C_{5+} was remarkably large for some alkalized and/or sulfided catalysts.

Table 16
Short Tests of Tungsten-Based Catalysts[a]

Operating conditions			*Ni, W on alumina*		*Ni, W on alumina*		*W on alumina*	
Temp., °C	*Feed H_2/CO*	*Space velocity, h^{-1}*	*K + R*[b]	*K + R + S1*[b]	*R*	*R + S2*[b]	*K + R*	*K + R + S3*[b]
350	1.97	200	3.4[c]	13.8	2.6	8.6	2.6	9.0
			(27.8)	(46.5)	(33.3)	(0)	(79.2)	(14.5)
400	1.94	200	6.1	19.6	3.7	16.2	2.6	3.6
			(8.0)	(12.5)	(36.5)	(0)	(58.2)	(14.9)
450	2.02	200	15.6	33.2	3.9	29.5	2.2	3.8
			(0.01)	(13.0)	(0)	(21)	(11.2)	(0)
450	1.00	200	9.7	23.0	6.7	22.1	6.2	7.8
			(2.7)	(27.1)	(26.6)	(2.5)	(71.4)	(62.8)
450	0.99	115	16.1	40.7	7.8	27.2	10.9	11.0
			(1.5)	(31.6)	(0)	(0)	(69.0)	(37.1)
450	3.01	200	(4.0)	22.5	4.2	10.4	2.5	3.7
			(0)	(30.3)	(0)	(0)	(38.4)	(30.0)
350	1.98	100	2.9	12.1	3.2	5.7	3.1	3.4
			(60.5)	(77.0)	(58.2)	(15.1)	(90.8)	(83.9)

[a] Pressure, 20 atm. Adapted from Madon *et al.* (*24*).
[b] K, Alkalized to ~3% K_2O; R, reduced in H_2 at 450°C; S1, presulfided with H_2S in H_2 to 0.71% S; S2, presulfided to 0.93% S; S3, presulfided to 1.3% S.
[c] Upper number of pair denotes the conversion of H_2 + CO in percent. The lower number in parenthesis is the percent of CO converted to C_{5+}.

6.5 Concluding Statement

On catalysts containing iron, cobalt, nickel, and ruthenium, H_2S and most other sulfur compounds are dissociatively chemisorbed on the metal. The surface sulfur is held very much more strongly than the sulfur in bulk sulfides. Generally, it is impossible to remove sulfur from poisoned catalysts under conditions that will not destroy the catalyst. Apparently the poisoning by sulfur compounds occurs mainly by the sulfur atoms chemisorbing on the metal sites and eliminating them from the catalytic reaction.

Some promoters, particularly alkali, increase the resistance of the catalyst to poisoning, but these improvements are small. Often partly poisoned catalysts produce higher molecular weight products than the fresh catalyst. In some single-pass, fixed-bed reactors, the inlet end of the catalyst will serve as a purifier for the rest of the bed, and reasonably long lives may be obtained with some catalysts.

Nevertheless, most of the present catalysts of the group iron, cobalt, nickel, and ruthenium must be considered to be too sensitive to sulfur poisoning for use with sulfur-containing synthesis gas, and it is recommended that essentially all of the sulfur, to concentrations of a few parts per billion, be removed.

Catalysts of molybdenum and tungsten seem to have a potential for the development of sulfur-resistant catalysts; however, present catalysts generally have low activity and poor selectivity. Catalysts with high sulfur resistance capable of operating with synthesis gas containing large quantities of sulfur would probably not be very useful, because a substantial fraction of the sulfur would probably be incorporated in the synthesis products.

6.6 Addendum

Two pertinent papers were published after the manuscript for Chapter 6 was completed:

1. Fritzharris, W. D., Katzer, J. R., and Manogue, W. H., *J. Catal.* **76,** 369 (1982). This work is similar to reference *1* in some respects and considers poisoning by H_2S of nickel on nonporous alumina spheres and plates, nickel on porous α-alumina, and nickel on quartz.
2. Agrawal, P. K., Katzer, J. R., and Manogue, W. H., *J. Catal.* **74,** 332 (1982). This work describes poisoning by H_2S of ruthenium on fused alumina plates.

In addition, an extensive review of the interactions of sulfur compounds with metals and sulfur poisoning in the hydrogenation of CO, the ammonia synthesis, and other reactions has been published [Bartholomew, C. N., Agrawal, P. K., and Katzer, J. R., *Adv. Catal.* **31,** 135 (1982)].

References

1. Agrawal, P. K., Fitzharris, W. D., and Katzer, J. R., *in* "Catalyst Deactivation" (B. Delmon and G. F. Froment, eds.), p. 179. Am. Elsevier, New York, 1980.
2. Agrawal, P. K., Katzer, J. R., and Manogue, W. H., *J. Catal.* **69,** 327 (1981).
3. Anderson, R. B., *in* "Catalysis" (P. H. Emmett, ed.), Vol. 4, p. 242. Van Nostrand-Reinhold, Princeton, New Jersey, 1956.
4. Anderson, R. B., *in* "Experimental Methods in Catalytic Research," (R. B. Anderson, ed.), Vol. 1, pp. 34–40. Academic Press, New York, 1968.

5. Anderson, R. B., Karn, F. S., Kelly, R. E., and Shultz, J. F., *Bull.–U.S., Bur. Mines* **628,** 16 (1965).
6. Anderson, R. B., Karn, F. S., and Shultz, J. F., *J. Catal.* **4,** 56 (1965).
7. Anderson, R. B., and Whitehouse, A. M., *Ind. Eng. Chem.* **53,** 1011 (1961); Anderson, R. B., *J. Catal.* **4,** 293 (1956).
8. Balatnikova, Yu. I., Apel'baum, L. O., and Temkin, M. I., *Zh. Fiz. Khim.* **32,** 2717 (1958).
9. Bayer, J., Stein, K. C., Hofer, L. J. E., and Anderson, R. B., *J. Catal.* **3,** 145 (1964).
10. Benson, A. E., and Field, J. H., U.S. Patent 2,886,405 (1959).
11. Benson, H. E., Field, J. H., and Haynes, W. P., *Chem. Eng. Prog.,* **52,** 433 (1956); Benson, H. E., Field, J. H., and Jimeson, R. M., *ibid.* **50,** 356 (1954).
12. Brill, R., FIAT, Reel R-19, Frames 7088–7104 (1941).
13. Dalla Betta, R. A., Piken, A. G., and Shelef, M., *J. Catal.* **40,** 173 (1975).
14. den Besten, I. E., and Selwood, P. W., *J. Catal.* **1,** 93 (1962).
15. Dry, M. E., *in* "Catalysis—Science and Technology" (J. R. Anderson and M. Boudart, eds.), Vol. 1, pp. 201–202. Springer-Verlag, Berlin and New York, 1981.
16. Fischer, F., *Brennst.-Chem.* **16,** 1 (1935).
17. Frear, G. L., Shultz, J. F., and Elmore, K., *Tenn. Val. Intern. Rep.* **R545,** 1–22 (1950).
18. Herington, E. F. G., and Woodward, L. A., *Trans. Faraday Soc.* **35,** 958 (1939).
19. "Fuel Research Report," Fuel Research Board, Greenwich, England, 1955, p. 20; 1956, p. 22.
20. Karn, F. S., Shultz, J. F., Kelly, R. E., and Anderson, R. B., *Ind. Eng. Chem. Prod. Res. Dev.* **2,** 43 (1963).
21. Karn, F. S., Shultz, J. F., Kelly, R. E., and Anderson, R. B., *Ind. Eng. Chem. Prod. Res. Dev.* **3,** 33 (1964).
22. King, J. G., *J. Inst. Fuel* **11,** 484 (1938).
23. Klostermann, K., and Hobert, H., *J. Catal.* **63,** 355 (1980).
24. Madon, R. J., Bucker, E. R., and Taylor, W. F., *U.S. Energy Res. Dev. Admin. (Dep. Energy), Final Rep.* Contract No. E(46-1)-8008 (1977).
25. Madon, R. J., and Shaw, H., *Catal. Rev.–Sci. Eng.* **15,** 69 (1977).
26. Madon, R. J., and Tayor, W. F., *Adv. Chem. Ser.* **178,** 93 (1979).
27. Madon, R. J., and Taylor, W. F., *J. Catal.* **69,** 32–43 (1981).
28. McCarty, J. G., and Wise, H., *J. Chem. Phys.* **72,** 6332 (1980).
29. Mittasch, A., *Adv. Catal.* **2,** 81 (1950).
30. Murchison, C. B., *Pap.—Coal Technol. '80 [Eighty] Conf., 1980.*
31. Murchison, C. B., and Murdick, D. A., *Prepr. Pap., AIChE Meet., 1980.*
32. Myddleton, W. W., British Patent 509,325 (1939).
33. Ng., C. F., and Martin, G. A., *J. Catal.* **54,** 384 (1978).
34. Oliphant, J. L., Fowler, R. W., Pannell, R. B., and Bartholomew, C. H., *J. Catal.* **51,** 229 (1978).
35. Perdereau, M., and Ondar, J., *Surf. Sci.* **20,** 80 (1970).
36. Pichler, H., *Adv. Catal.* **4,** 322 (1952).
37. Rapoport, I. B., "Chemistry and Technology of Synthetic Fuels," 2nd ed. Gostoptekhizdat, Moscow, 1955 [translated into English by M. Shelef, for the National Science Foundation and the Dept. of Interior, Washington, D. C., Israel Program for Scientific Translations, Jerusalem, 1962].
38. Rapoport, I. B., and Muzovkaya, O. A., *Khim. Tekhnol. Topl. Masel* No. 2, pp. 18–24 (1957); No. 5, pp. 19–27 (1957).
39. Rostrup-Nielsen, J. R., *J. Catal.* **21,** 171 (1971).
40. Shultz, J. F., Abelson, M., Stein, K. C., and Anderson, R. B., *J. Phys. Chem.* **63,** 496 (1959).

41. Shultz, J. F., Hofer, L. J. E., Karn, F. S., and Anderson, R. B., *J. Phys. Chem.* **66,** 501 (1962).
42. Shultz, J. F., Karn, F. S., and Anderson, R. B., *Ind. Eng. Chem.* **54,** No. 5, 44 (1962).
43. Shultz, J. F., Karn, F. S., and Anderson, R. B., *Rep. Invest.—U.S., Bur. Mines* **6974,** 1–20 (1967).
44. Shultz, J. F., Karn, F. S., Anderson, R. B., and Hofer, L. J. E., *Fuel* **40,** 181 (1961).
45. Shultz, J. F., Karn, F. S., Bayer, J., and Anderson, R. B., *J. Catal.* **2,** 200 (1963).
46. Wheeler, A., *Adv. Catal.* **3,** 250–327 (1951) *in* "Catalysis" (P. H. Emmett, ed.), Vol. 2, pp. 105–166. Van Nostrand-Reinhold, Princeton, New Jersey, 1955.
47. Wheeler, A., and Robell, A. J., *J. Catal.* **13,** 299 (1969).

CHAPTER 7

The Kölbel–Engelhardt Synthesis

by H. Kölbel and M. Ralek**

7.1 Introduction

In the early Fischer–Tropsch (FT) synthesis work, it was found that the reaction on iron catalysts occurs in a different way from that on cobalt and nickel catalysts. The simplified net reaction for the cobalt catalysts is

$$CO + 2\,H_2 = \text{—}CH_2\text{—} + H_2O \qquad \Delta H\,(227°C) = -165\ \text{kJ} \qquad (1)$$

and for the iron,

$$2\,CO + H_2 = \text{—}CH_2\text{—} + CO_2 \qquad \Delta H_R\,(227°C) = -204.8\ \text{kJ} \qquad (2)$$

Although the Fischer school already had considered a secondary reaction of water,

$$CO + H_2O = CO_2 + H_2 \qquad \Delta H_R\,(227°C) = -39.8\ \text{kJ} \qquad (3)$$

this thought did not persist. Franz Fischer's idea, that the hydrocarbon synthesis must go through a carbide intermediate dominated the thinking (*1*). Thus, the oxygen must come from CO decomposition and, over cobalt catalysts, is reacted with hydrogen to form water; in contrast, over iron catalysts, it reacts with additional CO to form CO_2. Consequently, two different mechanisms for the FT synthesis must be assumed, although the hydrocarbon synthesis on both catalysts proceeds through hydrogenation of the carbide and leads to similar hydrocarbon-product distributions.

This contradiction, as well as the function of steam during CO hydrogenation (*2*) and the role of the conversion reaction [Eq. (3)] in the FT synthesis (*3*) on both catalysts, was up to 1948 the motive for many experiments. On the one hand, it could be shown that the FT reaction on cobalt is the same as that of iron [Eq. (2)] if low space velocities and CO-

* Institute for Technical Chemistry, Technical University, Berlin (West), Germany.

rich synthesis gas were used. Thus, Eq. (3) is favored by long contact times as well as excess CO. On the other hand, the FT reaction on iron is similar to that on cobalt [Eq. (1)], if the intermediate water is removed from the reactor, so that the secondary reaction [Eq. (3)] is hindered. Thus, it was proved that Eq. (1) is the primary synthesis reaction, and that Eq. (2) is the sum of the synthesis reaction [Eq. (1)] and the following water–gas shift reaction [Eq. (3)].

Systematic research on the role of the water–gas shift in the reaction led in 1949 to the development of a new hydrocarbon synthesis from CO and steam; this synthesis is referred to in the literature as the Kölbel–Engelhardt (KE) synthesis and is shown in Eq. (4), and occurs on iron (*4*),

$$3\ CO + H_2O = -CH_2- + 2\ CO_2 \qquad \Delta H_R\ (227°C) = -244.5\ kJ \tag{4}$$

cobalt (*5*), nickel (*6*), and ruthenium (*7*) catalysts. The reaction can be run under specific conditions and leads not only to hydrocarbons but also to a series of oxygen-containing compounds that are 10% of the total product (*8*).

The further development of the KE synthesis led in 1959 to direct syntheses of primary aliphatic amines from CO, steam, and ammonia (*9*) and, later, to direct syntheses of terminal aliphatic secondary and tertiary amines (*10, 11*), as well as cyclic alkylamines (*12*). In order to carry out the KE synthesis on a pilot-plant scale, the slurry-phase reactor, developed by Kölbel and Ackermann (*13, 14*) has been used.

The process makes possible hydrocarbon synthesis without the production of hydrogen, and is suited for use with hydrogen-free and hydrogen-poor, CO-rich gases, especially from industrial gases such as producer gas and blast-furnace gas. The primary products of the KE synthesis consist of mainly unbranched aliphatic hydrocarbons of all chain lengths that, depending on the operating conditions, can be primarily olefins in which the double bond is mainly terminal.

7.2 Reaction Mechanism

From kinetic experiments (see Section 7.6), it can be proved that the KE reaction runs through an intermediate hydrogen-producing step with its immediate consumption (*15, 16*). This intermediate hydrogen can come, in part, from the product water of the previous FT reaction through a water–gas shift reaction according to the scheme

$$\begin{aligned} CO + 2\ H_2 &= -CH_2- + H_2O \\ 2\ CO + 2\ H_2O &= 2\ H_2 + 2\ CO_2 \end{aligned} \tag{5}$$

Thus, the KE reaction consists of two common mechanisms, the FT reaction and the water–gas shift.

7.3 Stoichiometry

The overall net reactions for the formation of hydrocarbons are

Formation of olefins

$$3n\ CO + n\ H_2O = C_nH_{2n} + 2n\ CO_2 \tag{6}$$

Formation of paraffins

$$(3n + 1)\ CO + (n + 1)\ H_2O = C_nH_{2n+2} + (2n + 1)\ CO_2 \tag{7}$$

The oxygenates are formed by the following net reactions:

Alcohols

$$3n\ CO + (n + 1)\ H_2O = C_nH_{2n+1}OH + 2n\ CO_2 \tag{8}$$

Acids

$$(3n + 1)\ CO + (n + 1)\ H_2O = C_nH_{2n+1}COOH + 2n\ CO_2 \tag{9}$$

Carbonyl compounds

$$(3n + 2)\ CO + (n + 1)\ H_2O = C_nH_{2n+1}CHO + (2n + 1)\ CO_2 \tag{10}$$

Unwanted side reactions are

1. Methanation

$$4\ CO + 2\ H_2O = CH_4 + 3\ CO_2 \tag{11}$$

2. CO decomposition

$$2\ CO = C + CO_2 \tag{12}$$

with the following reaction to methane

$$2\ C + 2\ H_2O = CH_4 + CO_2$$

3. Oxidation of the catalyst metal by water and CO_2

The reaction equations of the KE synthesis can be formally regarded as the FT reaction and multiple additions of the water–gas shift reaction. Two examples are given:

Paraffin formation

FT reaction	$n\ CO + (2n + 1)\ H_2 = C_nH_{2n+2} + n\ H_2O$	
Water–gas shift	$(2n + 1)\ CO + (2n + 1)\ H_2O = (2n + 1)\ CO_2 + (2n + 1)\ H_2$	
KE reaction	$(3n + 1)\ CO + (n + 1)\ H_2O = C_nH_{2n+2} + (2n + 1)\ CO_2$	(14)

Olefin formation

FT reaction	$n\,CO + 2n\,H_2 = C_nH_{2n} + n\,H_2O$
Water–gas shift	$2n\,CO + 2n\,H_2O = 2n\,CO + 2n\,H_2$
KE reaction	$3n\,CO + n\,H_2O = C_nH_{2n} + 2n\,CO_2$ (15)

The theoretical yield of olefinic products, based on converted CO, is 208.5 g of hydrocarbons per normal (S.T.P.) cubic meter (hereafter designated m^3_n) CO.

7.4 Thermodynamics

The thermodynamic product-formation probabilities in a system of several parallel and consecutive reactions that are in equilibrium with each other can only be considered from the calculations of simultaneous equilibria. Calculations of this type for the FT reaction (*17, 18*) and for the KE synthesis (*18*) have been published.

The thermodynamic data for single reactions of the KE synthesis, which are considered independently, are easily calculated on the basis of the known thermodynamics of the FT reaction (*19, 20*), because the reaction enthalpies of the KE and FT reactions differ only by single or multiple values of the enthalpy of the water–gas shift reaction.

Thermodynamic calculations of the probabilities of formation of a series of oxygen-containing products are found in the literature (*21*). The calculations of the thermodynamic maximal conversion of several KE reactions as well as the related amine synthesis can be found in references *22–24*.

7.5 Catalyst

In principle, all catalysts that are active for the FT reaction and oxidation resistant under the KE reaction conditions are active for the KE synthesis (*25*). Iron catalysts are particularly valuable (*4*), but cobalt (*5*) and nickel (*6*) can also catalyze the reaction. Ruthenium catalysts (*7*) under high pressure produce longer hydrocarbon chains (polymethylenes).

The main objectives for catalyst development are, on the one hand, the highest possible activity and life, and on the other hand, the highest possible selectivity in terms of the synthesis reactions and suppression of the side reactions. These side reactions are carbon deposition from CO, methanation, the nonstoichiometric water–gas shift reaction, as well as oxidation of the catalyst by steam (*2*). These reactions are also not desired in the FT synthesis. Possible damage to the catalyst from oxidation can be suppressed by a certain ratio of CO to H_2O in the synthesis gas. This ratio for iron is ~2.6 (*4*) and corresponds to the general stoichiometry. An excess of CO leads to carbon deposition, whereas an excess of water causes oxidation.

Nitriding the iron catalyst hinders the CO decomposition. Addition of alkali salts suppresses methanation. Copper or silver promotes the reduction and makes possible the formation of metal at lower temperatures, which decreases sintering. Use of carriers and supports (dolomite or kieselguhr) is also advantageous, especially for the fixed-bed catalysts (*4*). It is not necessary to use a support in a slurry-phase synthesis with suspended catalysts.

7.6 Kinetics

The kinetic measurements of the KE synthesis were carried out in a fixed-bed reactor [catalyst depth 600 mm, diameter 20 mm, with five sampling valves to determine concentration profiles (*15, 16, 26, 27*)].

The iron catalyst used was from Rheinpreussen AG, the cobalt catalyst from Ruhrchemie AG, and the nickel catalyst was homemade. The iron catalyst was precipitated with hot ammonia from an iron nitrate solution; the dried iron hydroxide was supported on kieselguhr and had the following composition: 100 : 10 : 0.5 : 0.2 : 50 iron : magnesium : copper : K_2CO_3 : kieselguhr. Before use in the synthesis, the catalyst was reduced with CO and with hydrogen (*25*).

The cobalt catalyst used was the standard FT catalyst from Ruhrchemie AG (*25*) with the following composition: 30.3 wt % Co, 1.5 ThO_2, 3.3 Fe_2O_3, 2.1 MgO, and 62.1 SiO_2 (as kieselguhr). The catalysts were received from the manufacturer already reduced (*26*).

The nickel–kieselguhr catalyst, with the addition of manganese and aluminum oxide, was precipitated according to the method of Fischer and Meyer (*28*), and was reduced in hydrogen (*27*). The total nickel content in the reduced catalysts was 37.5 wt %, and the part of the nickel as metal was 74%.

Details of how the catalysts were activated in the reactor are published elsewhere (*26, 27*) (see also Section 7.7.1.3).

Experiments with iron catalysts were carried out in the pressure range 7–23 atm and the temperature range 232–286°C. As described in Section

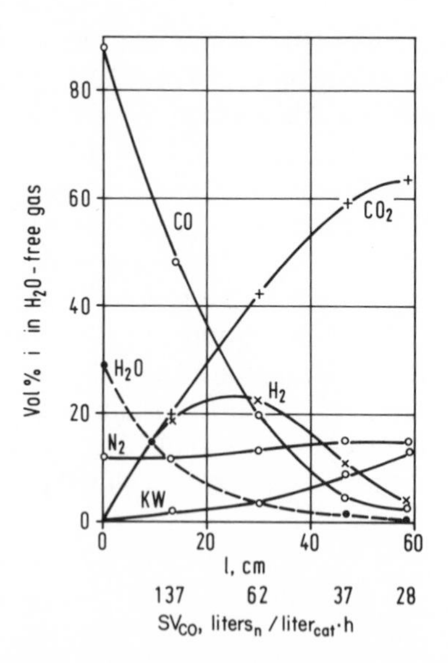

Fig. 1 Course of Kölbel–Engelhardt reaction on iron at 7 atm and 260°C, 33 h of operation. Fresh gas: CO, 4.7 atm; H_2O, 1.65 atm; N_2, 0.65 atm. From Kölbel and Hammer (*15*).

7.5, the CO/H_2O ratio of the synthesis gas had to be held within a narrow interval, 2.15–3.2. The cobalt catalyst was tested in the temperature range 200–270°C and pressure range 21–23 atm. Lower pressures led to an unfavorable KE reaction rate, whereas an increase above 30 atm favored the formation of cobalt carbonyls (*29*). The CO/H_2O ratio was between 2.1 and 3.9. Kinetic measurements with nickel catalysts were made in the temperature range 190–215°C and pressure range 1–2.5 atm. Higher pressure led to the formation of nickel carbonyls and to catalyst corrosion (*29*). The CO/H_2O ratio was 1.5–3.0.

Figure 1 shows typical concentration profiles of single reaction components on iron catalysts, measurement of which was made possible by the sampling valves. Underneath the abscissa (length of the catalyst bed) are the corresponding space velocities of the synthesis gas (thus accounting for the volume contractions). The curves for CO and H_2O decrease, and the curves for CO_2 and for hydrocarbons in the gas phase increase with increasing bed length. The curve for hydrogen shows a pronounced maximum.

The corresponding curves from cobalt catalysts differ from those of iron in that the intermediate hydrogen is distinctly smaller and has a maximum at smaller CO conversion, that is, toward the gas entrance of the bed. Other than that, the curves of both catalysts are similar. On nickel, the corresponding curves are also similar except that the hydrogen maximum is flatter than that determined for cobalt. Thus, over the entire catalyst bed, with the exception of a short initial portion of the bed, there is a small, constant partial pressure of hydrogen.

The hydrogen concentration profile is typical of a consecutive reaction. With relatively high partial pressures of CO and H_2O, hydrogen is formed faster than it is used. The hydrogen partial pressure thus rises. With decreasing partial pressure of CO and H_2O, the rate of formation of hydrogen decreases. Increasing hydrogen partial pressure increases its consumption in the hydrocarbon synthesis, until at the maximum, hydrogen is formed as it is used. After that, the partial pressures of CO and H_2O have decreased so much that hydrogen is used faster than it is formed. Thus, the hydrogen partial pressure declines.

The kinetics for such a reaction can only be described by a complicated rate law that takes into account the kinetics of hydrogen-producing as well as hydrogen-consuming reactions.

If the conversion of CO is split into the fraction that goes into the formation of CO_2, symbolized by $U_{CO \to CO_2}$, and the fraction that goes into the formation of hydrocarbons, symbolized by $U_{CO \to HC}$. The conversion of CO can be expressed as

$$r_{CO} = \frac{dU_{CO}}{d(1/SV_{CO})} = \frac{d(U_{CO \to CO_2} + U_{CO \to HC})}{d(1/SV_{CO})} = r_{CO \to CO_2} + r_{CO \to HC}. \quad (16)$$

Measurements made on the catalysts showed that the deposition of carbon on the catalyst and the subsequent formation of CO_2 by the Boudouard reaction were negligibly low. The catalyst reduction state remains constant during synthesis, which excludes the formation of CO_2 by reducing the catalyst with CO. Hence, the formation of CO_2 must take place strictly according to Eq. (3). This makes $U_{CO \to CO_2} = U_{CO \to HC}$. The rate of hydrogen formation is then

$$r_{CO \to CO_2} = \frac{dU_{CO \to CO_2}}{d(1/SV_{CO})} = k_1 p^a_{CO} p^b_{H_2O}, \quad (17)$$

The reverse reaction need not be taken into account because the equilibrium constant of the reaction [Eq. (3)] is between 50 and 100 under synthesis conditions.

The overall rate of all hydrogen-consuming reactions can be given as approximately

$$r_{CO \to HC} = \frac{dU_{CO \to HC}}{d(1/SV_{CO})} = k_2 p^c_{CO} p^d_{H_2}. \quad (17a)$$

For the KE reaction, the rate equation for all three systems of catalysts within the range of synthesis parameters investigated is

Table 1
Exponents and Rate Constants of Eq. (18)[a]

Catalyst	*a (CO)*	*b (H_2O)*	*c (CO)*	*d (H_2)*	k_1	k_2
Iron (233°C)	0.5	0.2	0.2	1.5	13	3
Cobalt (228°C)	1.0	0.3	0.3	1.0	0.9	1.7
Nickel (194°C)	0	0.6	0	0.6	4.5	11.2

[a] From Kölbel and Gaube (*6*), Kölbel and Hammer (*15*), Kölbel *et al.* (*16*), and Gaube (*27*).

$$r_{CO} = \frac{dU_{CO}}{d(1/SV_{CO})} = k_1 p_{CO}^a p_{H_2O}^b + k_2 p_{CO}^c p_{H_2}^d.$$

The reaction rate r_i (volumes of gas at S.T.P. converted per hour times volume of catalyst) was derived from the data typical of the technical syntheses in the following manner (*6*). The conversion component *i* was calculated by

$$U_i = \frac{x_i^0 - x_i(1 - k)}{x_i^0},$$

where x_i^0 is the molar fraction of component *i* entering the reactor, x_i the molar fraction of component *i* leaving the reactor, and $1 - k$ the ratio of total gas leaving the reactor to the total gas entering it ($1 - k = v_{in}/v_{out}$). If the feed of gas *i* entering the reactor is F_i (volumes at S.T.P. per hour), and the catalyst bulk volume *V*, assuming that the diffusional effects along the reaction tube can be neglected compared to the gas velocity, the following equation is valid:

$$F_i dU_i = r_i dV,$$

where dU_i is the change of conversion taking place in the differential volume dV at the rate r_i. The ratio F_i/V is the space velocity SV_i of the component *i*. It has the dimensions of volumes of gas (S.T.P.) entering the reactor per hour times the volume of catalyst. The reaction rate is then

$$r_i = \frac{dU_i}{d(1/SV_i)} \quad \frac{\text{volumes of gas } i \text{ (S.T.P.) converted}}{\text{hour} \times \text{volume of catalyst}};$$

this is equal to the slope of the function $U_i = f(1/SV_i)$.

The values of the exponents for each catalyst are different. In Table 1, the average values for a relatively large pressure range are given, as well as the values of both rate constants k_1 and k_2. These are, of course, dependent and bounded by the activity and condition of the catalyst, which in turn is a function of the time on stream. The net activation

Table 2
Influence of Pressure on Yield and Product Composition[a]

Pressure, atm	*CO Conversion, %*	*Yield, g/m_n^3 CO*	*Alcohol, wt %*	*Acid number, mg KOH per gram*
1	94.0	163.2	0.1	0.7
11	93.5	168.5	0.4	3.0
30	92.5	175.5	3.0	5.3
50	92.6	174.0	5.2	7.2
70	96.0	195.0	12.1	10.8
100	91.5	216.0	29.5	30.5

[a] Precipitated-iron catalyst, 235°C. From Kölbel and Engelhardt (*2*).

100%), portion of oxygenates only 2 wt %; space velocity of 160 h^{-1} (CO conversion 15%), oxygenate portion 12 wt %. Also, the portion of branched hydrocarbons rises with decreasing space velocities, which is a consequence of the isomerization reaction at longer contact times.

7.6.3 Influence of Pressure

The influence of pressure on the product composition is shown in Table 2 (*2*). The oxygen content of the products rises considerably with increasing pressure, especially between 30 and 100 atm, from 1 to 7.1%, corresponding to an alcohol content of 3 to nearly 30%. Also, the acid number rises from 5.3 to 30.5 mg KOH per gram. Corresponding to the incorporation of oxygen, the yield of total product at the same CO conversions rises from 163 to 216 g/m_n^3 CO. The olefin content of the total product was between 39 and 56 wt %.

7.7 Fixed-Bed Reactor Synthesis

In contrast to the FT synthesis, the KE synthesis has not been used on a pilot-plant scale. During the development of the KE synthesis (1948–1954), the main interest was the use of blast-furnace exit gas. Most of the work was done with CO-poor synthesis gas, 30–40% CO with the remainder nitrogen. The technical experiments were done with original blast-furnace gas.

7.7.1 Iron Catalysts

7.7.1.1 Reactor The experiments were run in a liquid-heated, laboratory tube reactor (bed length 700 mm, diameter 20 mm, throughput up

Table 3
Catalysts for Fixed-Bed Reactors

Supported catalysts
100 Fe, 0.2 Cu, 0.25 K_2CO_3, 100 kieselguhr
100 Fe, 10 Cu, 10 Mg, 50 kieselguhr with 2 wt % K_2CO_3
Unsupported catalysts
100 Fe, 10 Cu, 10 Mg, 2 wt % K_2CO_3
100 Fe, 0.2 Cu, 0.6 K_2CO_3

to 60 liters of gas per hour) without recycle. Additional experiments were carried out in a 6-m-long reactor with three catalyst tubes 19 mm in diameter. These three tubes had a capacity of 4 liters of catalyst and made possible extended tests with a gas throughput up to 2 m^3/h (*31*).

7.7.1.2 Catalysts Both supported and unsupported precipitated-iron catalysts with promoters (*25, 32*) were used for the fixed-bed synthesis (*14, 31*). Sodium carbonate or aqueous ammonia was used as the precipitating agent. The composition of several typical catalysts are given in Table 3.

7.7.1.3 Catalyst Activation The catalysts were activated prior to synthesis (*14*), that is, the oxidic iron compounds were changed to catalytically active iron phases. This activation was carried out with a $CO–H_2$ mixture ($CO/H_2 \sim 0.5$ at space velocities of 100 h^{-1} and 230°C for 70 h), or the catalyst was first carbided with CO and the carbide phase partially hydrogenated with hydrogen (270°C, 24 h, space velocities 100 h^{-1}). The degree of reduction was 60–70%, and the carbon in the catalyst was 6–10 wt %. Some experiments were also made in which the catalysts were activated directly with blast-furnace gas.

7.7.1.4 Synthesis Gas The synthesis gas used had an average composition of 34 vol % CO, 7% CO_2, 2% H_2, and 57% N_2 and corresponded to the blast-furnace gas available in 1952. Steam was then added to the gas so that the gas entering the reactor had a CO/H_2O ratio of 2.4–2.6. This corresponded to the optimal conditions of the synthesis (*3*).

7.7.1.5 Synthesis Conditions The activated catalysts were put into operation under pressure at the lowest possible temperature (200°C). Most of the syntheses were carried out at a pressure of 21 atm and an initial temperature of 220°C. In order to keep the CO conversion at 90%, the synthesis temperature was gradually raised, so that after an on-stream time of 1600 h, it reached 270°C. The space velocity for 90% CO conversion was in the range 100–50 $liters_n$ CO per liter of catalyst per hour.

Table 4
Conditions and Product Composition of a Fixed-Bed Synthesis[a]

Catalyst	100 Fe, 10 Cu, 10 Mg, 50 kieselguhr with 2% K_2CO_3	
Catalyst volume	0.18 liter	
Activation	48 h with gas (6.9% CO_2, 33.9% CO, 2.0% H_2, 57.2% N_2)	
Synthesis	240°C, 21 atm, SV_{CO} 100 h^{-1}, conversion = 94%, average time on stream 235 h, gas with CO/H_2O = 2.4, laboratory reactor	
Yield		
g Product per normal cubic meter blast-furnace gas		
Methane		6.3
Ethylene		1.7
Propane		8.2
Butane		11.5
Light liquid hydrocarbons		11.9
Oil and waxes		19.3
Total		
g Products per normal cubic meter blast-furnace gas		58.8
g Products per normal cubic meter CO		173.7 (i.e., 83.3% theory)

[a] From Kölbel and Engelhardt (*31*).

7.7.1.6 Product Composition The product composition is dependent on the catalyst and the synthesis conditions. Some examples are given for a fixed-bed synthesis in Table 4.

7.7.1.7 Regeneration Iron catalysts decline in activity because of poisoning, recrystallization, oxidation, and formation of stable iron carbides and elementary carbon. Under optimum synthesis conditions, 1 kg of catalyst can produce ~400 kg of hydrocarbon products. The used catalyst can be oxidatively regenerated many times to its previous activity. The oxygen content must be controlled such that the oxidation does not lead to local overheating, because this can damage the catalyst. The oxidized regenerated catalyst can then be activated and put to use (*14*).

7.7.2 Cobalt Catalysis

As is noted in the FT synthesis, cobalt catalysts (*5, 8, 29*) for the KE synthesis also have a longer catalyst life than the iron catalysts. In comparison to the FT synthesis, the KE synthesis leads under comparable conditions to products with a higher portion of olefins. Methanation is almost totally suppressed, its yield being about 2–5 g CH_4 per normal cubic meter CO [FT synthesis, 10–20 g CH_4 per normal cubic meter synthesis gas (*25*)]. The portion of oxygenates is strongly dependent on the synthesis pressure.

Table 5
Synthesis Conditions of Kölbel–Engelhardt Middle-Pressure Synthesis on Cobalt Catalysts

Composition of the dry synthesis gas, vol %	
(a) 89 CO, 0.3 H_2, 10 N_2	$CO/H_2O = 1.5$
(b) 42 CO, 0.8 H_2, 55 N_2	$CO/H_2O = 2$
Synthesis temperature	
Initial, 180°C	
After 835 h on stream, 232°C (*6*)	
Synthesis pressure, 11 atm	
CO Conversion	
(a) 85.8%, degree of liquifaction 98.5%	
(b) 78.6%	

7.7.2.1 Synthesis Conditions The KE reaction was studied in a fixed-bed reactor (bed length 600 mm, diameter 20 mm). The cobalt catalyst from Ruhrchemie AG (composition, see Section 7.6) was received already reduced. The activation conditions were 165–178°C with a nitrogen-rich synthesis gas (20.4% CO, 78.9% N_2 in 55 h). The synthesis conditions for the KE middle-pressure synthesis on cobalt are given in Table 5. These differ from those with iron catalysts mainly by lower synthesis temperatures and a lower CO/H_2O ratio in the synthesis gas.

For conditions (a) in Table 5, a yield of ~200 g of C_{2+} hydrocarbons was reached with an olefin portion of 70% and only 2 g of methane per normal cubic meter CO. For conditions (b), from every 1.0 m_n^3 CO, ~180 g of hydrocarbons and 20 g of oxygenated compounds were produced, including ~12 g of organic acids (including esterified components), 5 g of alcohols (including esterified components), ~0.5 g of aldehydes, and 2.5 g of ketones (*8*). The oxygenated products, 10% of the total, were produced on cobalt catalysts at a moderate conversion (for comparison, see Section 7.6.2). In Table 6, the product composition of the oxygenates is shown.

7.8 Comparison of Fischer–Tropsch and Kölbel–Engelhardt Reactions

Both syntheses have several common characteristics (catalysts, side reactions, etc.); however, they differ from each other in several ways (*33*):

Table 6
Oxygenated Compounds[a]

Alcohols		*Aldehydes*		*Ketones*		*Carboxylic acids*	
Methanol	50.7	Formaldehyde	4.8 (Formaldehyde, Acetaldehyde)	Acetone	17.7	Formic acid	4.2
Ethanol	26.8	Acetaldehyde		Diethylketone	2.0 (Diethylketone, Cyclohexanone)	Acetic acid	19.6
Propanol	19.4 (Propanol to *n*-Hexanol)	Propionaldehyde	0.4	Cyclohexanone		Propionic acid	8.0
n-Butanol		Valeraldehyde	12.6 (Valeraldehyde to *n*-Octanal)			*n*-Butyric acid	42.1
n-Pentanol		Butyraldehyde				*n*-Valeric acid	58.6 (*n*-Valeric acid to Lauric acid)
n-Hexanol		*n*-Hexanal				*n*-Caproic acid	
		n-Heptanal				*n*-Caprylic acid	
		n-Octanal				Lauric acid	
						Esters	24.30

[a] In moles per 1000 m_n^3 CO.

1. The kinetics, the course of reaction, and especially the partial pressures throughout the catalyst bed in both reactions are totally different. The hydrogen concentration in the gas phase in the KE reaction is smaller than that of the FT synthesis. This is noticeable in that the all hydrogen-consuming reactions for product formation are retarded in the KE synthesis. In contrast to the FT synthesis on cobalt, in which mainly paraffinic hydrocarbons are produced (*25*), the KE synthesis leads to a large amount of olefinic and oxygen-containing products. The small hydrogen partial pressure is not sufficient to hydrogenate these primary reaction products. Correspondingly, the unwanted methanation reaction over cobalt as well as on iron catalysts is almost suppressed, because this reaction is strongly dependent on hydrogen partial pressure. Even on nickel catalysts (*6*), methane is only 25% of the hydrocarbon product and hydrocarbons larger than C_{15} are produced (*27*). This is in strong contrast to the FT synthesis, in which methane is the main product in the same temperature range.

2. Because the hydrogen-consuming reaction steps are closely coupled with the hydrogen-producing steps, and the influence of the hydrogen-consuming reactions on the total CO conversion declines with shorter contact times, the olefins and oxygenates increase with shorter contact times. Thus, the KE synthesis produces a higher portion of unsaturated and oxygen-containing compounds than the FT synthesis.

3. The two syntheses differ also in the position of the catalyst zones endangered by oxidation in a fixed bed. In the FT synthesis, the oxidation zone is at the end of the catalyst bed, where much CO_2 and H_2O are present but little hydrogen and CO. In comparison, for the KE synthesis, this zone is at the inlet of the bed where a high steam partial pressure is dominant. Although high steam partial pressure is needed for kinetic rea-

sons, a certain CO/H_2O ratio must be maintained, according to the sensitivity of the catalyst to oxidation.

4. The relative rates of hydrogen-consuming and hydrogen-producing reactions over the iron catalysts can be influenced by addition of alkali promoters. This in turn influences the height of the intermediate hydrogen maximum, the consumption ratio CO/H_2, and finally, the product spectrum. Thus, the KE synthesis is quite flexible in changing the product distribution.

7.9 Conclusion for Large-Scale Operation

7.9.1 Fixed-Bed Synthesis

From the kinetics given, conclusions for carrying out the KE synthesis in fixed-bed reactors can be made. Operation of the FT synthesis on a large scale can be done better with recycle or in a staged process, such that heat transport can be controlled better. The secondary water–gas shift reaction can be suppressed by condensing the water in the recycle, thus controlling the consumption ratio CO/H_2. Removal of products can also suppress the secondary hydrogenation of olefins, the reduction of oxygen-containing compounds, as well as hydrogenation and cracking reactions. Removal of products and water increases the partial pressure of CO and hydrogen, and thus the rate of reaction.

Control of the KE reaction by analogous techniques is not possible, because water itself is a reactant. If products are taken out by condensation, then new water must be added to synthesis gas. Separation of the two reactions—the water–gas shift and FT reactions—in two reactors, in which the first reactor operates under optimal conditions for the water–gas shift reaction and the second for the FT reaction, is not possible. All the specific features of the KE synthesis disappear (*33*). In order to control heat transport, the KE synthesis can be carried out in either a staged reactor or a recycle reactor without removal of products. With staged reactors, fresh catalyst can be used in the second and following stages to achieve better conversion of the remaining CO.

Furthermore, when recycling, the back mixing of the recycle gas with fresh feed causes the hydrogen partial pressure to be constant along the bed. Thus, the yield, the average rate, is raised, and the catalyst life is lengthened.

7.9.2 Fluidized- and Entrained-Bed Techniques

The favorable effect of back mixing should also be evident by carrying out the synthesis in a fluidized bed or even an entrained bed. However, these two techniques work at higher temperatures (>300°C) to suppress the formation of wax and the risk of agglomeration of the catalyst particles and to achieve a reaction rate that is high enough to maintain fluidization of the bed or sufficient conversion in the short contact time in the entrained bed.

High synthesis temperatures are unfavorable in the KE synthesis, because it favors carbon deposition (Boudouard reaction) and thus catalyst deactivation (*34*). For these reasons, it appears that the fluidized bed and entrained bed are not suitable for the KE synthesis.

7.9.3 Liquid-Phase Synthesis

The use of the bubble-column reactor (*35, 36*) with suspended catalyst for the KE synthesis brings special advantages, as it does in the FT synthesis (*14, 36*). The use of a slurry reactor guarantees uniform temperature in the reactor and a high catalyst effectiveness factor of ~1. Furthermore, because of the excellent heat-transfer characteristics of the suspension, local overheating (hot spots) is reduced. With the liquid-phase reactor, temperatures can be increased 20–25°C above those of the fixed-bed reactor, without losing selectivity to hydrocarbons as a result of methanation or carbon deposition.

In addition to these effects, a bubble-column reactor also provides advantageous back mixing. The stationary intermediate hydrogen maximum that is common in fixed-bed reactors is leveled out. The catalyst is charged uniformly with the synthesis gas and its life is lengthened. The methanation is suppressed because it has a higher order of reaction for hydrogen than the main synthesis reactions.

The back mixing in the bubble column is not only favorable for mass transport but also for heat-evolution balance in the reactor. In contrast to the fixed-bed reactor, in which the heat evolution varies from phase to phase, the heat production is distributed evenly throughout the bubble column. Catalyst life is thus increased and the arrangement of cooling elements simplified.

7.9.3.1 Reactor Reactions of the liquid-phase synthesis were carried out in a bubble-column reactor (length 6 m, diameter 50 mm, effective reaction volume, 9 liters; Fig. 4).

7.9.3.2 Catalysts Unsupported precipitated-iron catalysts were used for the liquid-phase synthesis; the only cobalt catalyst tested was from Ruhrchemie AG (*25*).

The precipitated-iron catalysts contained 0.2–1.2 wt % K_2CO_3 promoter. The dried and tempered oxidized catalyst was ground in an oscillating ball mill with a little of the fluid phase. Dry grinding is not recom-

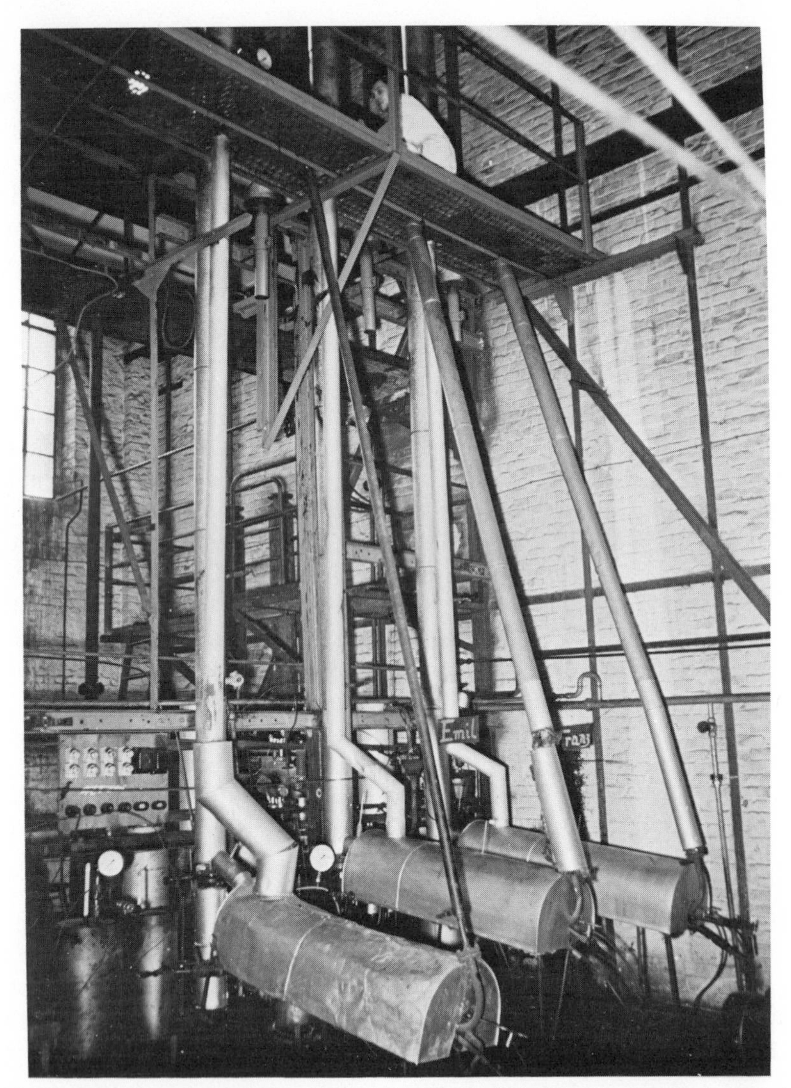

Fig. 4 Bubble-column reactor. Capacity, ~5000 g hydrocarbons per day. From Kölbel *et al.* (*13*).

mended because this tends to melt and smear phases together, particularly the dominant hexagonal crystalline platelets of α-Fe_2O_3. The catalyst particles after grinding must not exceed a diameter of 0.05 mm, in order to minimize sedimentation.

The catalyst concentration in the suspension has an optimum of ~10 wt % calculated on the basis of iron. Lower concentrations reduce the reactor efficiency whereas higher concentrations (up to 20–25 wt %) lead to a high suspension viscosity, which in turn diminishes the gas–liquid interface, leads to a lowering of mass transport, and results in a lower conversion. It is thus necessary to choose an optimum between catalyst particle size, concentration, and gas throughput (*14*).

7.9.3.3 Activation The concentrated, dried, and ground catalyst-containing suspension was poured into the bubble column, containing molten wax, which had been heated to 15–30°C above the operating temperature. The suspension was then fluidized and brought to the operating pressure (1–2 atm) by introduction of an inert gas (N_2, CO_2). After reaching the activation temperature, synthesis gas (CO/H_2 = 1.3–1.5) was suddenly introduced (SV ~ 2000 h^{-1}); after the first few minutes, an "activation wave" occurs that can be detected by a high CO_2 concentration in the exit gas. In this activation wave, the catalyst is reduced and simultaneously carbided by the Boudouard reaction. The CO_2 concentration in the exit gas declines very quickly. After the CO_2 concentration slowly and asymptotically approaches a limit, the activation is ended after several hours. The temperature is lowered to the synthesis operating temperature, depending on the desired throughput and conversion. The catalyst is then ready for operation.

The activation temperature depends on the catalyst activity and selectivity. It lies above the synthesis temperature. At too high an activation temperature, there is a danger of "overcoking," whereas at too low a temperature the activation does not "kick off". Also important is the heating time in inert gas, which must be as short as possible with the sudden introduction of the synthesis gas into the suspension. During activation, the fraction of the small catalyst particles is enlarged by carburization, which raises the dispersion of the solid phase in the suspension eventually to a quasi-colloidal state.

7.9.3.4 Operating Data for the Liquid-Phase Synthesis Representative operating data for hydrocarbon synthesis from blast-furnace gas are given in Table 7. The yield of hydrocarbons with three or more carbon atoms (C_{3+}) was 160 g per standard cubic meter of CO in the feed gas at a CO conversion of more than 90%.

7.9.3.5 Product Composition The liquid-phase synthesis is characterized by high flexibility of product composition. The product spectrum is particularly dependent on catalyst composition and synthesis temperature. The optimum temperature range is 250–300°C. Increasing the alkali content increases the hydrocarbon chain length with an increasing olefinic portion in the products. Higher synthesis temperatures favor the formation of shorter chain products.

Table 7
Operating Data and Results of Liquid-Phase Synthesis of Hydrocarbons from Exit Gas of Blast Furnace[a]

Operating Conditions	
Average composition of feed gas, %	7 CO_2, 2 H_2 34 CO, 57 N_2
Inlet gas, m_n^3/h	2.0
Effective reactor space (volume of suspension, including gas at operating condition), liters	9
Catalyst amount, g Fe	384
Synthesis pressure, atm	16
Synthesis temperature, °C	240–280
Catalyst loading, normal liters CO per gram Fe per hour	1.8
CO/H_2O ratio in inlet gas	2.6
Results	
CO Conversion, %	93
Yield of hydrocarbons, g/m_n^3 CO	182
C_{3+} Hydrocarbon yield, g/m_n^3 CO	160
Total hydrocarbon production (C_{3+}, total over the operating time), kg per kilogram Fe	405
Space–time yield of C_{3+} hydrocarbons, kg hydrocarbons per cubic meter reactor space in 24 h	272

[a] From Kölbel (*36a*).

In Table 8 are two product compositions from different promoted, precipitated-iron catalysts. Under condition (a), short-chain hydrocarbon formation results; with (b), higher molecular weight products. These products are particularly suitable as raw materials for petrochemicals

Table 8
Composition of Synthesis Products[a]

Fraction	*Carbon number*	*Boiling range, °C*	*Products, wt %* (*a*)	*Products, wt %* (*b*)
Methane–ethane	$C_1 + C_2$	—	8.6	8.1
Ethylene	C_2		3.5	2.1
Gasol	$C_3 + C_4$		29.8	15.1
Naphtha	C_5–C_{10}	to 190	43.1	17.8
Kogasin I	C_{11}–C_{14}	190–260	8.1	3.2
Kogasin II	C_{15}–C_{18}	260–320	4.0	9.7
Paraffin wax	C_{19}–C_{27}	320–450	1.7	15.0
Solid paraffins	$>C_{27}$	>450	1.2	29.0
Total yield of hydrocarbons, g m_n^3 CO			174	186

[a] From Kölbel (*36a*).

because 80–90% of gases and an average of 70% of the liquid products are olefinic. Improved selectivity of the KE synthesis is presumably possible only by using new selective catalysts. Such selective catalysts for the FT synthesis have been described in the literature (*34*).

7.10 Liquid-Phase Kölbel–Engelhardt Synthesis of High Molecular Weight Paraffinic Hydrocarbons

In this section a further development (*7, 37–39*) of the polymethylene synthesis on the ruthenium catalysts discovered by Pichler and Buffleb in 1940 (*40*) is described. When CO or CO-containing gases are fed into a suspension of metallic ruthenium in water at pressures above 50 atm and temperatures of 150–250°C, a layer of paraffins forms on the top of the water. The molecular weight of the paraffins is up to 7000 and with a melting range up to 130°C. Conversions up to 100% have been reached. The net reaction follows Eq. (4) and is strongly exothermic [ΔH_R (200°C) = −239 kJ per mole —CH_2 for molecular weights >300]. Side products are CO_2 and small amounts of CH_4 and H_2.

In this one-step synthesis the water acts as a reactant, as a suspension liquid, as the heat-exchange medium, and as a cooling agent by evaporation (*7, 37*). It can be inferred here that the hydrogen from the water–gas shift reaction is used directly at the synthesis site and is only in part desorbed from the catalyst.

The degree of liquefaction, that is, the ratio of CO that reacts to condensable hydrocarbons to the total CO converted, is dependent on temperature and pressure (Fig. 5). At higher temperatures and low pressure, methane is the only hydrocarbon. Its formation can be suppressed by use of higher pressures. The average molecular weight of hydrocarbons decreases with increasing temperature, as is shown in Table 9 (*39*).

The wax is mainly crystalline and shows the crystal structure of *n*-paraffins. Determination of acid and ester numbers shows one acid or ester group for about every 400 methylene groups. The products have properties similar to the polymethylenes synthesized from CO–H_2 mixtures (*40*).

7.10.1 Catalysts

The selectivity and activity of the ruthenium catalysts was dependent on the type of preparation. The starting material for these catalysts was

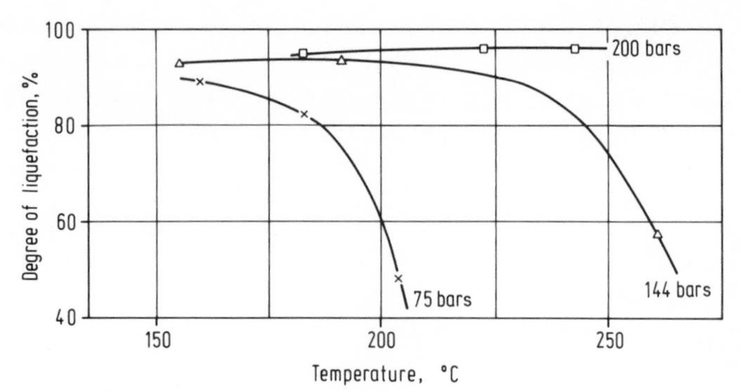

Fig. 5 Degree of liquefaction as a function of temperature at different pressures. From Kölbel *et al.* (*39*).

ruthenium(IV) oxide, which was precipitated from water-soluble potassium ruthenate with methanol at the boiling temperature of the solution. This precipitated ruthenium(IV) oxide was put directly into the reactor. Its reduction to metal by CO occurred in the first few minutes of the synthesis run (*7, 39, 40*).

The catalysts had a favorable life and its performance at a 90% CO conversion (120 atm, 210°C) was ~700 normal liters CO per hour per kilogram catalyst. From the experimental results it should be noted that raising the molecular weight of the products by increasing the pressure to high values is not possible because of catalyst corrosion by carbonyl formation.

Table 9
Dependence of the Upper Limit of Melting Range F_p of the Products on Operating Pressure p and Temperature T

p, atm	T, °C	F_p, °C
124	186	121
	211	117
	235	104
146	181	119
	229	116
	261	116
203	222	120

Table 10
Data for the Amine Synthesis in a Laboratory Fixed-Bed Reactor[a]

Precipitated-iron catalyst	100 Fe, 0.6 K_2CO_3, 0.2 Cu
Synthesis gas	52.6% CO, 20.6% H_2O, 0.9% NH_3, 23.7% Ar
Operating conditions	
Pressure, atm	11
Temperature, °C	219–235
CO Conversion, %	80
NH_3 Conversion, %	31–35
Portion of amines in product	

Average chain length	wt %
3	7.5
5	2.1
9	2.3
20	4.3
40	2.0
	20.2

[a] See Kölbel and Trapper (*9*) for the amine synthesis.

7.11 Direct Synthesis of Amines

The addition of nitrogenous bases [ammonia (*9*), primary (*10, 22*) and secondary (*10, 23*) aliphatic amines, and piperidine and pyrollidine (*12, 24*)] to the synthesis gas during KE synthesis leads to the formation of terminal alkylamines, whose portion of total product (hydrocarbons and amines) can reach up to 25 wt % of the total product (see also Chapter 5).

With a suitable choice of the bases and operating parameters, a high degree of selectivity in the alkylation can be achieved. Amines are formed with alkyl chain lengths over a broad range of carbon numbers. The technical data on the direct synthesis of amines from ammonia, CO, and water in a fixed-bed reactor (*9*) are given in Table 10.

Use of a slurry reactor (liquid-phase synthesis) is unfavorable. Conversion of ammonia is noticeably smaller than in the fixed-bed reactor. The reasons have not been explained (*41*).

Direct alkylation of ammonia in the FT (*42*) or KE synthesis is advantageous from a technical standpoint and the starting materials are inexpensive. The direct synthesis of amines can compete better with other technical amine syntheses if the conversion and selectivity relative to the carbon number in the alkyl chains can be increased.

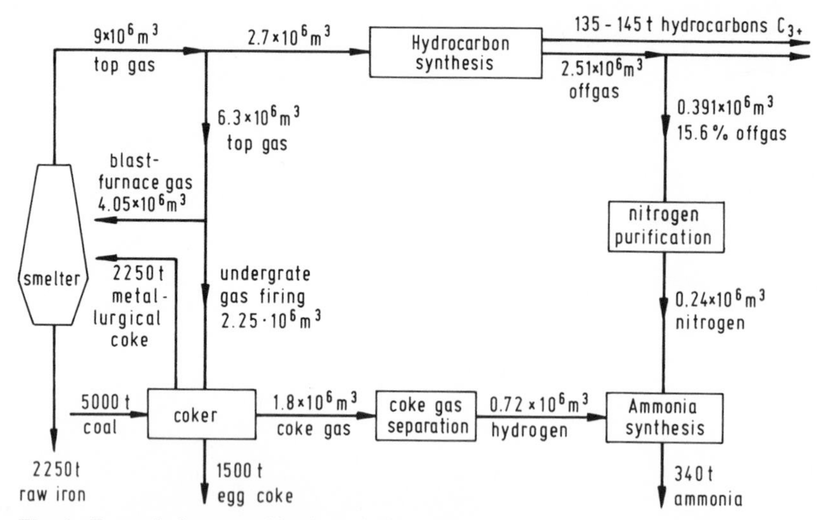

Fig. 6 Example for a combined steel plant, coke plant, hydrocarbon and ammonia synthesis (daily balance). t, Metric tons. From Kölbel (*43*).

7.12 Technical and Economic Prospects

7.12.1 Condition in the 1950s

At the time of the development of the KE synthesis (1948–1954), the main objective and interest was the use of CO-containing exit gases, especially blast-furnace gases, for the production of hydrocarbons (*43*). This was a particularly good source of CO for the synthesis, as well as a source of nitrogen for the ammonia synthesis, because it had a CO content of 32–35 vol %, 55–60 vol % N_2, and was almost sulfur free.

Figure 6 is a diagram of the then-planned combination steel plant, coke plant, KE synthesis, and ammonia synthesis. The diagram assumed a 1950–1955 daily production of 2250 [metric] tons of raw iron. We start with the amount of coke that comes from 5000 tons of coal per day. For the firing of the blast furnace, 2.25×10^6 m³ gas (=25% of the total blast-oven gas) is needed; 45% of the blast-furnace gas (4.05×10^6 m³ per 24 h) are necessary for the blast furnace, including 5% loss. The remaining 30% of the gas, 2.7×10^6 m³ per 24 h, becomes 135–144 tons of hydrocarbons per 24 h. About 15% of the synthesis offgas, that is, 391,000 m³ per 24 h, is freed of CO_2, after which the gas contains more than 90% nitrogen. After further purification, it is ready for ammonia synthesis. Separation of the coke-oven gas, 1.8×10^6 m³ per 24 h, gives 720,000 m³ per 24 h of hydrogen, which, when reacted with the nitrogen from the hydrocarbon

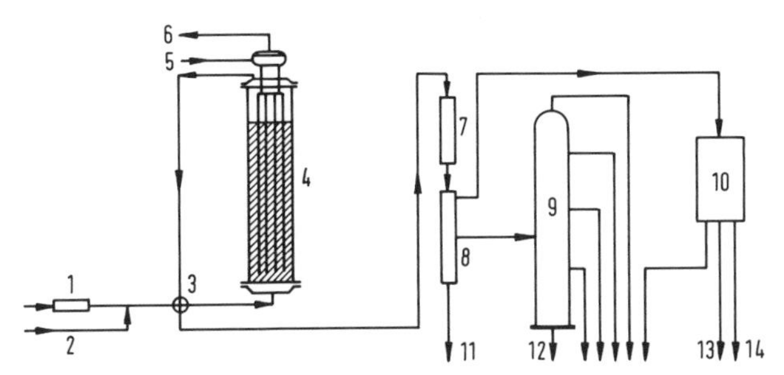

Fig. 7 Flow chart of a Kölbel–Engelhardt synthesis plant. 1, Compression of gases; 2, steam; 3, heat exchanger; 4, liquid-phase synthesis oven; 5, cooling water; 6, middle-pressure steam; 7, cooler (condenser); 8, separator; 9, distillation column; 10, CO_2 wash and activated-charcoal adsorber; 11, oxygen-containing products; 12, hydrocarbons at different boiling points; 13, CO_2; 14, nitrogen for ammonia synthesis. From Kölbel (*43*).

synthesis, gives ~340 tons NH_3 per 24 h. From the coke-oven gas separation comes 880,000 m^3 of rich gas for the Siemens–Martin steel plant, which has a high heating value. Additional lean gas is needed for the rolling mill under certain circumstances, but because it is too impure to have any chemical value, it can be produced from low-quality coal. From such an industrial complex it is possible to use ~40% of the actual coke for the chemical reactions of iron-ore reduction and hydrocarbon synthesis, not counting the possibility of ammonia synthesis. Industrial use of the KE slurry synthesis (Fig. 6) is shown in a flowchart in Fig. 7. After dust removal, the blast-furnace gas is compressed to an operating pressure of 16 atm, mixed with steam, led into the synthesis oven, and reacted to 220–270°C to hydrocarbons with a CO conversion of more than 90%. The products are led to a condenser and the lowest boiling hydrocarbons are adsorbed in an activated charcoal trap. Washing under pressure removes the CO_2. The remaining gas consists of more than 90% nitrogen and can be used for the ammonia synthesis. The economics of the process, based on 1950s prices (*44*), pointed to an expected profitability for such a combined plant, involving coal, chemistry, iron, and energy.

7.12.2 Present Condition

With present progress of the steel industry, the CO content of the exit gas has continually declined, so that today's exit gases contain only 15–18 vol % CO and thus are not directly usable for the KE synthesis for economic reasons. However, there are other possible uses for industrialization of the KE synthesis. Any review of the economic prospects of the KE synthesis must include the question of a necessary minimum required for an economical synthesis plant. This size lies in the area of 100,000 tons per year of primary products. This in turn requires a CO consumption of ~80,000 m_n^3 per hour. Furthermore, the question of a necessary minimum CO concentration must be considered, below which it would not be economical. Too high an inert portion of the gas would mean higher operating costs, especially energy costs for compression of the gas. This minimum CO concentration is ~30 vol %. Thus, the exit gas of a steel and smelter works does not have the required minimum CO concentration. A potential CO source for the KE synthesis could be a mixture of CO-poor blast-furnace gas and CO-rich Bessemer gas with the required 30% CO concentration. Other CO-containing exit gases, for example, from carbide or phosphorus production, do not contain the required CO amount.

An interesting development of the processing of cheap, CO-rich synthesis gas is the second generation of coal gasifiers ($CO/H_2 < 1.9$), which offer the combination of simultaneously (in one reactor) running the FT and KE reactions. Carbon monoxide–rich synthesis gas has a limited use in the FT synthesis, and therefore a water–gas shift reaction must be done in order to adjust the CO/H_2 ratio for the FT synthesis. According to Hildebrand and Joseph (*45*), it would be advantageous, for economic reasons, to add steam to CO-rich synthesis gas and lead it directly to simultaneously operating KE and FT reactors. The studies of Sapienza and coworkers (*46*) carry this thought further, in that instead of coal gasification to $CO–H_2$ synthesis gas, the partial coal oxidation to CO with an immediately following KE synthesis reactor for hydrocarbon synthesis is recommended. By flowing the CO_2 produced by the KE synthesis back into the coal reactor, the coal oxidation reaction can be better thermally controlled. The CO_2 can be reacted mostly to CO and, thus, the KE synthesis can be repeated. From other, up to now partly published reports, it is obvious that the KE synthesis is enjoying a resurgence of interest in Germany as well as in the United States, Canada (*47, 50*), and the Soviet Union (*51–56*).

References

1. Fischer, F., and Koch, H., *Gesammelte Abh. Kennt. Kohle* **10,** 575 (1930).
2. Kölbel, H., and Engelhardt, F., *Erdoel Kohle* **3,** 529 (1950); **5,** 1 (1952).
3. Kölbel, H., and Engelhardt, F. *Chem.-Ing.-Tech.* **22,** 97 (1950); **23,** 153 (1951).
4. Kölbel, H., and Engelhardt, F. *Brennst.-Chem.* **32,** 150 (1951); **33,** 13 (1952); *Angew. Chem.* **64,** 54 (1952).
5. Kölbel, H., and Vorwerk, E. *Brennst.-Chem.* **38,** 2 (1957).
6. Kölbel, H., and Gaube, J. *Brennst.-Chem.* **42,** 149 (1961); *Erdoel Kohle* **14,** 263 (1961).

7. Kölbel, H., and Bhattacharyya, K. K. *Liebigs Ann. Chem.* **618,** 67 (1958).
8. Kölbel, H., Kuschel, J., Hammer, H. *Justus Liebigs Ann. Chem.* **632,** 8 (1960).
9. Kölbel, H., and Trapper, J. *Angew. Chem.* **78,** 908 (1966).
10. Kölbel, H., Kanowski, S., Abdulahad, I., and Ralek, M. *React. Kinet. Catal. Lett.* **1,** 267 (1972).
11. Kölbel, H., Abdulahad, I., and Ralek, M. *Erdoel Kohle* **28,** 385 (1975).
12. Deckert, H., Kölbel, H., and Ralek, M. *Chem.-Ing.-Tech.* **47,** 1022 (1975).
13. Kölbel, H., Ackermann, P., and Engelhardt, F. *Erdoel Kohle* **9,** 303 (1956).
14. Kölbel, H., and Ralek, M. *Catal. Rev.—Sci. Eng.* **21**(2), 225 (1980).
15. Kölbel, H., and Hammer, H. *Z. Elektrochem.* **64,** 224 (1960).
16. Kölbel, H., Engelhardt, F., Hammer, H., and Gaube, J. *Actes Congr. Int. Catal., 2nd, 1960* p. 953 (1961).
17. Tillmetz, K.-D., *Chem.-Ing.-Tech.* **48,** 1065 (1976).
18. Anderson, R. B., Lee, C. B., and Machiels, J. C. *Can. J. Chem. Eng.* **54,** 590 (1976).
19. Anderson, R. B., *in* "Catalysis" (P. H. Emmett, ed.), Vol. 4, Chapter 1. Van Nostrand-Reinhold, Princeton, New Jersey, 1956.
20. Kölbel, H., *in* "Chemische Technologie" (K. Winnacker and L. Küchler, eds.), p. 439. Carl Hanser Verlag, Munich, 1959.
21. Kuschel, J., Ph.D. Thesis, Technical University, Berlin (1966).
22. Kanowski, S., Ph.D. Thesis, Technical University, Berlin (1973).
23. Abdulahad, I., Ph.D. Thesis, Technical University, Berlin (1974).
24. Deckert, H., Ph.D. Thesis, Technical University, Berlin (1977).
25. Storch, H., Golumbic, N., and Anderson, R. B. "The Fischer–Tropsch and Related Syntheses." Wiley, New York, 1951.
26. Hammer, H., Ph.D. Thesis, Technical University, Berlin (1959).
27. Gaube, J. W., Ph.D. Thesis, Technical University, Berlin (1960).
28. Fischer, F., and Meyer, K., *Brennst.-Chem.* **12,** 225 (1931); **14,** 47 (1933).
29. Kretschmann, S., Ph.D. Thesis, Technical University, Berlin (1958).
30. Schmidt, B., not published.
31. Kölbel, H., and Engelhardt, F., not published.
32. Frohning, C. D., Kölbel, H., Ralek, M., Rottig, W., Schnur, F., and Schulz, H., *in* "Chemical Feedstocks from Coal" (J. Falbe, ed.), p. 309. Wiley, New York, 1982.
33. Kölbel, H., Engelhardt, F., and Hammer, H., *Brennst.-Chem.* **42,** 149 (1961).
34. Kölbel, H., and Ralek, M., *Chem. Ind. (Frankfurt)* **31,** 700 (1979).
35. Kölbel, H., Ackermann, P., and Engelhardt, F., *Erdoel Kohle* **9,** 153, 225, 303 (1956).
36. Kölbel, H., Ackermann, P., Engelhardt, F. *Proc. World Pet. Congr.* **4c,** 227 (1956).
36a. Kölbel, H., *Chem.-Ing.-Tech.* **29,** 505 (1957).
37. Kölbel, H., and Hammer, H., *Chem. Process Eng.* **42,** 105 (1961).
38. Bhattacharyya, K. K., Ph.D. Thesis, Technical University, Berlin (1957).
39. Kölbel, H., Müller, W. H. E., and Hammer, H., *Makromol. Chem.* **70,** 1 (1964).
40. Schulz, H., *in* "Chemical Feedstocks from Coal" (J. Falbe, ed.), p. 482. Wiley, New York, 1982.
41. Winter, S., not published.
42. Bashkirov, A. N., Kagan, J. B., and Kliger, G. A., *Dokl. Akad. Nauk SSSR* **109,** 774 (1956).
43. Kölbel, H., *Stahl Eisen* **78,** 1165 (1958).
44. Schulze, J., Diplomarbeit, Institute for Technical Chemistry, Technical University, Berlin (1957).
45. Hildebrand, R. E., and Joseph, L. M., High quality transportation fuels from CO–H_2. *Methanol Symp., Middle Atl. Reg. Meet. Am. Chem. Soc., 13th, 23 March 1979* (1979).

46. Sapienza, R. S., Slegeir, W. A. R., Goldberg, R. I., and Easterling, B. Carbon monoxide—Resource of the future. BNL 28576. *Coal Technol.* Astrohall, Houston, Texas, 18–20 November 1980.
47. Maekawa, Y., Chakrabartty, S., and Berkowitz, N. Observations on Kölbel–Engelhardt hydrocarbon synthesis. *Can. Symp. Catal., 5th, Calgary, Alberta, 26–27 October 1977* (1977).
48. Gustafson, B. L., Ph.D. Thesis, Texas A&M University, College Station, Texas (1981).
49. Gustafson, B. L., and Lunsford, J. H., *J. Catal.* **74,** 393 (1982).
50. Niwa, M., Iizuka, T., and Lunsford, J. H., *J. Chem. Soc., Chem. Commun.* p. 684 (1979).
51. Golovko, A. K., Kravtsov, A. V., Maximov, Yu. V., Smolyaninov, S. I., and Lavrov, N. V., *Khim. Tverd. Topl.* (*Moscow*) **13,** 130 (1979).
52. Udrea, I., Moroianu, M., Udrea, M., and Nicolescu, I. V., *Heterog. Katal., 4th, 1979* Vol. 2 (1979).
53. Smolyaninov, S. I., and Kravtsov, A. V., *Izv. Tomsk. Politekh. Inst.* **214,** 3 (1977).
54. Smolyaninov, S. I., and Mironov, V. M., *Izv. Tomsk. Politekh. Inst.* **214,** 13 (1977).
55. Smolyaninov, S. I., Mironov, V. M., and Zilberman, A. G., *Izv. Tomsk. Politekh. Inst.* **214,** 23 (1977).
56. Mironov, V. M., and Smolyaninov, S. I., *Izv. Tomsk. Politekh. Inst.* **214,** 30 (1977).

Addendum

H. Schulz and H. Gökcebay (*Prepr. Conf. Catal. Org. React., 9th, Charleston, South Carolina, April, 1982*) have extended the work of Deckwer and Ralek (Chapter 4, p. 150) on precipitated Fe–Mn catalysts by increasing the ratio Mn/Fe to ~11. Products from these catalysts contain a large fraction of olefins, ~85%, at carbon numbers from 2 to 15. In each fraction 95% of the olefins are terminal. A typical catalyst contained, by weight, 100 Fe : 1080 Mn : 100 silica : 20 alumina : 5 K_2O, and it was operated at 250°C and 10 atm of 2 H_2 + 1 CO gas. These preparations are called "matrix" catalysts. H. Schulz (personal communication) postulates that the Fe is isolated by a matrix of manganese oxide so that secondary reactions do not occur. The high yields of α-olefins are believed to be the primary synthesis products.

L. E. McCandlish [*J. Catal.*, **83,** 362 (1983)] developed a chain-growth scheme for the FTS involving stepwise addition of methylene, CH_2. His mechanism seems consistent with known organometallic reactions, and for isomer distributions of Pichler and Schulz (*107*) (see Chapter 5) his method predicted methyl species better than Anderson's SCG (Chapter 5, p. 193) and also predicted the formation of ethyl-substituted molecules. A methylene adds to a "naked" surface carbon to produce a surface-bonded vinylidene $M{=}C{=}CH_2$ (**a**); this species and its higher homologs $M{=}C{=}CHR$ (**b**) and $M{=}C{=}C(CH_3)R$ (**c**) are important intermediates. Methylene now adds to the carbon–carbon double bond to form a surface-bonded cyclopropylidene, for example from **b,** $M{=}\overline{C{-}CHR{-}C}H_2$. This three-membered ring can fragment in two ways such that two products are obtained after hydrogenation. The intermediate produced from species **b** splits to yield a straight chain or one with a methyl on the second carbon. Intermediate **c** yields a cyclopropane that splits to produce an extension of the chain or a quaternary structure that is constrained to rearrange to a species that may become an ethyl or larger substituent.

Chain growth proceeds by the following steps:

```
                                 C        C
RCC——→RCC*C        a            RCC——→RCC*C        a
  \                               \
   \                               \
    \                               \       C
     ↘ C*                            ↘      C*
      RCC          af                 RCC          af
```

Here the constants *a* and *f* have the same significance as in SCG, and the asterisks show the last carbon added. Although not mentioned in the paper, the mechanism predicts that the C_2 intermediate has a growth constant 2*a* compared with *a* for end addition to other species. It may be noted that P. T. Dawson and W. K. Peng [*J. Chem. Phys.*, **48,** 623 (1968); **54,** 950 (1971)] postulated an analogous intermediate for the synthesis and decomposition of ammonia on tungsten, W_2N_3H.

Index

K

U

W

X